T0134443

# Studies in Big Data

Volume 74

**Series Editor**

Janusz Kacprzyk, Polish Academy of Sciences, Warsaw, Poland

The series "Studies in Big Data" (SBD) publishes new developments and advances in the various areas of Big Data- quickly and with a high quality. The intent is to cover the theory, research, development, and applications of Big Data, as embedded in the fields of engineering, computer science, physics, economics and life sciences. The books of the series refer to the analysis and understanding of large, complex, and/or distributed data sets generated from recent digital sources coming from sensors or other physical instruments as well as simulations, crowd sourcing, social networks or other internet transactions, such as emails or video click streams and other. The series contains monographs, lecture notes and edited volumes in Big Data spanning the areas of computational intelligence including neural networks, evolutionary computation, soft computing, fuzzy systems, as well as artificial intelligence, data mining, modern statistics and Operations research, as well as self-organizing systems. Of particular value to both the contributors and the readership are the short publication timeframe and the world-wide distribution, which enable both wide and rapid dissemination of research output.

** Indexing: The books of this series are submitted to ISI Web of Science, DBLP, Ulrichs, MathSciNet, Current Mathematical Publications, Mathematical Reviews, Zentralblatt Math: MetaPress and Springerlink.

More information about this series at http://www.springer.com/series/11970

Nhien-An Le-Khac · Kim-Kwang Raymond Choo
Editors

# Cyber and Digital Forensic Investigations

A Law Enforcement Practitioner's Perspective

Springer

*Editors*
Nhien-An Le-Khac
School of Computer Science
University College Dublin
Dublin, Ireland

Kim-Kwang Raymond Choo (iD)
The University of Texas at San Antonio
San Antonio, TX, USA

ISSN 2197-6503          ISSN 2197-6511   (electronic)
Studies in Big Data
ISBN 978-3-030-47133-0      ISBN 978-3-030-47131-6   (eBook)
https://doi.org/10.1007/978-3-030-47131-6

This Springer imprint is published by the registered company Springer Nature Switzerland AG
The registered company address is: Gewerbestrasse 11, 6330 Cham, Switzerland

# Foreword by David A. "Dave" Dampier

As a digital forensic trainer, educator, and researcher for the past 18 years, I have witnessed how much the field has progressed. When I first started working in the field, we were dealing with simple file system data and stand-alone computers or laptops. During the time since I started, file systems have gotten much more sophisticated, operating system data is now more plentiful, and mechanisms for storing it have gotten much better. Add to that is the onslaught on new digital devices that can now store data and be used to commit cybercrimes, and the world as we know it has exploded with new knowledge that we have to acquire and understand. We now, as digital forensic educators, researchers, and practitioners, have to constantly learn new consumer technologies, new file and operating systems, and new tools that can be used to acquire evidence of potential interest, and the list goes on.

I was excited to read this book, as it includes numerous contributions from digital forensic students, educators, researchers, and practitioners affiliated with the Master of Science (M.Sc.) in Forensic Computing and Cybercrime Investigation program at the University College Dublin, Ireland. The contributed chapters discussed the state-of-the-art advances on a broad range of topics, such as Internet of Things (IoT) forensics, malware (e.g., ransomware) forensics, CCTV forensics, cloud forensics, network forensics, and financial investigations. In a number of these chapters, challenges and limitations with existing tools are highlighted, which reinforces the importance of continuing research in digital forensics and the tools to do it.

In closing, the publication of this book is timely as it provides new insights into a broad range of digital forensic topics and addresses the growing challenge associated with digital forensics. I highly recommend it!

David A. "Dave" Dampier
Professor and Associate Dean for Research
College of Engineering and Computer Sciences
Marshall University
Huntington, WV, USA

# Foreword by Marcus K. Rogers

I am honored to write this Foreword as I have known Dr. Choo from the University of Texas, San Antonio, for many years and have continually been impressed with the efforts that he has undertaken to further and mature the scientific field of digital forensics and investigations. I have also had the pleasure of working with Dr. Le-Khac's world renowned M.Sc. program in Forensic Computing and Cybercrime Investigations at the University College Dublin. Some of the graduates from this program have spent time as visiting research scientists or my Ph.D. students at the Purdue University. It is great to see such excellent scholars bringing together other expert researchers and scholars to collaborate on such an important book.

As an ex-law enforcement officer, and currently the executive director of cybersecurity programs at the Purdue Polytechnic Institute, Purdue University, and chief scientist of the Tippecanoe County High Tech Crime Unit, I can truly appreciate the fact that this book takes a broad approach to various topics it covers. The audience for this book should be equally as broad as not only does it cover current technologies and forensic issues such as IoT device, ransomware, CCTV, PayPal, network and cloud forensics, Bitcoin, and ToR, it is written in a manner that is relevant to academics, practitioners, and law enforcement. The book's authors and editors are a heterogeneous group of experts ranging from academics to industry practitioners and law enforcement. This further adds to the breadth of coverage and gives it a unique approach to the topics.

In my opinion, this book will quickly become a standard reference source in the field and be adopted as required reading for investigators (practitioner and law enforcement) and in many digital forensic academic programs worldwide. I guarantee this will be on my students' required reading list!

I commend the editors and authors for their ongoing efforts in enhancing the digital forensics and investigation field's knowledge and understanding of current technologies and issues. I am eagerly anticipating updated versions of this book, as the technologies and topics it covers evolve and change.

Marcus K. Rogers
Professor/Executive Director
of Cybersecurity Programs
Purdue University
West Lafayette, IN, USA

# Acknowledgements

This book would not have been possible for the amazing students enrolled in the Master of Science (M.Sc.) in Forensic Computing and Cybercrime Investigation program at the University College Dublin, Ireland, as well as the research collaborators, who were willing to dedicate their time and efforts to work on the research and share their findings in this book.

We are also very fortunate to have extremely supportive colleagues (e.g., the instructors that taught the courses in the program), who go out of their way to support our efforts.

We are also extremely grateful to Springer and their staff for their support in this project. They have been most accommodating of our schedule and helping to keep us on track.

# Contents

# Editors and Contributors

## About the Editors

**Nhien-An Le-Khac** is a lecturer at the School of Computer Science (CS), University College Dublin (UCD), Ireland. He is currently the program director of MSc program in Forensic Computing and Cybercrime Investigation (FCCI)—an international program for the law enforcement officers specializing in cybercrime investigations. He is also the co-founder of UCD-GNECB Postgraduate Certificate in fraud and e-crime investigation. Since 2008, he is a research fellow in Citibank, Ireland (Citi). He obtained his Ph.D. in Computer Science in 2006 at the Institut National Polytechnique de Grenoble (INPG), France. His research interest spans the area of Cybersecurity and Digital Forensics, Data Mining/Distributed Data Mining for Security, and Fraud and Criminal Detection. Since 2013, he has collaborated on many research projects as a principal/co-PI/funded investigator. He has published more than 150 scientific papers in peer-reviewed journal and conferences in related research fields, and his recent edited book has been listed the Best New Digital Forensics Book according to the Book Authority.

**Kim-Kwang Raymond Choo** received the Ph.D. in Information Security in 2006 from the Queensland University of Technology, Australia. He currently holds the Cloud Technology Endowed Professorship at the University of Texas at San Antonio (UTSA). In 2015, he and his team won the Digital Forensics Research Challenge organized by the Germany's University of Erlangen-Nuremberg. He is the recipient of the 2019 IEEE Technical Committee on Scalable Computing (TCSC) Award for Excellence in Scalable Computing (Middle Career Researcher), 2018 UTSA College of Business Col. Jean Piccione and Lt. Col. Philip Piccione Endowed Research Award for Tenured Faculty, British Computer Society's 2019 Wilkes Award Runner-up, 2019 EURASIP Journal on Wireless Communications and Networking (JWCN) Best Paper Award, Korea Information Processing Society's Journal of Information Processing Systems (JIPS) Survey Paper Award (Gold) 2019, IEEE Blockchain 2019 Outstanding Paper Award, Inscrypt 2019 Best

Student Paper Award, IEEE TrustCom 2018 Best Paper Award, ESORICS 2015 Best Research Paper Award, 2014 Highly Commended Award by the Australia New Zealand Policing Advisory Agency, Fulbright Scholarship in 2009, 2008 Australia Day Achievement Medallion, and British Computer Society's Wilkes Award in 2008. He is also a fellow of the Australian Computer Society, an IEEE senior member, and co-chair of IEEE Multimedia Communications Technical Committee's Digital Rights Management for Multimedia Interest Group.

## Contributors

**Christopher Boyton** Irish Defence Forces, Newbridge, Ireland

**Ryan Brooks** Avon and Somerset Constabulary, Bristol, UK

**Kim-Kwang Raymond Choo** University of Texas at San Antonio, San Antonio, TX, USA

**Trung Q. Duong** Queen Belfast University, Belfast, UK

**Richard Gomm** Garda Síochana Ombudsman Commission, Dublin, Ireland; University College Dublin, Dublin, Ireland

**Darren Hayes** Pace University, New York, USA

**Kien Wooi Hew** Royal Canadian Mounted Police, Vancouver, BC, Canada

**Alexander Hilgenberg** State Police in Lower Saxony, Northeim, Germany

**Anca Jurcut** University College Dublin, Dublin, Ireland

**Andrew Kinder** National Crime Agency, London, UK

**Nhien-An Le-Khac** University College Dublin, Dublin, Ireland

**Madhusanka Liyanage** University College Dublin, Dublin, Ireland

**Robert McArdle** Trend Micro Ltd., Cork, Ireland

**William O'Sullivan** Irish Defence Forces, Newbridge, Ireland

**Neil Redmond** National Cyber Security Centre, Dublin, Ireland; University College Dublin, Dublin, Ireland

**Sebastian Schlepphorst** Federal Office for Information Security (BSI), Bonn, Germany; University College Dublin, Dublin, Ireland

**Lars Standare** Rhineland-Palatinate Police University, Buechenbeuren, Germany

**Le-Nam Tran** University College Dublin, Dublin, Ireland

**Cornelis Leendert (Eelco) van Veldhuizen** Team High Tech Crime, Dutch National Police, The Hague, The Netherlands

**Cian Young** Irish Defence Forces, Cork, Ireland

# The Increasing Importance of Digital Forensics and Investigations in Law Enforcement, Government and Commercial Sectors

**Nhien-An Le-Khac and Kim-Kwang Raymond Choo**

**Abstract** Digital forensics and investigations play an increasingly important role in a broad range of scenarios, such as those in pandemics (e.g. COVID-19 contact tracing), criminal cases, civil litigations, and national security cases. In this book, the editors documented their education and research activities with students enrolled in the Master of Science (MSc) in Forensic Computing and Cybercrime Investigation program at University College Dublin, Ireland, who are also from the law enforcement and government community in Canada, Ireland, Germany, The Netherlands, and United Kingdom. Collaboratively, the authors focused on topics ranging from Internet of Things (IoT) malware analysis, IoT testbed setup, malware analysis (e.g. ransomware), CCTV forensics, financial investigations (e.g. PayPal accounts and Bitcoins), cloud forensics, and network and ToR forensics.

**Keywords** Digital forensics · Digital investigations · Criminal investigations · Civil litigations · National security investigations

## 1 Introduction

The twenty-first century has seen significant advances in information and communications technologies (ICT) that were, perhaps, considered science fiction a decade or two ago. ICT now pervades many aspects of society today, and the number and range of digital device users are rapidly increasing. For example, if we take a stroll down a street in any city/country, we would likely encounter a broad range of digital devices, such as IP-based closed-circuit televisions (CCTVs), some Internet of Things (IoT) devices (e.g. temperature sensors, and remote terminal units—RTUs), autonomous vehicles, mobile devices (e.g. Android and iOS phones), as well as ICT that are embedded in human bodies such as Internet of Medical Things (IoMT; or

N.-A. Le-Khac (✉)
University College Dublin, Dublin, Ireland
e-mail: an.lekhac@ucd.ie

K.-K. R. Choo
University of Texas at San Antonio, San Antonio, TX, USA

© The Editor(s) (if applicable) and The Author(s), under exclusive license to Springer Nature Switzerland AG 2020
N.-A. Le-Khac and K.-K. R. Choo (eds.), *Cyber and Digital Forensic Investigations*, Studies in Big Data 74, https://doi.org/10.1007/978-3-030-47131-6_1

known as embedded medical devices). Increasingly, IoT devices are also found in military and battlefield settings (e.g. Internet of Military/Battlefield Things).

There are, however, implications with the increased digitalisation of our society. For example, such devices can be the target of cybercriminals and those with malicious intent. This reinforces the importance of law enforcement investigators, including those that are tasked with physical/conventional crime investigations (e.g. murder, drug trafficking, money laundering and terrorism financing), to keep pace with ICT advances, understand the role of technologies in their investigations, understand the type and range of forensic/evidential artefacts that could be obtained from such devices, and how to go about data acquisition in a forensically sound manner.

Digital forensics (DF) and digital investigations (DI) are not restricted to only law enforcement matters, as the use of ICTs is also prevalent in commercial and national security settings. In other words, DF and DI play a role in civil litigations and national security investigations, such as those involving nation states (e.g. espionage), or cyber warfare like activities (e.g. corrupting supply chains to inject hardware Trojans). They may also play a role in pandemics, such as COVID-19 contact tracing. Hence, there is also a need for ongoing education and research in DF and DI, involving and targeting a broad range of stakeholders.

In this book, both editors of this book documented their education and research activities with one such stakeholder group, namely: law enforcement and government community. The lead authors of the next ten chapters (i.e. Chaps. "Defending IoT Devices from Malware" to "The Bitcoin-Network Protocol from a Forensic Perspective") are law enforcement and government employees who are also enrolled in the Master of Science (MSc) in Forensic Computing and Cybercrime Investigation program at University College Dublin, Ireland, collaborated with both editors[1] and other researchers on a broad range of topics, described in the next section.

## 2 Organisation of This Book

Chapters "Defending IoT Devices from Malware" and "Digital Forensic Investigation of Internet of Thing Devices: A Proposed Model and Case Studies", respectively led by William O'Sullivan (Irish Defence Forces, Ireland) and Alexander Hilgenberg (State police in Lower Saxony, Germany), focus on IoT device forensics. In Chap. "Defending IoT Devices from Malware", the authors examined two popular IoT malware, namely: Mirai and Qbot, and proposed mitigation and detection strategies based on their analysis. In Chap. "Digital Forensic Investigation of Internet of Thing Devices: A Proposed Model and Case Studies", the authors explained how one can set up an IoT testbed/laboratory for training of future forensic investigators.

Chapter "Forensic Investigation of Ransomware Activities—Part 1", led by Young (Irish Defense Force, Ireland) and in collaboration with McArdle (Trend Micro Ltd.,

---

[1] Nhien-An Le-Khac is the Director of the MSc in Forensic Computing and Cybercrime Investigation program, and Kim-Kwang Raymond Choo is the external examiner of the program.

Ireland) and both editors, focuses on ransomware analysis. Similarly, Chap. "Forensic Investigation of Ransomware Activities—Part 2" led by Boyton (Irish Defence Forces, Ireland) focuses on ransomware forensics.

CCTVs are found not only in public places (e.g. streets and public transportation venues), but also at private homes and offices. Hence, they are a potential source of evidence (e.g. video recordings). However, the variety of CCTV systems and the significant volume of data that can potentially be extracted from CCTV video can be challenging for forensic examiners. Chapter "CCTV Forensics in the Big Data Era: Challenges and Approaches", authored by Gomm (Garda Síochana Ombudsman Commission, Ireland), Brooks (Avon and Somerset Constabulary, UK), and Hew (Royal Canadian Mounted Police, Canada) in collaboration with both editors, focuses on CCTV forensics.

In Chap. "Forensic Investigation of PayPal Accounts", Standare (German Police, Germany) collaborated with Hayes (Pace University, USA) and both editors to study financial fraud investigation, using PayPal as a case study.

Cloud forensics is the focus of Chap. "Digital Forensic Approaches for Cloud Service Models: A Survey" led by Schlepphorst (Federal Office for Information Security, Germany), where we give an overview of digital forensic approaches for different cloud service models.

The forensic investigation of mobile networks is always a challenge for law enforcement due to their ad hoc nature, hence in Chap. "Long Term Evolution Network Security and Real-Time Data Extraction" led by Redmond (Deloitte LLP Ireland, also a former member of Irish National Cyber Security Centre) we describe the security issues of LTE and GSM networks as well as how to extract artefacts from these networks.

Chapter "Towards an Automated Process to Categorise Tor's Hidden Services", led by Kinder (National Crime Agency, UK), examines at how ToR hidden services are set up, and presents an approach to successfully identify criminal sites without manual interaction. Finally, Chap. "The Bitcoin-Network Protocol from a Forensic Perspective", led by Veldhuizen (Team High Tech Crime, Dutch National Police, The Netherlands), forensically examines the Bitcoin-network protocol.

**Nhien-An Le-Khac** is a lecturer at the School of Computer Science (CS), University College Dublin (UCD), Ireland. He is currently the program director of MSc program in Forensic Computing and Cybercrime Investigation (FCCI)—an international program for the law enforcement officers specializing in cybercrime investigations. He is also the co-founder of UCD-GNECB Postgraduate Certificate in fraud and e-crime investigation. Since 2008, he is a research fellow in Citibank, Ireland (Citi). He obtained his Ph.D. in Computer Science in 2006 at the Institut National Polytechnique de Grenoble (INPG), France. His research interest spans the area of Cybersecurity and Digital Forensics, Data Mining/Distributed Data Mining for Security, and Fraud and Criminal Detection. Since 2013, he has collaborated on many research projects as a principal/co-PI/funded investigator. He has published more than 150 scientific papers in peer-reviewed journal and conferences in related research fields, and his recent edited book has been listed the Best New Digital Forensics Book according to the Book Authority.

**Kim-Kwang Raymond Choo** received the Ph.D. in Information Security in 2006 from the Queensland University of Technology, Australia. He currently holds the Cloud Technology Endowed Professorship at the University of Texas at San Antonio (UTSA). In 2015, he and his team won the Digital Forensics Research Challenge organized by the Germany's University of Erlangen-Nuremberg. He is the recipient of the 2019 IEEE Technical Committee on Scalable Computing (TCSC) Award for Excellence in Scalable Computing (Middle Career Researcher), 2018 UTSA College of Business Col. Jean Piccione and Lt. Col. Philip Piccione Endowed Research Award for Tenured Faculty, British Computer Society's 2019 Wilkes Award Runner-up, 2019 EURASIP Journal on Wireless Communications and Networking (JWCN) Best Paper Award, Korea Information Processing Society's Journal of Information Processing Systems (JIPS) Survey Paper Award (Gold) 2019, IEEE Blockchain 2019 Outstanding Paper Award, Inscrypt 2019 Best Student Paper Award, IEEE TrustCom 2018 Best Paper Award, ESORICS 2015 Best Research Paper Award, 2014 Highly Commended Award by the Australia New Zealand Policing Advisory Agency, Fulbright Scholarship in 2009, 2008 Australia Day Achievement Medallion, and British Computer Society's Wilkes Award in 2008. He is also a fellow of the Australian Computer Society, an IEEE senior member, and co-chair of IEEE Multimedia Communications Technical Committee's Digital Rights Management for Multimedia Interest Group.

# Defending IoT Devices from Malware

**William O'Sullivan, Kim-Kwang Raymond Choo, and Nhien-An Le-Khac**

**Abstract** As the number of internet users continues to grow, so do the numbers and types of devices people connect to; hence, a larger attack surface. For example, the Qbot and Mirai botnet malware are capable of infecting devices across different chipset architectures, and both malware were reportedly responsible for a number of high profile DDoS attacks in recent times. These two malware families (and many others) generally affect a broad range of consumer grade appliances, and many of these appliances (also referred to as devices) are insecure or not designed with security in mind. While researchers have focused on areas such as attacking the botnet owner's payment infrastructure, reversing the botnet and using it as a countermeasure in grey-hat counterattack, etc., there are many more questions that have not been addressed. For example, are users putting too much trust in manufacturers and failing to take adequate measures to protect their own networks? Hence, in this paper we investigate two most popular families of Internet of Things (IoT) malware, Mirai and Qbot, to understand how they spread, what attacks they are capable of, who could be responsible, and what are the motivations of the threat actors. We also propose an efficient solution to scan for Mirai- and Qbot-related vulnerabilities in IoT devices and systems. We then study what companies can do to help protect themselves from attacks. Simple steps such as correctly configuring appliances, carrying out risk assessments and creating an action plan are discussed as proactive measures that could be taken to facilitate threat reduction and incident response.

**Keywords** Malware analysis · IoT devices · Mirai · Obot

W. O'Sullivan (✉)
Irish Defence Forces, Newbridge, Ireland
e-mail: billy.osullivan@hotmail.com

K.-K. R. Choo
University of Texas at San Antonio, San Antonio, TX, USA

N.-A. Le-Khac
University College Dublin, Dublin, Ireland

© The Editor(s) (if applicable) and The Author(s), under exclusive license to Springer Nature Switzerland AG 2020
N.-A. Le-Khac and K.-K. R. Choo (eds.), *Cyber and Digital Forensic Investigations*, Studies in Big Data 74, https://doi.org/10.1007/978-3-030-47131-6_2

# 1 Introduction

Internet of Things (IoT) botnets have made big headlines in recent times, from the attacks on Sony and Microsoft's online services in December 2014 [1] to the attacks that brought down French hosting provider OVH and Dyn (a DNS provider who provide services to websites such as PayPal, Twitter, Spotify and Netflix) in late 2016 [2].

These attacks can prove devastating to those on the receiving end. On the 3rd October 2016, MalwareTech estimated that a massive 120,000 unique devices were infected with Mirai malware [3]. Mirai was also responsible for the largest denial of service attack ever seen with a 1.1 Tbps attack on hosting company OVH [3].

These botnets also have an effect on the user whose equipment was infected, using extra bandwidth and possibly causing devices to operate poorly. Indeed it seems that the only ones to benefit from these botnets are the threat actors who create and use them.

Solutions already exist to bring down botnets, but sometimes these can be messy and require interaction between law enforcement officials from different jurisdictions. The politics and bureaucracy of such interactions can often hinder the effectiveness of these types of operations.

Online tutorials exist on YouTube [4] for botnet code, which has been dumped online. With this easily locatable information, anyone motivated enough can attempt to start their own botnet. The motivation behind this paper is to find a method to prevent devices from becoming infected in the first place.

Today, a problem exists with IoT devices used in Smart Home [5] or wearable devices [6] becoming easily infected with malware. The IoT devices used in Qbot and Mirai attacks include devices that an average home user might not think need protecting. Devices such as routers, DVR's and cameras were the bots in these attacks. The chances are that the owners of the devices infected were not aware that a vulnerability even existed with their devices.

This poses a big problem for the future of undermining the threat of botnets. How can we stop these devices from becoming infected if the owners are not aware that a vulnerability exists? Vulnerabilities may always be a factor in today's world, but most vulnerabilities can be acted on once known about. Indeed, if the owners of devices were aware of the vulnerabilities residing in their devices and of a solution to fix them back in 2014, Qbot and Mirai might not even exist today. Hence, in this chapter, we are looking at (i) Examining Mirai and Qbot and find what vulnerabilities they exploit in order to spread; (ii) Design a vulnerability scanner based off of the study of Mirai and Qbot that will find any vulnerable devices on a network; (iii) Presenting the user of the scanner with measures to be taken to patch up these vulnerabilities.

## 2 Related Work

In [7], the authors examines the Mirai source code that was dumped online. They looked at the source code and operation. They noted that Mirai spreads by brute forcing IOT devices that use 'Busybox' via telnet. To this end Mirai uses a list of 60 hardcoded usernames and passwords. Once access has been gained, the host IP address and credentials used are reported to the C&C server for exploitation. Once a host becomes infected, it in turn starts scanning the internet looking for new hosts to infect.

With this in mind they take note of the ramp in telnet activity on the internet observed by the UCSD Network Telescope starting in June 2016, which is a good indication of when Mirai started spreading.

They also take note of the software used to run the C&C server. The C&C uses a CLI to manage the botnet. As previously stated, when the bots find a host that can be accessed, the IP address and credentials are reported back to the C&C.

Also of note was the fact that the Mirai source code included a program to encrypt strings for use in the bot software. Encrypted strings include the IP addresses of the C&C server. The use of this encryption and also the removal of binaries after infection is seen as an attempt to frustrate those who attempt to reverse the bots code, but since the source code is now available reversing Mirai is no longer necessary.

The author's solution to the problem was to modify Mirai and deploy it in such a way as to remove its attack ability, still lock the host down to stop future infection and to send an alert with the data sent back to the C&C to the relevant network authority. Given the legality of this method they recommend that this solution is only used by law enforcement agencies and/or government bodies.

The authors of the paper do recognize the legal and ethical issues with their method to resolve the problem. The heavy handedness of their approach, the fact that they themselves would be responsible for using other people's equipment without their consent, along with the fact that the author put no provision into locking down possible infection to their own legal jurisdiction raises legal issues not just in their own country, but could cause an international incident if this solution was adapted with a unilateral approach.

While this was only tested by the author in a private test environment, the results were positive and there is no technical reason that this could not succeed on a large scale. The main conclusion that the nt has no incentive for ISP's, manufacturers or owners of IoT devices to properly secure these devices. The author sees the way forward as government intervention, a supporting legal framework and using their approach with the Mirai botnet to induce action from ISP's, manufacturers and home users.

In [8], authors examined how DDoS for hire services work, specifically those that were offered by Lizard Stresser, Asylum Stresser and VDO. Their research demonstrated that these three services alone had over 6000 paid subscribers and that they have launched over 600,000 attacks.

Perhaps more interestingly they investigate the payment method used when paying for these services. At the time their paper was written, they discovered that there were two main methods of paying for these services—via PayPal or via Bitcoin. They found that Lizard Squad, which only offered Bitcoin as a payment method, had a 2% sign up to subscription conversion rate. They compared this to Asylum Stresser which had 15% and VDO which had 23%, both of which accepted PayPal and Bitcoin as payment methods.

With this information they set up a payment intervention with the help of the FBI and PayPal. They found that responsive payment service providers such as PayPal aided greatly in limiting the ability of the attackers and increasing the risk to attackers in accepting payment through these methods.

While the authors' solution seems good initially, the final result will be that digital currencies will be used increasingly in the future as a method of avoiding this method of take down, and to better hide the identities of the threat actors.

Ryan's research [9] had the aim of studying the characteristics of compromised devices belonging to botnets. Several passive detection techniques were used to this aim including honeypots, VirusTotal and Shodan.

In Ryan's work he focused on the routers vulnerable to CVE-2014-9222 which is a vulnerability by which threat actors could gain privileges via a crafted cookie which triggers memory corruption, known as the "Misfortune Cookie" [10]. It was discovered over 200 brands of routers could have been affected.

Using Shodan, it was possible to show the extent of the issue surrounding this vulnerability. What this paper demonstrated was how widespread vulnerabilities can be and the power of Shodan when it comes to finding them.

Another research on malware analysis by using classification approaches discussed in [11]. These approaches can be considered when enough samples of malware on IoT devices are collected.

## 3 Analysing of IoT Malware

The source code for both Qbot and Mirai have been released online. This removes the need to try and reverse engineer the malware from pre-compiled samples, and instead allows the higher level language of each to be examined to see how the code works. As such the model chosen to examine these malware samples will take the following form: (i) Code Analysis; (ii) Internet Search; (iii) Results.

Following the analysis of the malware samples separately, both samples have been compared. The information retrieved in this section will then be used in designing a vulnerability scanner for use in home networks.

### 3.1 Mirai Malware Analysis

The Mirai code was obtained from—https://github.com/jgamblin/Mirai-Source-Code [12].

### 3.1.1   Code Analysis of Mirai Bot Component

The relevant Botnet Files examined are listed as follows: attack.c, attack.h, attack_app.c, attack_gre.c, attack_tcp.c, attack_upd.c, checksum.h, includes.h, killer.c, killer.h, main.c, protocol.h, rand.c, rand.h, resolv.c, resolv.h, scanner.c, scanner.h, table.c, table.h, util.c, util.h.

The structure of the files before compilation gives a clear indication that this application is meant to be used in some form of network attack. For instance the file names: attack_udp.c, attack_tcp.c, attack_gre.c. Also when the strings in the code are examined for these files references to well-known denial of service attack types can be seen such as: attack_udp_dns, attack_udp_plain, attack_tcp_syn, attack_tcp_ack, attack_tcp_stomp, attack_gre_ip, attack_gre_eth.

Also in the file labelled scanner.c is a list of 60 unique username and password combinations. That in combination with other strings in the file such as those below seems to indicate that the application attempts to spread by brute forcing its way onto systems. 'finished telnet negotiation, received username prompt, received password prompt, received shell prompt, retrying with different auth combo!, Found verified working telnet'.

There is nothing in the code to suggest persistence of this bot on the victims device rebooting.

The bot seems to spread by brute forcing its way onto systems of various architectures such as x86, ARM, ARM7, MIPS, etc. Each sample is an ELF 32 bit executable. Its purposes seem to be three fold:

1. To spread itself onto various systems, but not onto a specific list of IP ranges to avoid gaining attention. The list of excluded ranges includes—DOD of the United States, US Postal Service, HP, General Electric Company and Internal IP ranges.
2. To shut down vulnerabilities that other bots could use to gain access to the system and also hunt for then remove qbot and zollard malware.
3. To perform denial of service attacks in the form of TCP based attacks, UDP based attacks, GRE based attacks or HTTP based attacks. Also statistics of the attack seem to be stored such as attack duration and targets.

### 3.1.2   Internet Search of Mirai

In [13], the author concludes that the Mirai malware's main target is ARM chips, although it can target several different architectures also. They also see Mirai as the next iteration of the Qbot malware (also known by the following names: GayFgt, LizKebab, Torlus, Bash0day, Bashdoor, Bashlite).

Another research of Mirai [3] concludes that attacks such as those seen by Mirai will become more common due to the numbers of IoT devices and also sees a large threat, if a way is found to detect vulnerable devices hidden behind NAT. They say that the time has come for manufacturers to stop shipping devices with global default passwords and switch to randomly generated passwords displayed on the bottom of devices.

In [14], the authors concluded that the biggest issue with devices being vulnerable to Mirai is that IoT devices tend to have weak passwords by default. They also state that the best way to keep your devices safe is to constantly monitor your network for unsecured devices.

### 3.1.3   Result of Mirai Analysis

**Who is responsible?**

While no one knows for certain, it seems likely from the code that this was made by a script kiddie. The fact that it has a prank in it, as well as the ability to target game servers does not suggest that this was a nation state threat actor. It also would seem that this is not made by organized crime as when it was used, it was quite noisy and attracted a lot of attention.

It also does not seem to be the work of Hacktivists. While a hacking group calling themselves New World Hackers did try to take responsibility for some attacks, the general consensus is that it is unlikely them. That, coupled with the lack of messages or demands makes it seem like a hacktivist group is unlikely responsible.

This seems much more likely to be the work of a talented but inexperienced individual. The reason that they probably fit into this bracket is for the following reasons:

1. The list of IP ranges included in their code to be off limits. Why have this list at all? The attacks are so noisy that excluding these IP ranges would not help hide the source. Perhaps the list is there because they don't want to harm US government and large American corporation's devices to be viewed as a patriot? But this would then mean the author is probably American and not some other nationality that would not care about these ranges.
2. The rick-roll in the code seems juvenile and unprofessional.
3. The attacks seemed spur of the moment and more to bring prestige than for any political reason.
4. The attacker did not seem prepared for the amount of attention these large attacks would bring, meaning they had to dump the code prematurely.
5. Once dumped the code would not be very useful after a time without serious revision as its predatory nature locks down the device on infection. This is because too many people using the code would mean much smaller botnets that would not be capable of the attacks seen.

**What the does the malware do?**

This botnet purpose is threefold:

1. Scan the internet for other vulnerable hosts and report their IP addresses back to a central server.

2. Carry out attacks on targets as assigned by a central server. Infected devices will listen for instructions on port 48101.
3. Lock down the infected device to prevent further infection and also remove any other infections found on the device.

Its attack capabilities are comprised of the following denial of service methods:

1. UDP Flood which works by sending a large number of UDP packets to random ports on a remote host. As a result, the distant host will: Check for the application listening at that port; See that no application listens at that port; Reply with an ICMP Destination Unreachable packet [15].
2. UDP Flood optimized for speed.
3. UDP attack on valve gaming servers is an UDP (amplification) attack designed to use all available resources on a server. The attack is designed to send TSource Engine Query requests to a gaming server, so many requests that server cannot process all of them and creates a denial of the gaming service [16].
4. UDP DNS based water torture attack caused by sending spurious searches to a victims DNS servers, for example: DNS water torture is an attack by which a random string is appended to a victims domain, such as wefcsd.www.google.com, this query is then sent to the service providers DNS, who will attempt to contact the authoritative name server to find the answer. If no answer is found there it will contact the next name server and so on. When a fail response is sent back to the host, a new query is sent with a different string appended to the start of the URL [17].
5. TCP SYN attack where A SYN flood is a form of denial-of-service attack in which an attacker sends a succession of SYN [18].
6. TCP ACK attack where the victim server attacked by an ACK flood receives fake ACK packets that do not belong to any of the sessions on the server's list of transmissions. The server under attack then wastes all its system resources trying to define where the fabricated packets belong. This results in productivity loss and partial server unavailability [19].
7. TCP STOMP attack where the bot opens a full TCP connection and then continues flooding with ACK packets that have legitimate sequence numbers, in order to hold the connection open [20].
8. GRE based attack where GRE (Generic Routing Encapsulation) is a tunneling protocol that can encapsulate a wide variety of network-layer protocols inside virtual point-to-point links over an IP network. Ironically, GRE tunnels are often used by DDoS scrubbing providers as part of the mitigation architecture to return clean traffic directly to the protected target [20].
9. HTTP layer 7 attack where a flood GET/POST requests are sent [21].
10. CFNULL layer 7 attack where a large POST is sent to a server carrying a payload of junk data. This attack is designed to consume a web servers resources [21].

**What does it require to function?**

In order to successfully infect a host, the host must meet certain criteria as follows:

1. The host must have a public IP address (and not be in the excluded IP ranges). This means that devices with a private IP address are safe from this variant.
2. Telnet port 23 must be accessible.
3. The device must have a chipset in one of the following architectures—arm, arm7, mips, ppc, sh4, sparc, x86, mpsl, m68k.
4. The device must have one of the 60 sets of login credentials mentioned in the code review.
5. The device must be running Linux Busy Box instruction set.

**What is vulnerable to infection?**
Any system running busy box with telnet exposed to the internet, default credentials and with a public IP address not in the exclusion list should be seen as extremely vulnerable.

**What is the infection vector?**
Infection occurs when a bot finds a vulnerable device; it sends the details back to a central server which in-turn logs into the device and uses a wget command to pull down the malware from a server hosting the malware.

**Where the victims/attacker is located?**
Infected hosts are located all around the globe (cf. MalwareTech map, June 8th 2017).

**When most of the attacks occurred?**
The largest attacks occurred during September and October of 2016 before the source code was dumped.

**Why the attacks occurred?**
At present the only reason appears to be vandalism.

**How it can be stopped/controlled?**
To remove the malware from a device the device just needs to be reset. The best way to stop an attack like this from happening again would be a multi-tier approach:

1. More regulation from government bodies requiring that both ISP's and manufacturers of internet capable devices to force security standards on them. For instance, ISP's could be forced to scan their IP ranges for vulnerable devices. On finding one they could notify their customer about the find and potentially kill any traffic to the devices after a certain amount of time. Manufacturers could be forced to adhere to standards such as not having passwords hard coded into the firmware of their devices, and to move away from outdated unsecure protocols like Telnet.
2. Education for the home user who might not understand that the implications of leaving unsecured devices open to the internet is a bad idea. This could come in the form of a warning from the manufacturer on the packaging of any device purchased, ISP's warning their customers that an unsecured device has been detected or even a public service announcement in the form of a television commercial/pamphlet outlining best security practices for internet connected devices in the home.

## 3.2  Qbot Malware Analysis

The Qbot code was obtained from—https://github.com/gh0std4ncer/lizkebab/blob/master/server.c [22].

### 3.2.1  Code Analysis of Qbot Bot Component

This malware has code built in for numerous attacks including Hold Flooding, Junk Flooding, TCP Flooding and UDP Flooding. There is also a whole section of code dedicated to sending emails for an attack to spam/DDoS a victim's email, but this section is all commented out. Perhaps the bot author could not get it working? Clearly this is designed to carry out denial of service attacks.

It also has a section of usernames and passwords. This is different than Mirai in that it appears any combination of usernames and passwords can be used together as they are defined in separate lists instead of pairs defined together as with Mirai. These appear to be used to perform a telnet scan for hosts with default telnet credentials exposed to the internet.

The author of the bot takes the effort at the start of the code to write a comment that this software is an IRC bot and also says that the bot will run on nearly any system. Indeed when the code is checked we can see a section dedicated to sending and receiving data over IRC as a method of controlling the bot.

By default, the code checks for the following chipset architectures on the victim device: Arm, Mips, Mpsl, PPC and x86.

It also has several attack methods available. There is nothing in the code to suggest persistence of this bot on the victims device rebooting. The bot skips the following IP ranges while scanning for hosts: 0.x.x.x, 10.x.x.x, 100.64.x.x to 100.127.x.x, 127.x.x.x, 169.254.x.x, 172.16.x.x to 172.31.x.x, 192.0.2.x, 192.88.99.x, 192.168.x.x, 198.18.x.x to 198.19.x.x, 198.51.100.x, 203.0.113.x, 224.x.x.x and greater.

This bots' purposes seem to be twofold:

1. To spread itself onto vulnerable systems via telnet using a list of default credentials.
2. To carry out denial of service attacks in the form of Hold, Junk, TCP and UDP floods. It should be noted that the code suggests that the duration of these attacks is customizable and also the ability to stop an attack also exists.

One other interesting aspect of this code is that the authors appear to name themselves in a comment at the start of the code, "Chippy1337 and @packetprophet present:" and they even go as far as to include a license agreement where users can send Bitcoin if they like the software—LICENSE AGREEMENT: If you lulz'd, you must sent BTC to 121cywjXYCUSL2qN7MnQAzSHNsWotUrea7.

### 3.2.2  Internet Search of Qbot

On searching Qbot initially it became apparent of the many different aliases that the malware goes by, including: Bashlite, Gafgyt, Lizkebab, Qbot, Torlus and Lizard Stresser [23].

A detailed analysis of various versions of Qbot can be found in [24]. In their analysis they suggest that the reason the malware has so many names is to try and throw security researchers off their trail. They also suggest that although this malware can target a variety of devices connected to the internet such as IoT devices and routers, its main target is unsecured routers.

Journalist Brian Krebs notes that home users can help fight this malware threat by logging on their internet connected devices such as routers and change the default passwords to something more secure [25]. He also makes the suggestion of changing the default DNS settings of home routers to a custom DNS server that can be set up for free at OpenDNS.com as this would help filter out known malicious sites and phishing scams.

In addition, on researching the Qbot botnet the following link was found containing set up guides for various versions of Qbot at: https://pastebin.com/u/Jihadi4Prez. The same individual has their own YouTube channel where they have various video tutorials in place to set up a few different botnets including Mirai and Qbot at: https://www.youtube.com/channel/UCXM4xUOmJk3Px2qiG9x1ygg.

### 3.2.3  Results of Qbot Analysis

**Who is responsible?**
The group calling themselves Lizard Squad are responsible for the initial strain of Qbot. The group were trying to set up a booter service online. Several individuals were arrested and questioned in connection with the initial attacks, including Vinnie Omari and Julius Kivimäki.

**What it does?**
The botnets purpose is twofold:

1. To spread itself onto vulnerable systems via telnet using a list of default credentials.
2. To carry out denial of service attacks in the form of Hold, Junk, TCP and UDP floods. It should be noted that the code suggests that the duration of these attacks is customizable and also the ability to stop an attack also exists.

Its attack capabilities are comprised of the following denial of service methods:

1. Hold Flood—This attack holds open TCP connections in an effort to exhaust resources.
2. Junk Flood—This attack sends a random string of junk characters to a TCP port.

3. TCP flood—Sends a flood of TCP segments with the flag specified when sending the attack command.
4. UDP flood—Sends a random string of junk characters to a UDP port.

## What does it require to function?

In order to successfully infect a host, the host must meet certain criteria as follows:

1. The host must have a public IP address (and not be in the excluded IP ranges). This means that devices with a private IP address are safe from this variant.
2. Telnet port 23 must be accessible.
3. The device must have a chipset in one of the following architectures—Arm, Mips, Mpsl, PPC and x86, although the code could be easily compiled for other architectures.
4. The device must have one of the sets of login credentials mentioned in the code review.

## What is vulnerable to infection?

Any system running busy box with telnet exposed to the internet, default credentials and with a public IP address not in the exclusion list should be seen as extremely vulnerable.

## What is the infection vector?

Infection occurs when a bot finds a vulnerable device. It sends the details back to a central server which in-turn logs into the device and uses a 'wget' command to pull down the malware from a server hosting the malware.

## Where the victims/attacker is located?

Infected hosts were global (Source: Level 3 Threat Research Labs late 2016).

## When most of the attacks occurred?

The largest attacks related to Qbot were 25th December 2014 and again in September 2016, in conjunction with Mirai. Other attacks continued in between the two dates (and probably still happen).

## Why the attacks occurred?

Early attacks were a part of the launch of a booter service launched by Lizard Squad. The attack in September 2016 appears to be for both prestige and vandalism.

## How it can be stopped/controlled?

To remove the malware from a device, the device just needs to be reset.

The best way to stop an attack like this from happening again would be a multitier approach:

- More regulation from government bodies requiring that both ISP's and manufacturers of internet capable devices to force security standards on them. For instance, ISP's could be forced to scan their IP ranges for vulnerable devices. On finding one they could notify their customer about the find and potentially kill any traffic to the devices after a certain amount of time. Manufacturers could be forced to adhere to standards such as not having passwords hard coded into the firmware of their devices and moving away from outdated unsecure protocols like Telnet.
- Education for the home user who might not understand that the implications of leaving unsecured devices open to the internet is a bad idea. This could come in the form of a warning from the manufacturer on the packaging of any device purchased, ISP's warning their customers that an unsecured device has been detected or even a public service announcement in the form of a television commercial/pamphlet outlining best security practices for internet connected devices in the home.

## 3.3   Comparison of Mirai and Qbot

In this section, a Mirai botnet will be compared to a Qbot botnet. The bot binaries themselves will be discussed, as will the command and control methods.

### 3.3.1   Attack Methods

As seen earlier in the code review, Mirai has more attack options than Qbot available. Mirai's attack options seem to have built upon Qbot's and expanded to include the following attacks:

1. UDP attack to target Valve gaming servers.
2. UDP DNS based attack.
3. GRE based attacks.
4. HTTP layer 7 attack where a flood GET/POST requests are sent.

### 3.3.2   Method of Propagation

Both pieces of malware seem to propagate in similar fashions:

1. An infected bot scans the internet for random IP addresses with telnet exposed.
2. Once a device is found the infected bot tries to brute force access with a list of default credentials.
3. If access is gained Mirai and Qbot both act a little differently:

   - For Mirai: it reports back to the command and control with the necessary information—IP address and credentials required for access. From here a loader server is used to deliver a script which downloads the malware from a malicious domain.

- For Qbot the bot forces a shell script to run which downloads the malware.

4. For Qbot this is it, the new bot checks into the C&C and continues the process until an order is received. For Mirai, the bot searches for any competing malware on the system and shuts it down once found. It also locks down SSH and telnet so other malware can't do the same to it. Once the device is locked down, it repeats the cycle until a command is received.

### 3.3.3  Method of Command and Control and Offering of Services

Qbot has a relatively simple C&C structure. It is a program uses IRC to send and receive data to and from bots under its control. The threat actors, in this case Lizard Squad, did the following to set up their botnet and get customers.

**Botnet set up**—(Fig. 1)
**Botnet growth**—(Fig. 2)
**Advertise services**—(Fig. 3)
**Carry out attacks for profit**—(Fig. 4)

The data for the numbers of customers, subscribers, attacks, targets and the money earned have been retrieved from the research paper "Stress Testing the Booters: Understanding and Undermining the Business of DDoS Services" [8].

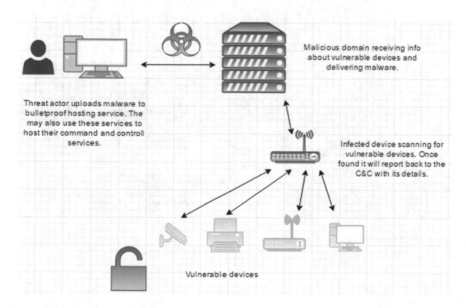

**Fig. 1** Qbot set up

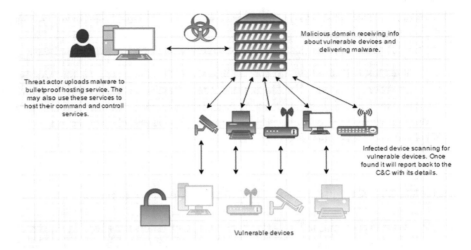

**Fig. 2** Qbot growth

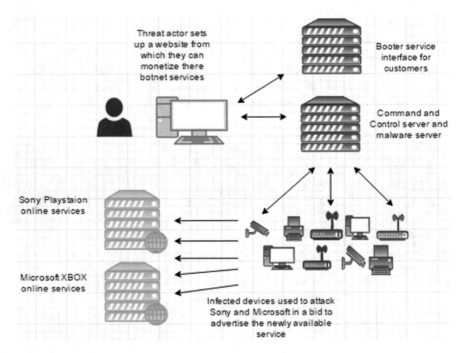

**Fig. 3** Qbot advertisement

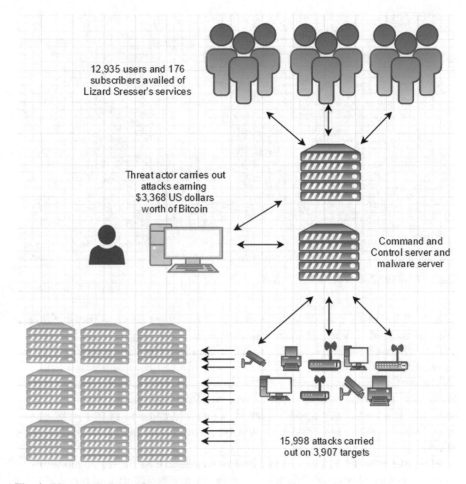

12,935 users and 176
subscribers availed of
Lizard Sresser's services

Threat actor carries out
attacks earning
$3,368 US dollars
worth of Bitcoin

Command and
Control server and
malware server

15,998 attacks carried
out on 3,907 targets

**Fig. 4** Qbot attacks for profit

Mirai on the other hand has its C&C written in 'GO'. It makes use of a C&C and a loader. These share data through a database such as MySQL. The C&C details are stored in the bot as a domain rather than just an IP address. To understand how these parts fit together, a diagram was drawn to better demonstrate (Fig. 5).

## 3.4   Conclusion of Mirai and Qbot Research

Mirai appears to build upon the capabilities and techniques used in Qbot. It has more attack options available than Qbot and due to its command and control structure is more resilient to being shut down (if the C&C domain is taken down, the threat actor could just change the IP address of the domain to bring it back up).

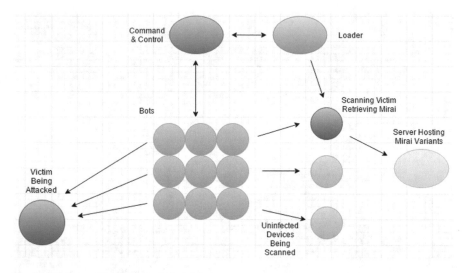

**Fig. 5** Mirai command and control structure

In addition Mirai seemed to be able to pull off larger attacks. This could be a result of its predatory nature by which it removes competing malware and secures devices against further attacks. This would have resulted in large botnets before the source code was released.

The chances are that a similar attack with Mirai as it exists would not be possible due to both its predatory nature and the release of the source code. When the source code was released, many unskilled threat actors attempted to take advantage and use it, unaware that they were diluting the maximum size of botnets achievable with Mirai.

Most likely the next iterations of Mirai will include binaries for infecting 64 bit Linux systems and possible Windows systems in an effort to try and infect a wider range of devices. It also would not be surprising if variants were made to perform digital currency mining, such as bitcoin mining, as due to the nature of these activities they could easily go unnoticed by victims.

What both Mirai and Qbot has shown is that security concerns need to be taken more seriously for every device connected to the internet. The days of antivirus being enough for residential users to secure their home networks have long since passed. It is no longer just about protecting laptops and servers from malware; in an age where even televisions and refrigerators can connect to the internet, another solution is required.

# 4 Adopted Approach for IoT Malware Analysis

As was seen in the literary survey, a number of solutions exist—network scanners can perform deep packet inspection on network traffic and alert when it finds something suspicious, software such as NMap and ZenMap can scan network ranges for open ports, and there are even more comprehensive scanners such as Nessus that can find a range of vulnerabilities.

The problem with those is that the only ones designed to pro-actively hunt for vulnerabilities are designed for security professionals. This research paper aims to make a vulnerability scanner that is simple to use and aimed at home users rather than security professionals. Also when a vulnerability is found it should guide the user to fix it as easily as possible.

The scanner has been designed with information taken from the operation of two different IOT botnet samples, Qbot and Mirai. Specifically the following information was used in the design:

- The method used to scan for devices which in both samples is telnet over port 23.
- The credentials used to attempt brute forcing access to the device.

## 4.1 Process

With this in mind the following design brief was adopted to design the solution (Figs. 6 and 7):

1. The core solution should have two main purposes:

    (a) Carry out scans on a user endpoint (Windows 10 OS) on demand via a GUI interface.
    (b) Carry out continuous vulnerability scanning on a locally accessible web server (Raspberry Pi) and display results in a secure fashion.

2. Both solutions should be simple to use.
3. Where possible python should be used to program the solution and external library requirements should be kept to a minimum.
4. The scanner would scan a network range for devices responding to a PING.
5. On finding devices it would then scan for open ports, specifically port ranges 1–6000 (most important is that all well-known ports from 1 to 1023 are scanned, 6000 ports were picked in order to speed up the scans).
6. If port 23 was found to be open, an attempt to access the device should be made using the credentials found while investigating the malware samples.
7. If a successful connection is made the user should be alerted to the fact and guided to change the login password.
8. An external scan should also be carried out by checking Shodan for any information on the external IP address.

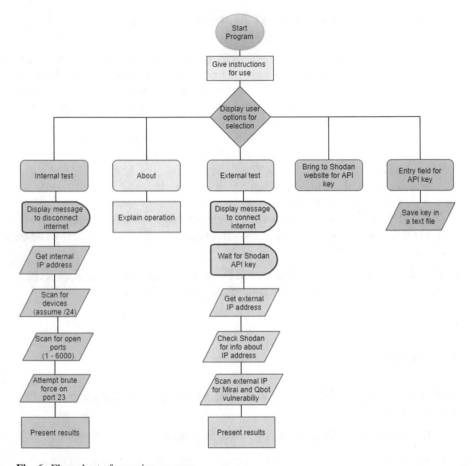

**Fig. 6** Flow chart of scanning process

9. The external scan should also scan the routers external IP for the Mirai and Qbot vulnerability.
10. The web server version of the solution should offer downloads of the locally run scan software, a telnet client and versions of putty.
11. The web server should offer guidance and ideas to protect your home network such as using OpenDNS, displaying examples of commercially available products and switching to using SSH over Telnet where possible.
12. The web server should also instruct the user how to change login credentials with the telnet client available for download.

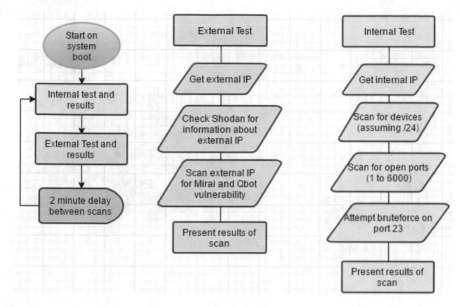

**Fig. 7** Flow chart of web browser scanner

## 4.2  Securing the Webserver

In order to secure the Raspberry Pi webserver, both hardware and software were used. This solution makes use of an LCD and a switch connected to the Raspberry Pi (Fig. 8).

The LCD is used by default to display the date and time in UTC format, as well as the current IP address of the server. The button is used to control access to the webserver. When the button is pressed, the LCD displays a new message:

'Webservice ON For: 300 seconds'.

Also the script activates the apache service with the following command:

os.system("sudo/home/pi/apachestart.sh")

Once the 300 s are up the script disables the apache service again and continues scanning for a button press.

## 5  Experiments and Discussion

Both the vulnerability scanner and the continuous vulnerability scanner successfully detected a vulnerable device amongst other devices on a home network. They also display the results so the user can correct the issue. Both scanners also successfully scan the external IP address for vulnerabilities related to Mirai and Qbot, as well as

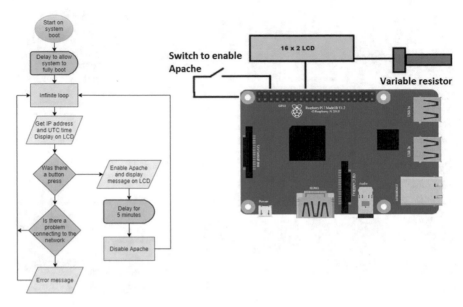

**Fig. 8** Solution flow chart and Raspberry Pi block diagram

checking Shodan for information. Here is a sample of the output of scanning for both the internal and external scans (Fig. 9).

The approach taken in this chapter was to create two separate vulnerability scanning systems:

```
*****************
* Internal Scan *
*****************

Now scanning your internal network to check for
vulnerabilities.
This scan checks your local network for open ports
and checks any device open for telnet connections
for Mirai vulnerability.


Proceeding with scan ...

*******************
* IP & Ports Scan *
*******************

IP Address: 192.168.0.16
Open Ports: [80, 139, 443, 445, 631, 3910, 3911]

IP Address: 192.168.0.1
Open Ports: [53, 80, 1990, 5431, 5916]

IP Address: 192.168.0.37
Open Ports: [22, 23]

IP Address: 192.168.0.13
Open Ports: [135, 139, 443, 445, 554, 902, 912, 1536, 1537]

***************************
* Mirai & QBOT Telnet Scan *
***************************

Scanning 192.168.0.37 for Mirai and QBOT vulnerability
Host: 192.168.0.37 compromised. The user name is: root The password is: toor

Scanning Finnished At 22:14:02...
Scanning Duration: 0:14:22.163000...

Testing finished, your modem can be plugged back into the internet.
```

```
*****************
* External Scan *
*****************

Scanning your external network to check for
vulnerabilities.
This scan checks Shodan's online database
for any information already available about your
network.
It also scans your external interface to check
if it is vulnerable to known Mirai intrusion
methods.


Proceeding with scan ...

Started At: 22:18:09

***************
* Shodan Scan *
***************

Your external IP is: 78.19.166.153
Checking Shodan for your external address...
No information available for your IP address on Shodan

*************************************
* Scanning External IP For Known   *
*    Mirai & QBOT Vulnerabilities   *
*************************************

Scanning your routers external IP78.19.166.153
for Mirai & QBOT vulnerability

Device not vulnerable to Mirai or QBOT attack

Scanning Finnished At 22:19:49...
Scanning Duration: 0:01:40.486000...
```

**Fig. 9** Results of internal and external scan

1. The first one a program to be run by a user when needed on their endpoint.
2. The second one runs continuously on a webserver and serves the results on a secure locally accessible webpage.

These two solutions were designed to be easy to use, with simple advice given to fix the problem. Also a simple python telnet client was made to allow the user to make necessary changes.

These solutions will expose Mirai and Qbot vulnerabilities in the home network and also display any open ports that could potentially be a risk. When tested on a home network with a vulnerable device both solutions successfully detected the vulnerable device and presented the IP address and credentials used to gain access.

This chapter looked at the current state of botnet research and found that many solutions exist, ranging from the methods of securing networks, fighting back by targeting the money, using honeypots to gather intelligence and even using botnets in a morally questionable way to gather information about unsecured devices to alert the owners.

One area that seemed lacking was the use of vulnerability scans in the home network. This fact, along with recent high profile botnets such as Mirai and Qbot presented an interesting research opportunity:

- To study Mirai and Qbot, find the motivations behind the attacks, how the malware spreads and what devices they commonly infect.
- To use the information gained in the study of Qbot and Mirai to come up with a design brief for a home vulnerability scanning solution to check for Qbot and Mirai vulnerabilities.

It was found that both Qbot and Mirai spread via port 23 using the telnet protocol. They both had a list of commonly used default credentials which were used to try and bruteforce access to devices which had port 23 exposed. They were also both ELF type malware, written in C and designed to be easily compiled for various architectures such as Arm, Mips and x86 chipsets. Qbot's main target was routers and Mirai's was CCTV cameras.

Due to the similarities, it seems that Mirai is the next step in the progression of Qbot. Mirai brings added functionality that Qbot lacked such as the ability to remove competing malware and the ability to lock down devices. It also has a larger variety of attacks available then Qbot had.

With regards to the attacks each performed, there does appear to be differences in the threat actors involved. Qbot was initially used by Lizard Squad in order to generate revenue in exchange for carrying out attacks via their booter service. Mirai was used to attack websites such as Krebs on Security and Dyn DNS, which did not seem to have any monetary reward.

Two systems were designed with the information gathered through the analysis of Qbot and Mirai:

- A vulnerability scanner that can be run on a Windows 10 system. It requires Python 2.7 and also the Shodan API library to be installed. It has a graphical user interface to allow easy interaction and uses multi-threading in order to speed up scanning. It does make the assumption of a /24 network being used on the internal network. It can run a scan on the internal network and external IP address of the Internet connection to check for Mirai and Qbot vulnerabilities.

- The webserver is a continuous vulnerability scanner which carries out the same scans as vulnerability scanner but on an ongoing basis. A webpage is used to display results of the scan. The webpage can only be accessed when a button is pressed on the webserver. This button enables the apache service for 5 min, after which access is stopped by disabling apache. An $16 \times 2$ LCD is used to display information such as the IP address used to access the webpage, and the time remaining when the apache service is enabled in seconds.

When tested, both solutions successfully detected vulnerably devices on the network and alerted the credentials used to gain access. A simple telnet client was also made that allows one to access telnet enabled devices once they have the correct credentials.

With regards to companies being prepared it was found that there are many steps they can take to being ready for an incident. Companies should make sure their existing infrastructure is configured correctly to help minimise risk. A risk assessment should be carried out to ensure that appropriate investment is being made in the infrastructures security. Documentation should also be created with a plan for responding to attacks.

## 6   Conclusion and Future Work

Telnet botnets such as Mirai and Qbot take advantage of the low hanging fruit of badly secured and exposed telnet devices. They use these IoT devices, which seem to have been forgotten when it comes to securing home networks, to great effect in cyber-attacks from the Sony and Microsoft attacks to the attack that brought down Dyn.

As the numbers of devices connecting to the internet grow, it is reasonable to assume that unless something is done more of these style of attacks will occur. The solution for this problem will not be easy, and will require action from multiple parties:

- Manufacturers need to take more responsibility when designing devices. Hard coded default passwords should be dropped for a password that needs to be changed on first use. Outdated protocols such as telnet should be swapped for more secure protocols such as SSH. Firmware should be easily updateable in order to fix any zero day vulnerabilities which become apparent too.
- Governments need to monitor the types of devices consumers are buying and connecting to the internet. If trends occur such as devices continuing to be sold without proper security precautions taken on the device, regulations should be forced on the manufacturers for the good of all internet users.

- ISP's need to take a more active role in spotting malicious traffic. While not many people would advocate deep packet inspections on all network traffic, simply spotting patterns of where traffic is going could be enough to thwart some attacks.
- Home users carry perhaps the biggest burden in stopping these attacks in the future. A layered approach needs to be taken when securing home networks. A network monitor should be used to spot malicious traffic coming and going from the network. OpenDNS should be used to mitigate some of the threats that exist. Also continuous vulnerability scanning should be performed in order to spot vulnerabilities before they can be exploited.
- Companies need to take charge of their networks security. They need to be sure that every device is configured correctly in order to help deal with an attack. Risk assessments should be carried out to ensure that the correct investment is being made to protect their infrastructure. Documentation should be created to make sure everyone knows what to do when an incident does occur.

The solution offered in this chapter is only part of the overall solution. It is hoped that at the very least finding vulnerabilities before they can be exploited should allow measures to be taken to correct any issues that exist on a network, thus eliminating the threat that these devices pose.

If the opportunity presents itself to carry out more work in this area, it would be interesting to extend this research on Smart Vehicle security and forensics [26]. In addition, it would be beneficial to focus on a more rounded vulnerability scanner that would scan for a wider range of vulnerabilities than just Qbot and Mirai.

# References

1. Krebs on Security: Cowards Attack Sony PlayStation, Microsoft xBox Networks. http://krebsonsecurity.com/2014/12/cowards-attack-sony-playstation-microsoft-xbox-networks/ (2014). Accessed 27 June 2019
2. HackRead: The Mirai botnet: what it is, what it has done, and how to find out if you're part of it. https://www.hackread.com/mirai-botnet-ddos-attacks-brief/ (2016). Accessed 27 June 2019
3. MalwareTech: Mapping Mirai: a botnet case study. https://www.malwaretech.com/2016/10/mapping-mirai-a-botnet-case-study.html (2016). Accessed 27 June 2019
4. YouTube:       Jihadi       x       tutorials.       https://www.youtube.com/channel/UCXM4xUOmJk3Px2qiG9x1ygg (2017). Accessed 27 June 2019
5. Goudbeek, A., Choo, K.-K.R., Le-Khac, N.-A.: A forensic investigation framework for smart home environment. In: 17th IEEE International Conference on Trust, Security and Privacy in Computing and Communications (IEEE TrustCom-18), New York, Aug 2018. https://doi.org/10.1109/TrustCom/BigDataSE.2018.00201
6. Alabdulsalam, S., Schaefer, K., Kechadi, M.-T., Le-Khac, N.-A.: Internet of things forensics: challenges and case study. In: Gilbert, P., Sujeet, S. (eds.) Advances in Digital Forensics XIV. Springer Berlin Heidelberg, New York (2018). https://doi.org/10.1007/978-3-319-99277-8_3
7. Jerkins, J.A.: Motivating a market or regulatory solution to IoT insecurity with the Mirai botnet code. In: 2017 IEEE 7th Annual Computing and Communication Workshop and Conference (CCWC), Las Vegas, 9–11 Jan 2017
8. Stress Testing the Booters: Understanding and Undermining the Business of DDoS Services. http://damonmccoy.com/papers/www2016-booter.pdf (2015). Accessed 27 June 2019

9. Botnet Detection: Honeypots and the Internet of Things. https://msmis.eller.arizona.edu/sites/msmis/files/documents/sfs_papers/ryan_chinn_sfs_masters_paper_0.pdf (2015). Accessed 27 June 2019
10. CVE: CVE-2014-9222. https://cve.mitre.org/cgi-bin/cvename.cgi?name=CVE-2014-9222 (2014). Accessed 27 June 2019
11. Linke, A., Le-Khac, N.-A.: Control flow change in assembly as a classifier in malware analysis. In: 4th IEEE International Symposium on Digital Forensics and Security, Arkansas, Apr 2016. https://doi.org/10.1109/ISDFS.2016.7473514
12. Github: Mirai Source Code. https://github.com/jgamblin/Mirai-Source-Code (2016). Accessed 27 June 2019
13. Malware Must Die: MMD-0056-2016—Linux/Mirai, how an old ELF malcode is recycled. http://blog.malwaremustdie.org/2016/08/mmd-0056-2016-linuxmirai-just.html (2016). Accessed 27 June 2019
14. Splunk: Analyzing the Mirai Botnet with Splunk. https://www.splunk.com/blog/2016/10/07/analyzing-the-mirai-botnet-with-splunk/ (2016). Accessed 27 June 2019
15. Bijalwan, A., Wazid, M., Pilli, E.S., Joshi, R.C.: Forensics of random-UDP flooding attacks. J. Netw. **10**(5), 287 (2015). https://doi.org/10.4304/jnw.10.5.287-293
16. Hot Hardware: Mirai IoT DDoS Botnet Source Code Reveals Specific Targeting of Valve Source Engine Games on Steam. https://hothardware.com/news/mirai-iot-ddos-botnet-source-code-targets-valve-source-engine#WvZOQVKi252ACL1t.99 (2016). Accessed 27 June 2019
17. Secure64: Water Torture: A Slow Drip DNS DDoS Attack. https://secure64.com/water-torture-slow-drip-dns-ddos-attack/ (2014). Accessed 27 June 2017
18. Bogdanoski, M., Shuminoski, T., Risteski, A.: Analysis of the SYN flood DoS attack. J. Comput. Netw. Inf. Secur. **8**, 1–11 (2013). https://doi.org/10.5815/ijcnis.2013.08.01
19. DDOS-GAURD, ACK & Push ACK Flood: https://ddos-guard.net/en/terminology/ack-push-ack-flood. Accessed 27 June 2019
20. Security Week: What's the Fix for IoT DDoS Attacks? http://www.securityweek.com/whats-fix-iot-ddos-attacks (2016). Accessed 27 June 2019
21. F5 Labs: Mirai: The IoT Bot That Took Down Krebs and Launched a TBPS Attack on OVH. https://f5.com/labs/articles/threat-intelligence/ddos/mirai-the-iot-bot-that-took-down-krebs-and-launched-a-tbps-attack-on-ovh-22422 (2016). Accessed 27 June 2019
22. Github: Qbot Source Code. https://github.com/gh0std4ncer/lizkebab/blob/master/server.c (2015). Accessed 27 June 2019
23. Eduard, K.: BASHLITE Botnets Ensnare 1 Million IoT Devices. Security Week, 31 Aug 2016. http://www.securityweek.com/bashlite-botnets-ensnare-1-million-iot-devices (2016). Accessed 27 June 2019
24. Malware Must Die: MMD-0052-2016—Overview of "SkidDDoS" ELF++ IRC Botnet. http://blog.malwaremustdie.org/2016/02/mmd-0052-2016-skidddos-elf-distribution.html#gayfgt (2016). Accessed 27 June 2019
25. Krebs on Security: Lizard Stresser Runs on Hacked Home Routers. https://krebsonsecurity.com/2015/01/lizard-stresser-runs-on-hacked-home-routers/#more-29431 (2015). Accessed 27 June 2019
26. Le-Khac, N.-A., Jacobs, D., Nijhoff, J., Bertens, K., Choo, K.-K.R.: Smart Vehicle Forensics: Challenges and Case Study. Future Gener. Comput. Syst. (2018). https://doi.org/10.1016/j.future.2018.05.081
27. Efa: Raspberry Pi Image: Changes Made to Original Diagram Include the Addition of a Flow Chart and a Switch with LCD Screen. https://en.wikipedia.org/wiki/File:RaspberryPi_3B.svg, https://creativecommons.org/licenses/by/3.0/ (2016). Accessed 24 April 2020

**William O'Sullivan** is currently a technical support engineer at FireEye and a reservist for the Irish Defence forces. In 2017 he received an MSc for studies completed from UCD in the Forensic Computing and Cybercrime Investigation (FCCI) program. In 2011 he received an ordinary degree in Electronic Engineering and Military Communications Systems from Institute of Technology Carlow. He is currently studying another Masters program at NUIG titled Software Engineering & Database Technologies due to conclude in 2020 on successful completion. During his service as a full time member of the Irish Defence Forces he worked as a systems administrator as well as a member of the Cyber Incidence Response Team. His research interests include study of IoT vulnerabilities, machine learning techniques for malware detection, reverse engineering of malware and techniques for analysing data/data mining.

**Kim-Kwang Raymond Choo** received the Ph.D. in Information Security in 2006 from the Queensland University of Technology, Australia. He currently holds the Cloud Technology Endowed Professorship at the University of Texas at San Antonio (UTSA). In 2015, he and his team won the Digital Forensics Research Challenge organized by the Germany's University of Erlangen-Nuremberg. He is the recipient of the 2019 IEEE Technical Committee on Scalable Computing (TCSC) Award for Excellence in Scalable Computing (Middle Career Researcher), 2018 UTSA College of Business Col. Jean Piccione and Lt. Col. Philip Piccione Endowed Research Award for Tenured Faculty, British Computer Society's 2019 Wilkes Award Runner-up, 2019 EURASIP Journal on Wireless Communications and Networking (JWCN) Best Paper Award, Korea Information Processing Society's Journal of Information Processing Systems (JIPS) Survey Paper Award (Gold) 2019, IEEE Blockchain 2019 Outstanding Paper Award, Inscrypt 2019 Best Student Paper Award, IEEE TrustCom 2018 Best Paper Award, ESORICS 2015 Best Research Paper Award, 2014 Highly Commended Award by the Australia New Zealand Policing Advisory Agency, Fulbright Scholarship in 2009, 2008 Australia Day Achievement Medallion, and British Computer Society's Wilkes Award in 2008. He is also a fellow of the Australian Computer Society, an IEEE senior member, and co-chair of IEEE Multimedia Communications Technical Committee's Digital Rights Management for Multimedia Interest Group.

**Nhien-An Le-Khac** is a lecturer at the School of Computer Science (CS), University College Dublin (UCD), Ireland. He is currently the program director of MSc program in Forensic Computing and Cybercrime Investigation (FCCI)—an international program for the law enforcement officers specializing in cybercrime investigations. He is also the co-founder of UCD-GNECB Postgraduate Certificate in fraud and e-crime investigation. Since 2008, he is a research fellow in Citibank, Ireland (Citi). He obtained his Ph.D. in Computer Science in 2006 at the Institut National Polytechnique de Grenoble (INPG), France. His research interest spans the area of Cybersecurity and Digital Forensics, Data Mining/Distributed Data Mining for Security, and Fraud and Criminal Detection. Since 2013, he has collaborated on many research projects as a principal/co-PI/funded investigator. He has published more than 150 scientific papers in peer-reviewed journal and conferences in related research fields, and his recent edited book has been listed the Best New Digital Forensics Book according to the Book Authority.

# Digital Forensic Investigation of Internet of Thing Devices: A Proposed Model and Case Studies

**Alexander Hilgenberg, Trung Q. Duong, Nhien-An Le-Khac, and Kim-Kwang Raymond Choo**

**Abstract** Internet of Things (IoT) forensics is challenging, partly due to constant and rapid developments in the hardware and supporting software, as well as the underpinning infrastructure. This necessitates the development of a model that can be used to guide digital forensic investigations of IoT devices, while allowing flexibility to incorporate potential differences in (legal) requirements between jurisdictions. In this paper, we present one such forensic model, and describe how to set up an IoT testbed/lab to train new or inexperienced forensic investigators to examine devices and potential evidential sources. Finally, we evaluate the utility of our model using two case studies.

**Keywords** IoT forensics · Forensic model · IoT lab · Digital forensics

## 1 Introduction

Internet of Things (IoT) devices are deployed in applications such as home automation, security/surveillance, weather forecasting, and Industry 4.0 (also collectively referred to as Industrial IoT—IIoT). Generally, IoT devices are poorly configured (e.g. with default password and setting) and can be taken control of by attackers, for example by identifying and exploiting vulnerabilities in such devices or their underpinning infrastructure.

Forensically sound methods for data acquisition and data extraction from IoT devices are challenging and increasing so, due to the complexity and range of IoT

A. Hilgenberg (✉)
State Police in Lower Saxony, Northeim, Germany
e-mail: alexander.hilgenberg@ucdconnect.ie

T. Q. Duong
Queen Belfast University, Belfast, UK

N.-A. Le-Khac
University College Dublin, Dublin, Ireland

K.-K. R. Choo
University of Texas at San Antonio, San Antonio, USA

© The Editor(s) (if applicable) and The Author(s), under exclusive license to Springer
Nature Switzerland AG 2020
N.-A. Le-Khac and K.-K. R. Choo (eds.), *Cyber and Digital Forensic Investigations*,
Studies in Big Data 74, https://doi.org/10.1007/978-3-030-47131-6_3

devices. For example, not all IoT devices operate on a certain standard, and there is no user manual to guide the forensic investigation of such devices. Data acquisition and extraction approaches may depend on the nature of the offence, and can vary from device to device.

There is no one digital forensic model dedicated to the investigation of IoT devices, at the time of this research. In this paper, we propose an IoT based forensic model, which builds on existing models in the literature. We then present different scenarios to show the varieties of devices and how one can use the proposed model to conduct an IoT forensic investigation. In addition, we describe and explain in details how one can build an IoT laboratory to facilitate the research of future IoT devices, for example as part of the forensic lab's effort to keep abreast of the latest trend. We also evaluate our proposed model using two case studies.

The rest of this chapter is organized as follows: In the next section, we will briefly describe the extant literature. Section 3 is dealing with the important challenges of IoT forensics. We describe our approach in Sect. 4. We evaluate the proposed model in Sect. 5 with case studies. Finally, we conclude and discuss on future work in Sect. 7.

## 2   Related Work

There have been a number of digital forensic (investigation) models proposed in the literature. Digital forensic models are generally designed to facilitate or guide the investigation of digital devices, with the aim of acquiring and analyzing the digital evidence. Some earlier forensic models such as DFIM (digital forensic investigation model) focus on the physically accessing the digital devices of interest in order to conduct data acquisition [1]. However, such models are not suitable in environment where data may not be stored or available locally, for example data being stored or disseminated from IoT devices to the cloud or some fog or edge devices. The hybrid model in [2], on the other hand, was developed to facilitate one in acquiring access to evidence, either remotely or physically.

In [3], Oriwoh et al. proposed a model that takes a zone-based approach in the investigation of IoT devices and networks. Zone 1 focuses on the internal network of devices, which could be home devices and applications. These connections do not leave the network and are only found in the local area (e.g. Bluetooth and Wi-Fi). All hardware, software and networks that relate to a crime scene are catalogued and a decision is made about what is relevant to the case and what may hold evidence that will be useful to the case. Zone 2 includes all devices and software that are at the border of the network and that provide a communication medium between the internal and external networks. It works as a gateway for the internal IoT devices to outgoing services like cloud or remote access. Zone 3 covers all hardware and software that is outside of the network in question. This could be evidence from all cloud, social network, ISP (internet service provider) and mobile network provider's data. The advantage of this model is that all those elements can be undertaken in parallel or a Zone with a higher priority can be identified and investigators could prioritize their focus, say on a specific topic first. This model also reduces the complexity that could eventually be encountered in IoT environments, and ensures that investigators can focus on the identified areas and objects in preparation for investigations. The authors in [3] also proposed the Next-Best-Thing Triage (NBT) model. The forensic approach to such a model would have to recognize IoTware as they approach and join networks, and recognize when they leave. Identifying these devices and locating them would be a part of an ongoing investigation. Network movements need to be considered and monitored to a specific level. However, movements of these devices from one network to another can complicate a forensic investigation due to the need to obtain permission for other networks and perimeters of these disparate networks.

Perumal et al. [4], on the other hand, showed that the 1–3 Zone approach is a complex approach for IoT based forensic cases, and it does not have a clear direction on how to conduct the analysis and investigation. In response, they developed their own model, which starts from the investigation commencement to when the evidence is archived. This model is also based on machine to machine (M2M) communication, which is divided by their geo location and communication medium such as ZWave [5], Zigbee [5], LTE, Wi-Fi, Ethernet and other upcoming communication methods. Once located (the suspected medium that has communicated with the target device), the forensic investigator could proceed with triage examination. After these "IoT steps", the model returns into a common forensic procedure which would be the chain of custody, lab analysis, result, proof and defense, and archive and storage. However, modern IoT devices not only depend on M2M communications but can also send their data (or can be controlled) via cloud services. Different companies build cloud platforms in order to control hundreds of IoT devices. To exclude them in an ongoing investigation would lead to insufficient data. Now not only IoT becomes a challenge to investigate, cloud services add an additional layer of challenges to the investigation model.

Authors in [6] presented challenges in IoT forensics. Some of major challenges can be listed as the distribution of data, the life span of digital media, the requirements of cloud services, the varied format of data, device types and the lack of security mechanisms. They also pointed out the limitations of current forensic tools. However, authors did not propose any forensic model for IoT platforms.

Simplest models, like the DFIM [1] and hybrid model [2], are inadequate to deal with every aspect of IoT forensic investigations. These models either concentrate on the network aspect or do not consider parallel investigations (either crime scene or laboratory). The NBT approach can filter out unnecessary IoTware so that the investigator can only focus on central IoT control units or cloud platforms in order to minimize the workload (e.g., without examining every non-relevant IoT device). This triage examination is vital and should be one of the main stages in IoT forensic investigations. However, 1–3 Zones model could become complex dealing with a wide range of IoT devices. The upcoming model, therefore, needs to be very flexible and effective in an ongoing investigation. For example, how can an investigator tell which device is important/relevant to the investigation, and where can the investigator get most of the forensic evidence data.

## 3  Problem Statement

The digital forensic investigation in the future will likely evolve as technologies advance (e.g., in terms of how data is stored, and processed). One such example is the shift from traditional evidence sources, such as computers, to smartphones [7], cloud servers [8], unmanned ground/aerial vehicles [9], IoT devices in a smart home environment [5].

Therefore, we need to understand the various challenges underpinning IoT forensic investigations, such as the following:

- Different IoT devices may require different ways of acquiring information/artefacts of forensic interest/relevance, and these mechanics may also differ from case to case (e.g., due to the nature of the offense). It is also not realistic to expect that every forensic investigator/team has the expertise and/or resource to deal with all, or even most popular, IoT devices and systems. Hence, this may result in insufficient data handling and consequently legal complications. Hence, in this paper, we present a forensic model designed to assist the investigator to identify relevant artefacts when little or no information is given about the IoT data types and IoT devices. For example, how does one determine which data type is important, and which can be ignored or deferred for latter investigation?
- In addition, the volatility of IoT data can also be an issue due to the limitation of built-in memory on most inexpensive IoT devices. The investigators also have to determine at what point data is no longer relevant to the investigations.

## 4  Proposed IoT Forensic Model

To overcome the challenges of IoT based investigations, the goal here is to create a flexible model that is easy to follow and implement in an ongoing investigation. In other words, we need a module-based model that can be customized according to the case. Figure 1 describes our proposed IoT forensic model comprising preparation, identification, acquisition, analysis and reporting phases (see Sects. 4.1–4.5). The model is designed to work with both structured and unstructured data.

### 4.1  Planning/Preparation

In this phase, the present information can provide useful initial insights of the upcoming tasks. In the context of IoT forensics, more specific questions can be explored and examples include (i) What devices are affected? (ii) What is the nature of the offense (an advanced persistent threat (APT) case involving some technically skilled

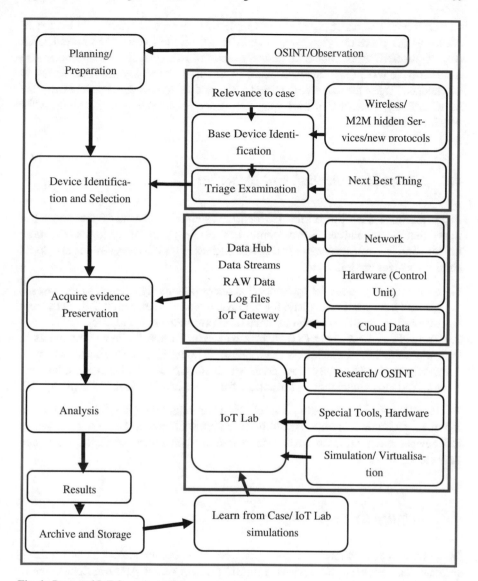

**Fig. 1** Proposed IoT forensic model

criminals, or a physical crime such as a wearable device found on the murder victim)? (iii) Is an observation necessary? If so, what are the preliminary results? (iv) Is the investigation time critical, and how it affects the process to obtain required authorization (e.g., court warrants)?

In this phase, open source intelligence (OSINT) can also provide relevant useful information, such as basic information about the suspect (e.g., is he/she technically advanced?). Using tools such as Shodan.io may also provide information about the suspect's environment (e.g., what services/hardware are associated with his/her IP addresses?), and scanning tools may reveal the presence of IoT based communication (like Zigbee/XBee) or Bluetooth connections. For example, investigators can also use hardware tools (e.g., beam antenna) to obtain a list of devices that needed to be checked. These devices are identified and listed with their unique addresses, like MAC address. Such information may help uncover devices that are used and/or possibly hidden.

## 4.2 Device Identification and Selection

In this phase, the investigators need to identify potentially relevant devices and determine their functionalities. For example, IoT gateways are likely to contain more information than an inexpensive IoT device such as a sensor. Therefore, investigators should consider the following.

1. Relevance to case: Is the located IoT devices potentially relevant to the investigation? It may not be realistic for the investigators to seize all IoT devices from the premise, and thus the importance of appropriate device filtering.
2. Triage/Next Best Thing (NBT): It is necessary to find the key components in order to gather as much data as efficiently as possible. In time-sensitive cases, the focus should be on central control units (e.g., Amazon Echo or router) that are likely to contain more information, for example using the NBT approach [3].

Overall, device identification and selection is crucial for successful acquisition of IoT data, and thus it is important to train frontline officers to be able to recognize IoT devices and get relevant assistance to acquire such devices and data, since such data can be volatile.

## 4.3 Acquire Evidence and Preservation

In contrast to conventional computing devices (e.g., computers), evidence acquisition can vary between IoT devices (e.g., 3D printers versus Amazon Echo), due to differences in storage, communication and other features. Therefore, the IoT based investigation model needs to sufficiently robust. We recommend having a data hub, where all acquired information can be stored and saved for analysis (locally or in the lab). The data hub can be expanded and upgraded, as necessary [10], and it is a collection of all data streams (compressed, raw data or just log files). The data hub can also be a collection of tools and devices that the investigator uses to collect as much information as possible. This, for example, can be done by extracting log files from a Bluetooth control unit or gathering login data from cloud based IoT devices. Smart locks, for instance, can be controlled via a smartphone application (app). Thus, the smartphone and the app are also relevant evidence sources.

Having gathered as much as possible relevant data from IoTware helps one to create a "digital story/timeline".

## 4.4 Analysis

The first question is: Can a preliminary analysis be performed at the crime scene, if required? For example, by conducting a preliminary analysis of a Bluetooth-enabled smart lock at the crime scene, the investigators can determine when the suspect returned home [11].

IoT Lab: Here, the IoT lab can determine the quality of the outcome and results from IoT devices under investigation. For example, specialized tools and hardware (e.g., using JTAG or chip off method) can be used to extract more data. Simulations may be performed in some cases, in order to understand how particular IoT devices behave or interact with other devices or systems. A test environment can collect additional data streams that can be analyzed, without the risk of data contamination. Reverse engineering of devices may also be undertaken, if expertise and resources are available. Clearly, the IoT lab has to also meet forensic standards.

## 4.5 Report

Due to the potential number of IoT devices involved in an investigation, the investigator may need to include a network layout of the environment and explain the data flow, etc.

Archive and Storage: Vital data also needs to be stored for later use (trial), and/or used for subsequent experiments for learning.

## 5 Experiment Platforms—IoT Lab

As discussed in the preceding section, having an IoT laboratory to simulating and secure acquired evidential data is beneficial. A specialized IoT lab allows one to test out new devices, create test scenarios and prepare the investigators in handling different tools.

## 5.1 IoT Hardware

Given the rapid developments of IoT devices, the equipment in the lab also needs to keep pace with the current market situation. For instance, currently one of the most widely used IoT development based boards is the Raspberry Pi and its most used OS platform Raspbian.

Linux OS and ARM CPU Architecture are also commonly used on IoT devices. There are different operating systems for the Pi hardware, such as Windows 10 IoT Core, Ubuntu Mate and Android Things. However, Raspberry Pi is not the only board that needs to be available in an IoT lab for testing purposes. A Pi is also suitable for creating IoT based investigation scenarios, train investigators and perform investigations on future threads on IoT hardware and software. Below are some examples of hardware to be included in addition to the Pi [12]:

- Arduino Uno R3 (easy to develop, a lot of hardware extensions, like radio modules)
- Intel Joule 570X (very powerful board, higher performance for video and machine vision applications)
- Thunderboard React (offers cloud-connected, Bluetooth Smart enabled solution for sensor data and prototype IoT applications. It comes with a free mobile app)
- NRF52-DK (NFC tag functionality, compatible with Arduino Uno R3, 3rd party Arduino shields usable).

We also need to understand how IoT devices communicate (e.g., using different protocols) and where data may be potentially stored or are available. Thus, the lab should also include communication devices such as Wi-Fi cards or USB dongles, GPS modules, GSM/3G, 4G/LTE and 5G based modem USB dongles, NFC module development boards, Z-Wave Me RaZberry Modul for Raspberry Pi (Z-Wave Transceiver), RaspBee ZigBee Modul for Raspberry Pi, and Freakduino 2.4 GHz Arduino Zigbee (XBee) board. At present, sensors such as camera-sensor, infrared sensor, and temperature sensor are one common source of evidential data. However, the data of those sensors and the sensor itself can be tampered with/modified. Thus, investigators should test this by checking the integrated sensor of their functionality and accuracy. For instance, the server room has a device that monitors temperature and humidity, and understanding how a sensor works on a particular device can enhance the detection of faulty data streams/inputs.

Having worked out what interfaces the IoT devices have would facilitate the extraction of data (e.g., using JTAG [13]).

## 5.2 Network Sniffing and Data Communication Acquiring

There are many ways to "sniff" network or wireless data from IoT devices, such as using a Raspberry Pi as the network router; thus, allowing vital information to be obtained. Hence, an IoT lab should consider building an IoT network to understand how information can be obtained from network communications.

Some IoT devices do not have connection ports like Ethernet or USB, as they may only use, say an Amazon Dash button. The button only communicates through Bluetooth and Wi-Fi. Only by breaking the device, the investigator can use the JTAG

[13] method to acquire data. Hence, it may be easier to intercept the Wi-Fi transmission to facilitate data acquisition, for example using a man-in-the-middle attack for services or applications that do not deploy certificate and public key pinning.

It may also be necessary to obtain an additional device that can handle multiple communication methods and capture the received signals. Such IoT sniffers can be executed on IoT development boards. At present, the Raspberry Pi is one good candidate due its ability to obtain different radio modules and run a full Linux operating system. In addition to the IoT sniffer, it is also recommended to use a network simulator/analyser.

## 5.3   IoT Software

Developments in IoT software and hardware are fast paced. For example, the size of IoT devices is getting smaller and its computational capabilities getting more powerful (e.g. 'bloated' operating systems can also run on some small IoT devices). At present, popular IoT based Linux OS that should be considered in an IoT lab include Raspbian OS [14], Windows IoT [15], Riot OS [16], Google Brillo/Android Things [17], ARM Mbed OS [18], and Nucleus RTOS [19].

Based on the authors' collective research and practitioner experience, it is also recommended that the IoT lab designer considers emulating the range of popular IoT operating systems on a normal computer, if practical. VM software can be used to emulate the ARM architecture, which allows one to test the various possibilities without the need for a dedicated hardware platform. In addition to running IoT operating software under VMs, we can also use Contiki OS [20] or other appropriate operating system (at the time of building the IoT lab) to simulate various communications technologies and device behavior. Having secure the vital data streams and raw data, the next step is to analyze the stream/data using appropriate forensic tools, such as Autopsy [21], X-Ways, EnCase [22], FTK Forensic [23], XRY [24] and Magnet AXIOM [25].

We also remark that pictures, videos and audio files are less commonly found on sensor-based devices like smart locks and smart vacuum cleaners. For devices that contain such content (pictures, videos and audio files), the IoT lab designer will need to ensure that existing commercial tools are compatible/capable of acquiring data from such devices, and whether other non-conventional tools may be required to deal with less common IoT devices.

## 6   Example Use Cases

We will now demonstrate how our proposed model (see Sect. 4) can be used to investigate a Raspberry Pi (use case 1) and other common IoT devices (use case 2).

## 6.1   Scenario 1: Raspberry PI

Raspberry Pi can be used for a broad range of applications, ranging from home automation to gaming console to building a mini server for site hosting or other services, and so on. In our example use case, we assume the suspect used the Raspberry Pi for the following purposes (Fig. 2):

- Kodi Media Player that is plugged to a TV
- OwnCloud service only accessible via some Wi-Fi interface
- Mini Router with multiple tools
- Storing pictures of planes on our system (simulating illegal materials such as child abuse materials and terrorism related materials); and
- Amazon Dash button used as an anti-forensic measure (e.g. terminate the system and send a SMS).

**Step 1: Planning/Preparation**
In this use case, we used a Wi-Fi jammer to prevent any communication between the

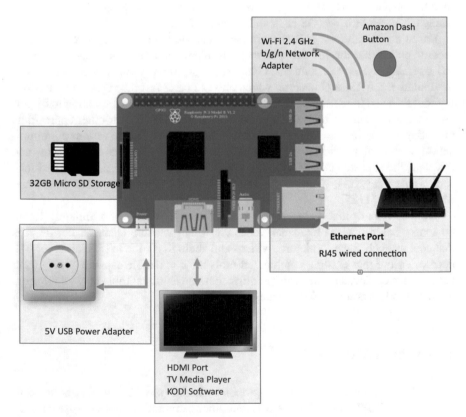

**Fig. 2** Hardware and connections used in scenario 1

Dash button and the Raspberry Pi. After the crime scene has been secured, the jammer could be deactivated and while an onsite wireless scan can reveal the estimated position of the devices (potential source of evidence). After acquiring the estimated position of the devices, the positions and other metadata can be saved as a CSV file, and inspection can then be taken to prioritize such devices (e.g. base stations versus devices connected to which base stations). The results can be imported to existing applications, such as Maltego, to identify and filter computing devices/systems of no or less interest. Depending on the network configuration, such an activity can provide a network map of potential source of evidence (see Fig. 3) and be used to inform subsequent evidential data acquisition strategy.

### Step 2: IoT Identification and Selection

Relevance to case: The Raspberry Pi, in this use case, is a hidden wireless access point and disguised as a Kodi media player. Hence, less experienced or time-poor investigators may not be able to identify/determine the relevance of this particular device. Hence, having a network map such as Fig. 3 could be useful to help determine its relevance.

Base Device Identification: A suspect may attempt to 'hide' the Raspberry Pi from a wireless scan. Recall that a Raspberry Pi has other communication ports that could be used, for example the RJ45 Ethernet port. The investigator on site could try to scan for other devices, for example using Zenmap network tool. Scanning the local area network of the suspect could still reveal the presence of a Raspberry Pi (or

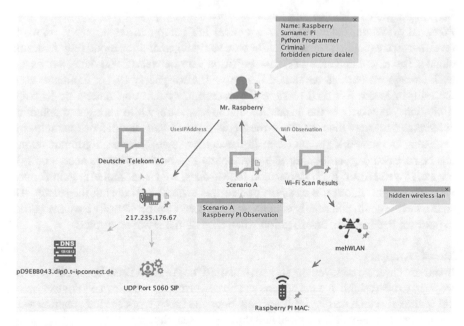

**Fig. 3** Maltego overview of scenario 1

other devices) in the network area. Also, tools such as Airodump-ng could be useful (assuming that no sophisticated anti-forensic measure is deployed).

Triage Examination: Other than standard computing equipment (e.g. notebook, personal computer and smartphones), Raspberry Pi (and other devices such as Amazon Echo) should be included for triage examination.

In this use case, as part of the IoT Device Identification and Selection,

- We identified the Raspberry PI as a key device of interest,
- The Dash button was activated and a SMS was sent to an unknown person (identified via Airodump-ng), and
- multiple computing devices were connected to the hidden wireless access point.

In other words, the importance of pre-planning cannot be understated. Having an IoT testbed can equip new/inexperienced investigators with some relevant knowledge that can help them prepare before going to the crime scene, for example understanding what equipment to bring (e.g. Wi-Fi jammer), what are the potential evidential sources (e.g. smart toys may also be a potential evidence source in domestic abuse or murder investigation, as the smart toys may capture images, videos or audio recordings relevant to the case), etc. For example in our use case, the investigator should be able to detect the estimated position of the hidden "mehWLAN" wireless access point with the right tool, such as Airodump-ng and a directional antenna (Fig. 4). With the integrated MAC-address vendor list, one can then determine the hidden wireless LAN to be the Raspberry Pi device.

**Step 3: Acquire Evidence and Preservation**
Potential evidential data sources in a typical IoT setup can reside at the network level, hardware (e.g. IoT devices and/or base stations), and/or the cloud (e.g. Amazon cloud). The extent and ease of data acquisition vary between these data sources, as each storage medium stores data differently. The Raspberry Pi, for example, uses the built-in Micro SD card to store the operating system and data. Due to page limitation, we will focus the discussion on the hardware, while noting that potential evidential data may also reside at the network level and cloud. For example, by analyzing the network stream (seen in the earlier stage), hidden communications can be revealed (e.g. certain devices connecting to a hidden wireless access point). Hence, if we perform live forensics on IoT devices, we can potentially gain further valuable insight. In our use case, the device was switched off during the search. As such a device could potentially contain an encrypted volume, therefore we may need to perform live forensic investigation when the volume is not encrypted.

**Step 4: Analysis**
In our use case, we believed the suspect had used the Amazon Dash button. However, we need to also establish what actions can be performed by this device? For example, did it actually run a script and if so, how does the script work? What function was opened and was any data manipulated during the search?

For the purpose of this use case, we focused on self-written scripts used on IoT development boards. We need to extract and understand their purpose and how they

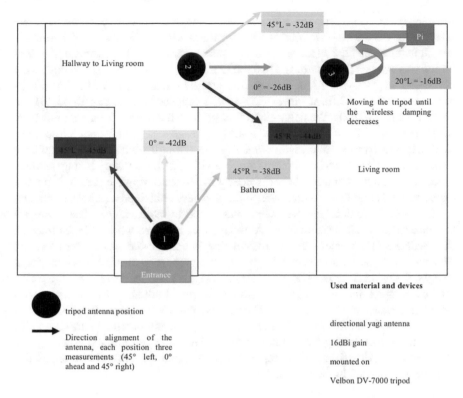

**Fig. 4** Detecting the raspberry Pi wireless access point

can be implemented. First, the internal Micro SD card was forensically acquired (read-only, write blocker) using the FTK Imager. Autopsy 4.3.0 was then used to mount the acquired image and investigate further traces and possible hidden files the suspect might use. For IoT devices, at this stage, the investigator should ask questions such as "What is the primary goal to retrieve from the device?", "How was the device used?", and "What did the suspect do in order to prevent the investigators from finding incriminating data?" Assuming that in the process of our investigation (earlier stages), we determined that the suspect used a Dash button. Hence, we need to obtain more information about the device that was found during the search and to understand how it operates. In particular, we are interested to know how the device communicates, etc. For example, after locating the MAC address of the device, can this address be used in some script to call home the Dash button? Therefore, we defined the first search patterns (or regular expressions) to search the image for evidential material on how the dash button was used. The MAC address can be obtained by simply using Airodump-ng again in scanning for wireless signals. Pushing the button, the device appears in the list. We also had to ensure that no other Wi-Fi signal appears during the scanning. This could be done by using just a very small Wi-Fi antenna that only reaches nearby Wi-Fi signals. We then translated the MAC address (e.g.: AC:63:BE:F8:B7:6F) for the vendor ID. Next, with the acquired MAC address it is now possible to detect whether the device had any contact with the Raspberry Pi. We can use the Autopsy search function and the MAC address as a search parameter. For instance, the dash button was used on the system and had the IP address 192.168.99.131. So now, what else might be processed through around the search? Using Autopsy Timeline (available since Version 4.1.0), we investigated what happened on the day 15.03.2017 at around 9 a.m. (+0100). We determined that prior to the device been shut down by the Python script, it did not only send a short message to a "friend", but it also modified data under/var/www/owncloud/data/hiddenadmin/files/Photos. The Raspberry Pi had an OwnCloud service running some activities. In the directory listing window, it became apparent that the "Photos" folder was deleted during the search. A natural question is whether the Python script could also have contained a remove command for this specific folder. However, what was stored in the folder? In fact, there are two possibilities at this point: (i) As Autopsy processed through the file, it created a $CarvedFiles folder where deleted files or files in unallocated spaces were located. This may vary and the results may be too much to process on the site; (ii) Concerning images, OwnCloud services create for each image a thumbnail to reduce the traffic load by compressing the images to a smaller format. Looking for these directories can lead to traces of stored images. The disadvantage here is that vital EXIF metadata is not available from the original source image. Other potential data could also be extracted, like the log folder under/var/log. The latter (folder) should always be considered for Unix/Linux devices, as it can potentially reveal what computer(s) was/were connected to the device.

**Step 5: Report**
Most forensic tools include a feature to create a report file with the previously bookmarked content. For this example, we used the "Report" function of Autopsy and the created report can be presented to other investigating team, for court purposes, or archived for future usage.

The last step could also be to create a manual on how to process through devices used by more experienced/technically inclined individuals who created tools/scripts to avoid files/data/devices from been found/restored.

## 6.2   Scenario 2: Smart House

In this use case, we use a smart home where there are a number of connected IoT devices (Fig. 5). By using multiple IoT devices, the goal here is to create an event based timeline that can tell a "story" of various events based on the sensors and data recordings. In other words, evidential data from multiple IoT devices can potentially facilitate the recreation of the chronology of events, and help the investigation (Fig. 6). In this case, we assumed that the smart home consists of smartphones and Bluetooth smart lock. Therefore, acquiring information such as timestamps from the smart lock is a starting point, as we would be able to identify who had used their credentials to

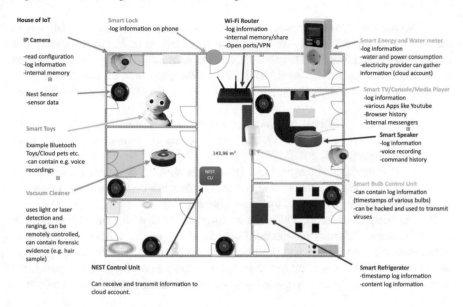

**Fig. 5** Scenario 2—smart house

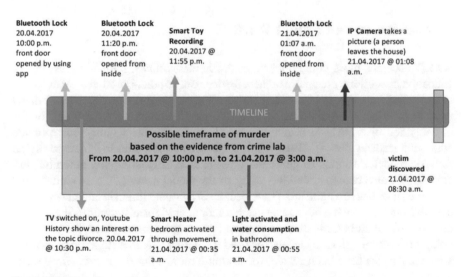

**Fig. 6** Timeline of scenario 2

enter the premises during a certain time period. Smart locks vary in sophistication and prices, and the Bluetooth smart lock we used for example starts around 70 Euros. A companion app is necessary to control the locking mechanism, and these apps generally provide a logging ability.

TV/Multimedia/Consoles: Smart TVs are also a common household item, where data can be stored on memory or memory/storage cards (e.g. Micro SD cards). For example, we determined that the smart TV in our use case, the suspect had browsed on the YouTube app for videos with a specific topic/interest. Consoles for instance do offer messaging functions, and hence consoles should also be seized for further investigation.

Smart Toys: The Teddy is watching you. With toys becoming more 'intelligent', they are a potential source of evidential data (e.g. incident involving the hack on CloudPet's unsecure MongoDB [26]).

Smart Heating/Light/Energy: Data from such smart IoT devices that control heater/light and measure energy and water consumption can potentially be useful in reconstructing the timeline of the household occupants.

IP Cameras: IP cameras, including a smart doorbell with camera feature, are another valuable information source. Such cameras generally have the ability to detect motion in the picture and send the picture/video sequence to a specific mail address or a cloud account provided by the manufacturers. Some IP cameras, for instance, can be configured to operate only on certain timeframes and have a built-in timer.

## 7    Conclusion and Future Work

As IoT becomes more commonplace in our homes, offices, public spaces, higher education institutions, etc., more of these devices can be potential sources of evidence in a criminal investigation (e.g. in cases where some system was compromised and used to facilitate a physical home invasion, physical stalking, or murder, or the investigation of a physical crime where IoT devices in the vicinity may provide additional leads/evidence). Thus, in this paper we presented an IoT focused digital forensic model, designed to guide the investigation of existing and emerging IoT devices. We also demonstrated the utility of the model using two case studies, and explained how one can build an IoT testbed lab to facilitate forensic understanding of new IoT devices. Such a lab can be easily reconfigured for different settings, based on the needs of an investigation, to train and prepare the investigators prior to them going to the crime scene. It can also be used to facilitate the investigation of the case after the devices have been seized, for example mimicking the actual crime scene layout.

When performing IoT forensics, it is important to note that anti-forensic measures may have been attempted by the suspect to circumvent or prevent certain data or devices from been discovered. For example, in our use case, we demonstrated that it can be easily done using simple Python scripts. Clearly, more sophisticated approaches, such as implementing a small Arduino board with the ability to destroy an internal hard disk drive by setting up a trap inside a computer case, can be deployed.

Thus, the lab can also include common anti-forensic approaches for training purposes. There is also a need to keep pace with developments in anti-forensics, in order to maximize the potential of evidence acquisition.

Future research also includes collaborating with other like-minded digital forensic researchers to use the proposed model in examining a broader range of IoT devices and share the findings of the forensic examination. The findings from different researchers and using different IoT devices may identify areas in the proposed model or testbed setup that need further refinement. Hence, we are also looking at extending our model to assist the current challenges in retrieving and analyzing artefacts for smart vehicle forensics [27].

# References

1. Kruse, W.G., Heiser, J.G.: Computer Forensics: Incident Response Essentials. Addison-Wesley (2002)
2. Lee, I., et al.: Challenges and research directions in medical cyber-physical systems. INSPEC Accession Numb. 12425479, 75–90 (2012)
3. Oriwoh, E., et al.: University of Bedforshire internet of things forensics: challenges and approaches (2013). https://doi.org/10.4108/icst.collaboratecom.2013.254159
4. Perumal, S., Norwawi, N.M., Raman, V.: Internet of things (IoT) digital forensic investigation model: top-down forensic approach methodology, pp. 1–5 (2015)
5. Goudbeek, A., Choo, K.-K.R., Le-Khac, N.-A.: A forensic investigation framework for smart home environment. In: 17th IEEE international conference on trust, security and privacy in computing and communications (IEEE TrustCom-18). New York, USA (August 2018). https://doi.org/10.1109/TrustCom/BigDataSE.2018.00201
6. Alabdulsalam, S., Schaefer, K., Kechadi, M.-T., Le-Khac, N.-A.: Internet of things forensics: challenges and case study. In: Peterson, G., Shenoi, S. (eds.) Advances in Digital Forensics XIV. Springer, New York, Berlin, Heidelberg. https://doi.org/10.1007/978-3-319-99277-8_3
7. Faheem, M., Le-Khac, N.-A., Kechadi, M.-T.: Smartphone forensics analysis: a case study for obtaining root access of an android samsung S3 device and analyse the image without an expensive commercial tool. J. Inf. Secur. 5(3), 83–90 (8 pages) (2014). http://dx.doi.org/10.4236/jis.2014.53009
8. Chen, L., Le-Khac, N.-A., Schlepphorst, S., Xu, L.: Cloud forensics: model, challenges, and approaches. In: Chen, L., Takabi, H., Le-Khac, N.-A. (eds.) Security, Privacy, and Digital Forensics in the Cloud. High Education Press, Wiley Inc. (April 2019). https://doi.org/10.1002/9781119053385.ch10
9. Roder, A., Choo, K.-K.R., Le-Khac, N.-A.: Unmanned aerial vehicle forensic investigation process: Dji Phantom 3 drone as a case study. In: 13th annual ADFSL conference on digital forensics, security and law. Texas, USA (May 2018)
10. Lea, R., Blackstock, M.: City hub: a cloud-based IoT platform for smart cities. In: 2014 IEEE 6th international conference on cloud computing technology and science (CloudCom) (CLOUDCOM). Singapore, pp. 799–804 (2014)
11. TEDx Talks: How the IoT is making cybercrime investigation easier|Jonathan Rajewski|TEDxBuffalo (2016). Available online: https://trvision.net/detail/how-the-iot-is-making-cybercrime-investigation-easier-jonathan-rajewski-tedxbuffalo-9CemONO6vrY.html. Accessed on 12 Dec 2019
12. Introducing the Top Ten Dev Boards of 2017. Available online: https://www.arrow.com/en/research-and-events/articles/the-top-ten-development-platforms-dev-kits-for-2017. Accessed on 3 Nov 2019

13. JTAG Explained (finally!): Why "IoT" makers, software security folks, and device manufacturers should care-senrio (2018). Available online: http://blog.senr.io/blog/jtag-explained. Accessed on 3 Feb 2019
14. Raspbian. Available online: https://www.raspbian.org. Accessed on 2 Feb 2019
15. Windows 10 IoT Core. Available online: https://developer.microsoft.com/de-de/windows/iot. Accessed on 2 Feb 2019
16. RIOT-The friendly operating system for the internet of things. Available online: https://riot-os.org/. Accessed on Nov 2019
17. Android Things. Available online: https://developer.android.com/things/index.html. Accessed on Nov 2019
18. mbed, device to data platform. Available online: www.mbed.com/en. Accessed on Nov 2019
19. Nucleus RTOS. Available online: https://www.mentor.com/embedded-software/nucleus/. Accessed on 2 Feb 2019
20. Contiki: the open source operating system for the internet of things. Available online: http://www.contiki-os.org/. Accessed on June 2019
21. Autopsy. Available online: https://www.sleuthkit.org/autopsy/. Accessed on Nov 2019
22. EnCase forensic software—top digital investigations solution. Available online: https://www.guidancesoftware.com/encase-forensic. Accessed on Nov 2019
23. Forensic Toolkit (FTK)|AccessData. Available online: http://accessdata.com/solutions/digital-forensics/forensic-toolkit-ftk. Accessed on Nov 2019
24. MSAB—the pioneers of mobile forensics. Available online: https://www.msab.com/. Accessed on Nov 2019
25. Magnet Forensics AXIOM. Available online: https://www.magnetforensics.com/magnet-axiom/. Accessed on Dec 2019
26. Cloud Pets, attack on the mongo database. Available online: https://www.bbc.com/news/technology-39115001. Accessed on 27 Dec 2019
27. Le-Khac, N.-A., Jacobs, D., Nijhoff, J., Bertens, K., Choo, K.-K.R.: Smart vehicle forensics: challenges and case study. Future Generation of Computer Systems, Elsevier (July 2018). https://doi.org/10.1016/j.future.2018.05.081

**Alexander Hilgenberg** received M.Sc. in Forensic Computing and Cybercrime Investigations in 2017 from UCD, Dublin. He is currently active as a computer forensic investigator at the police inspection Northeim in Germany.

**Trung Q. Duong** is a Reader in Telecommunications at Queen's University Belfast, U. His current research interests include IoT (applied to disaster management, smart agriculture, air-quality, smart cities), 5G networks (small-cell networks, ultra-dense networks, HetNets, physical layer security), machine learning and big data analytics (applied to environment and healthcare). He has authored/co-authored of 350+ papers including 220 ISI journal articles with approximately 10000 citation and h-index 54 in Google Scholar. He was awarded the Best Paper Award at the IEEE Vehicular Technology Conference (VTC-Spring) in 2013, the IEEE International Conference on Communications (ICC) in 2014, the IEEE Global Communications Conference (GLOBECOM) in 2016, IEEE Digital Signal Processing (DSP) in 2017, IEEE Wireless Communications & Mobile Computing in 2019 (IWCMC 2019) and IEEE GLOBECOM in 2019. He is a recipient of prestigious Royal Academy of Engineering Research Fellowship from 2016 to 2020 and has won the prestigious Newton Prize 2017.

**Nhien-An Le-Khac** is a lecturer at the School of Computer Science (CS), University College Dublin (UCD), Ireland. He is currently the program director of MSc program in Forensic Computing and Cybercrime Investigation (FCCI)—an international program for the law enforcement officers specializing in cybercrime investigations. He is also the co-founder of UCD-GNECB Postgraduate Certificate in fraud and e-crime investigation. Since 2008, he is a research fellow in Citibank, Ireland (Citi). He obtained his Ph.D. in Computer Science in 2006 at the Institut National Polytechnique de Grenoble (INPG), France. His research interest spans the area of Cybersecurity and Digital Forensics, Data Mining/Distributed Data Mining for Security, and Fraud and Criminal Detection. Since 2013, he has collaborated on many research projects as a principal/co-PI/funded investigator. He has published more than 150 scientific papers in peer-reviewed journal and conferences in related research fields, and his recent edited book has been listed the Best New Digital Forensics Book according to the Book Authority.

**Kim-Kwang Raymond Choo** received the Ph.D. in Information Security in 2006 from the Queensland University of Technology, Australia. He currently holds the Cloud Technology Endowed Professorship at the University of Texas at San Antonio (UTSA). In 2015, he and his team won the Digital Forensics Research Challenge organized by the Germany's University of Erlangen-Nuremberg. He is the recipient of the 2019 IEEE Technical Committee on Scalable Computing (TCSC) Award for Excellence in Scalable Computing (Middle Career Researcher), 2018 UTSA College of Business Col. Jean Piccione and Lt. Col. Philip Piccione Endowed Research Award for Tenured Faculty, British Computer Society's 2019 Wilkes Award Runner-up, 2019 EURASIP Journal on Wireless Communications and Networking (JWCN) Best Paper Award, Korea Information Processing Society's Journal of Information Processing Systems (JIPS) Survey Paper Award (Gold) 2019, IEEE Blockchain 2019 Outstanding Paper Award, Inscrypt 2019 Best Student Paper Award, IEEE TrustCom 2018 Best Paper Award, ESORICS 2015 Best Research Paper Award, 2014 Highly Commended Award by the Australia New Zealand Policing Advisory Agency, Fulbright Scholarship in 2009, 2008 Australia Day Achievement Medallion, and British Computer Society's Wilkes Award in 2008. He is also a fellow of the Australian Computer Society, an IEEE senior member, and co-chair of IEEE Multimedia Communications Technical Committee's Digital Rights Management for Multimedia Interest Group.

# Forensic Investigation of Ransomware Activities—Part 1

Cian Young, Robert McArdle, Nhien-An Le-Khac,
and Kim-Kwang Raymond Choo

**Abstract** Techniques employed by malware authors evolve and become more
advanced each day in an effort to bypass defences and evade detection. From 2013
to the present, a type of malware known as ransomware has increased exponentially in popularity with cyber criminals. Ransomware encrypts files on a victim's
filesystem and subsequently demands a ransom payment to release the files. The
exponential growth of ransomware poses a serious and real threat to end-users and
organisations worldwide. The exponential growth also poses serious challenges to the
security industry, such as the need to analyse and study the large volume of emerging
ransomware families. A problem exists in that new ransomware families may use
previously unseen techniques to evade detection and detonate successfully. A second
problem exists for security analysts when it comes to analysing the ever increasing
volume of emerging ransomware families. Malware analysis generally falls into one
of two categories: static and dynamic analysis. Dynamic analysis is effective at clas-
sifying malware, however it's ineffective at discovering newly developed techniques
or functionality. On the contrary static analysis is effective at discovering newly
developed techniques and functionality, however it requires significantly more time
to complete than dynamic analysis. The information gathered from static analysis
is essential to enable organisations better defend against these new attacks. The
information obtained from this research can be used to help defend against future
threats using similar techniques and highlight the effectiveness of manual analysis
to discover new and advanced techniques.

**Keywords** Ransomware · Forensic analysis · Reverse engineering · Malware

C. Young (✉)
Irish Defence Forces, Cork, Ireland
e-mail: cianjpyoung@gmail.com

R. McArdle
Trend Micro Ltd., Cork, Ireland

N.-A. Le-Khac
University College Dublin, Dublin, Ireland

K.-K. R. Choo
University of Texas at San Antonio, San Antonio, USA

N.-A. Le-Khac and K.-K. R. Choo (eds.), *Cyber and Digital Forensic Investigations*,
Studies in Big Data 74, https://doi.org/10.1007/978-3-030-47131-6_4

# 1   Introduction

Malware authors have total freedom when developing new pieces of malware. Developing new malware variants is trivial and the financial reward can be immediate. In 2014 almost one million new malware threats were released each day [1]. Symantec discovered more than 430 million new unique pieces of malware in 2015, up 36% from the previous year [2]. There have been many researches on malware analysis in literature so far [3, 4] but not specifically for ransomware. In terms of ransomware, one family was discovered in 2012. In 2015, 35 ransomware families were discovered. In 2016, this increased by 451% to 193 newly discovered families [5]. The year-on-year trend suggest a high likelihood that new families of ransomware will continue to emerge, incorporating never before seen techniques.

Analysing malware falls into one of two categories: dynamic and static analysis. Dynamic analysis is effective at identifying the purpose of malware but can fail to discover new techniques. Malware authors are mindful of malware analysis and often develop techniques to evade dynamic analysis. An alternative approach is an in-depth static analysis examination, which is highly effective at discovering advanced techniques and obtaining conclusive results.

Ransomware attacks are becoming increasingly common, causing significant damage to both end users and organisations worldwide. Modern ransomware families are well-developed, and implement strong encryption such as the Advanced Encryption Standard (AES), making brute force recovery computationally infeasible. Only when a victim pays the ransom demand will the decryption key(s) be provided, allowing the victim to regain access to their files and/or system.

Recently a new type of hybrid version of ransomware incorporating worm functionality has emerged. A worm is a type of malware that has the inherent ability to propagate and replicate from computer to computer without human interaction by exploiting computer vulnerabilities. The emerging ransomworm threat, incorporating both worm and encryption functionality, is a major hazard facing organisations worldwide; Affording cybercriminals a new opportunity for financial gain through a highly contagious and destructive threat.

This chapter adopts an in-depth manual analysis with the aim of discovering new techniques used in ransomworms and targeted ransomware attacks against enterprise networks. The knowledge gained will help defend against future attacks by identifying exploited security weaknesses and documenting the latest adversary techniques.

# 2 Background

## 2.1 Evolution of Ransomware

Ransomware attacks fall into one of two types:

(a) Locker ransomware attacks completely lock a victim's machine, preventing them from using it.
(b) Crypto ransomware attacks encrypt files on a victim's filesystem to render them inaccessible. Generally, this involves encrypting specific extensions such as .doc and .jpg rather than encrypting every file on disk. The targeting of specific file extensions is done to maximise impact resulting in a higher probability of the victim paying. The machine is otherwise operational.

The AIDS Trojan (1989) is the first recorded instance of ransomware. It was spread via floppy disk and used symmetric encryption to encrypt the name of all files on the C Drive. Victims were instructed to pay the ransom demand to a post office box in Panama [6]. The second recorded instance of ransomware was in 1996 when two researches released a paper titled "Cryptovirology" [7]. The research created a proof-of-concept ransomware capable of encrypting files on a victim's filesystem using RSA and TEA encryption. Both of these early events demonstrated the ransomware threat, however, 2005 would be the year in which the ransonware threat began to appear in the wild.

2012 is widely regarded as the year ransomware took off with the emergence Reveton [8], a type of locker ransomware. Reveton used social engineering techniques, displaying warning messages to victims purporting to be from law enforcement, stating their computer had been used in illegal activities, such as downloading child pornography. At its peak, Reveton activity resulted in 500,000 infection attempts over an 18 day period. A combined effort from international law enforcement agencies brought about the arrest of the cyber-criminals [1].

In 2013 the volume of ransomware attacks worldwide increased exponentially, with an increase of 500% compared to the number of attacks in 2012 [1]. Threat intelligence company SonicWall detected an increase from 3.2 million ransomware attack attempts in 2014 to 3.8 million in 2015, and an astounding 638 million in 2016 [9]. Since January 2016, US CERT detected an average of 4,000 ransomware attacks daily, a significant increase on the 1,000 daily attacks in 2015 [10].

The instigator to this sudden rise is the now notorious Cryptolocker (2013). Cryptolocker is widely accredited with defining a successful business model for ransomware attacks—implementing strong cryptography, correctly, and using the digital currency Bitcoin for ransom payment. The use of Bitcoin afforded the authors anonymity and the ability to scale operations easily. The adoption of this innovative business model brought about a significant increase in attacks worldwide, resulting in unprecendented levels of financial gain for cybercriminals. CryptoWall version 3, a ransomware family originating in 2013, accounted for an estimated total $325M

USD in damages in the US alone [11]. The rising attacks cumulated with the ransomware family Locky becoming the first family to make $1M USD per month in Q1 2016 [12].

## 2.2   Stages of a Ransomware Attack

The stages of a ransomware attack follow the Lockheed Martin Cyber Kill-Chain.

### 2.2.1   Infection Vector

- Phishing/Social Engineering: e-mails containing a malicious link or an attachment serving as an initial dropper [13].
- Compromised site: client-side script downloading initial payload resulting in ransomware infection [14].
- Drive-by Download: A compromised website delivering ransomware, for example Malvertising [15].
- Infection via Botnets [16].

### 2.2.2   Delivery and Execution

The ransomware executable is delivered to the victim's machine and executed. Execution usually includes fingerprinting the operating system (OS), establishing persistence, encrypting files, and deleting the original and unencrypted files.

### 2.2.3   Encryption

Ransomware encrypts files matching specific file extensions such as .doc and .jpg. Encryption and key management methods vary from family to family, with some families using the Windows cryptographic libraries and others writing custom cryptographic libraries and functions.

### 2.2.4   Backup Removal

Ransomware families thwart recovery by deleting backups and restore points on a victim's machine. For example, several ransomware families remove local backups using the Microsoft Windows tool, vssadmin. Segregated offsite backups can provide a useful restore solution following an infection.

### 2.2.5   User Notification

A ransom note notifies the user of the ransomware infection and provides instructions on how to pay for file recovery. A deadline of several days is often given.

### 2.2.6   Payment

Scalable payment processing afforded by digital cryptocurrencies has provided a platform for ransomware revenue to grow significantly over recent years. Bitcoin is one popular cryptocurrency that enables people to purchase goods digitally. Over 900 cyptocurrencies are available online [17].

### 2.2.7   Worm Functionality

A computer worm has the ability to self-propagate from host to host without human interaction. History shows how difficult it is to stop a resilient worm. Two classic examples of robust worms are Slammer and Conficker. Conficker, first detected in 2008, infected computers in over 190 countries by exploiting flaws in Windows software and propagated by using dictionary attacks on system administrator passwords [18]. Slammer, first detected in 2003, set a record by infecting over 75,000 servers in ten minutes, despite a patch (MS-02-039) [19] being available for six months.

   ZCryptor, a ransomware that emerged in early 2016, incorporated a ransomware module for file encryption and a second module for self-propagation. The worm module dropped autorun.inf on removable media attached to the victim's machine, thus creating a copy of itself before invoking its encryption routine [20].

## 3   SamSam Analysis

The SamSam ransomware campaign began targeting organisations in April 2016. SamSam is a multi-phased attack that included reconnaissance and lateral movement, before manually deploying the ransomware binary within the target network, similar to a modern intrusion following the cyber-kill chain [21, 22]. Notably, SamSam attackers gained their initial foothold on target networks through external web compromise as opposed to phishing or drive-by exploits.

   The ransomware targets Active-Directory Domain Environments and leverages Bitcoin as the cryptocurrency for ransom payment. The autonomy of SamSam attack is akin to a modern-day intrusion where the attackers follow a set of steps to achieve their objective.

1. Reconnaissance: Attackers scan in the internet facing servers to identify vulnerable hosts.

2. Delivery and Exploitation: Attackers launch the attack at targeted organisation by exploiting a vulnerability in the JMX-Console Web-App in JBoss, CVE-2010-0738 [23].
3. Installation: Establish persistence, escalate privileges and acquire credentials.
4. Lateral Movement: Enumerate the network, pivot internally and acquire more credentials.
5. Actions on Target: Leverage administrator tools, such as Sysinternals PS Exec, to deploy the ransomware to the Windows machines on the internal network.

## 3.1 Infection Vector and Preamble to Deploying Ransomworm

The initial server-side attack exploits an unpatched vulnerability in JBoss Application Servers [24]. The vulnerability, CVE-2010-0738, allows a remote user to submit specifically crafted HTTP requests to bypass JMX Console authentication as the application only enforces session protection for GET and POST requests, thus enabling an attacker to upload malicious files through other HTTP methods such as HTTP HEAD requests.

The initial host compromised acts as a beachhead facility for internal targeting. The purpose of this phase is to map the network and identify possible targets for infection. One investigation into the attacker's methodology uncovered the use of several tools and scripts to map the victim's network [25]. Authors indicate use of the credential stealing tool Mimikatz [26] to harvest credentials from compromised hosts.

## 3.2 The Ransomware Component

Table 1 describes a sample of SamSam components. After identifying systems of interest, the attackers deploy the ransomware component. RSA and AES cryptography are used to encrypt victim files. The ransomware is a .NET compiled binary

**Table 1** SamSam component

| Filename | endeavor2.exe |
|---|---|
| Internal name | endeavor2.exe |
| Description | .NET ransomware component of SamSam |
| Size | 176 KB |
| SHA1 | 1eb97c7ca98e75d64ad2d7b1ec5d5f6a67bb5c30 |
| Compile time stamp | 13/06/2017 20:07:36 UTC |

requiring Microsoft's .NET framework to execute on a victim's machine. Represented by the Program Database (PDB), the project name of the binary is endeavor2.pdb. Earlier versions of the ransomware had the project name "SamSam" which is where the ransomware campaign got its name. The open-source tool "de4dot" [27] enables de-obfuscation of strings in the binary.

The .NET decompiler ILSpy [28] is used to decompile the .NET binary. Figure 1 displays the encrypted strings in the decompiled binary. At the beginning the key "SALT" is present and each encrypted string is proceeded by the function *encc.myff11*. To enable analysis of the encrypted content, the code is compiled in Microsoft Visual Studio and subsequently run in debug mode.

The contents of the encrypted expression "e17" during debugging, reveals the TOR site used for payment help (Fig. 2).

```
{
    private static string e1 = "SALT";
    private static string e2 = encc.myff11("EAAAAImT1sZBSRCFQ7nMEMlT4pHnl6kIaubatyS/ZgjQJ6vw7BvLeLws8cryaW5xDKl9b954Ni65ABVPXivwLkOAUSFFQlWtlkO/tjE
    private static string e3 = encc.myff11("EAAAAORy8/rM976/bGRFgncjWaYTQK7YQDaRoFw7f5HMTg16", Program.e1);
    private static string[] e4 = Program.e2.Split(',');
    private static string[] e5 = Program.e3.Split(',');
    private static List<List<string>> e6 = new List<List<string>>();
    private static string e7 = Environment.MachineName + "<br><br>";
    private static string e8 = Directory.GetCurrentDirectory();
    private static string e9 = Environment.GetFolderPath(Environment.SpecialFolder.CommonApplicationData) + "\\" + encc.myff11("EAAAAMX+e6YD0ucAvSH
    private static string e10 = "";
    private static string e11 = encc.myff11("EAAAAF/rebF4vt3ScDLek314X/6TV3DubR1w8tLraPTsMzSntxin+nQmh5yE8ZQfKrx/fg==", Program.e1);
    private static string e12 = encc.myff11("EAAAAOwMlZxCSUQvYGKwF+U67EWoGgDfDQ88FtiBaxpG9Jf51", Program.e1);
    private static string e13 = encc.myff11("EAAAAMMOwbOuT4kLBaL8md1P0MuSh193WOtPmqnOim5LiPeb", Program.e1);
    private static string e14 = "1.7";
    private static string e15 = "12";
    private static string e16 = "6";
    private static string e17 = encc.myff11("EAAAACHZRSD/4CWviK9mcD2dyugCMX1qXCk8Qbt1iEF+duJ/M9V6B32E1hu2F+Dqez0R0sgf76z1AKge4sXy8kIf4E+2r2OoylicYE
    private static string e18 = encc.myff11("EAAAAJiSKY7YE/nqRLcCJhjlAWNKPo+2BB5I1mNnQJzUiZude6+gI8zWitWyluPjYrIxlxh0iidBxCsSTkg/IPnk0=", Program
    private static string e19 = "</html>\r\n<body style='background-color:lightgrey;'>\r\n<pre>\r\n<font color="Red"><center><h3>&#35;&#87;&#104;&#
;&#97;&#115;&#32;&#100;&#101;&#109;&#46;&#32;&#32;<br><h3><b>&#73;&#102;&#32;&#121;&#111;&#117;&#32;&#97;&#114;&#101;&#32;&#119;&#114
    private static string e20 = Path.GetPathRoot(Environment.SystemDirectory);
    private static string e21 = Process.GetCurrentProcess().ProcessName + ".exe";
    private static string e22 = encc.myff11("EAAAAPy5bjfoSQSjVPm262UhQL7c0IHqksbXo2BnPALCs5F+kmc60HFMnaFzNMal4SimQNr58pPNTJQEQqTdkyuoFcQ=", Program
    private static string e23 = encc.myff11("EAAAAEL1xd77lIVe9C0TG8HkleKPlT05VmhARyY4WBboM1DF", Program.e1);
    private static string e24 = Program.e9 + "\\" + Program.e23;
    private static string w1 = encc.myff11("EAAAAkeh5APALVqxq/UYwyAcFT1zhr8IA+mF/ROFma4UEWG", Program.e1);
    private static string w2 = encc.myff11("EAAAAPuN69xn3UGzk43lgHcHms2aEFilIxghoF5qApHlVBIx", Program.e1);
    private static string w3 = encc.myff11("EAAAALjLXuiBxifH2aSTXCLvmUDAxFM6UUGgre9TPDi0ZfRt1YSRyyh0lEFfSWKlOlEEag==", Program.e1);
    private static string w4 = encc.myff11("EAAAAP7WgNdexpV2NYmVa82TXfQ2wkwdkHg91UcVARIkR6N", Program.e1);
    private static string w5 = encc.myff11("EAAAAF0L4p1vlYBuLHKrW95diBqYZvPddnyj5tRlaDP0UH8y", Program.e1);
    private static string w6 = encc.myff11("EAAAAB0NcVk0d+vT6j6Rr0SdJxglDHF1kDojiDrd1ygF3gWF/4BRKViEuA/d7JlmIokXTcHzR7BA1jl3NjcaShSn01U=", Program.
    private static string w7 = encc.myff11("EAAAAJQViS7iXteGq8mVSPqLOng0Ex80FeIU9NglszmhOv6M", Program.e1);
    private static string w8 = encc.myff11("EAAAAFr56FHZzgjUceFq4/lLST9QsLAnOZ0qz0VZV1/QK8bt", Program.e1);
```

**Fig. 1** SamSam encrypted strings

**Fig. 2** Expression "e17" reveals an onion site used for ransomware payment

## Encryption

The ransomware component first performs a conditional check on the number of command line arguments and fails to execute if not provided with a single argument: a public RSA key in XML format (Fig. 3).

Next, the ransomware component creates the directory "greewin" within C:\ProgramData and drops the batch script, *startinfo.bat.* (Fig. 4). Following execution of the batch script, the ransomware binary will start encrypting files on the file system.

Prior to encrypting a file, the ransomware binary, endeavour2.exe, first checks the amount of space on disk to ensure adequate free space is available for the encrypted version of the file. Two files exist during the encryption process: the original *<file>.ext* and the encrypted data *<file>.ext.mention9823*. Once encrypted, the original file is deleted.

---

```
RSA4 - Notepad
File   Edit   Format   View   Help
<RSAKeyValue><Modulus>mzfIMBGEitQ2T3w0T5Ol4rfXIYBlvq2ssx/mcmvxzew0fcxnd/Cvzb
97hFRCfOH+Jj2+1CIzccAzmvkuI6ZgBQraHQcjqzk3eDIj6vzPNOekJGOFKxQEexzdNNKbWzNJTbw
RQwynJRQORC4XW5LpR9QpzdwQasss/2DSdaNnjdAD21DcScnksNIPPXx2dj1/lwOD5h9lNPBZz8zk
8kILD7Pt6rGBPNDgNm4MJSL/pJLZ5lAAsxwwwp9mLsafnfjnAFSasf3231AFADvxsSSSasfs2f/k3a
rOOTESHYbgwyQ4e8X9Jkdnsnds1nfsaa1w2asnmk9HD1JDBs/ghFERS==</Modulus><Exponent
>AQAB</Exponent></RSAKeyValue>
```

**Fig. 3** Public key in XML format

---

```
startinfo.bat - Notepad
File   Edit   Format   View   Help
@echo off
SETLOCAL EnableExtensions
set "EXE=SamSam2.exe"
set "PEXE=C:\Users\dan\Desktop":loopFOR /F %%x IN ('tasklist /NH /FI "IMAGENAME eq %EXE%"')
DO IF %%x == %EXE%
goto FOUND
goto END
:FOUND
ping 127.0.0.1 -n 5 > NUL
goto loop
:END
DEL "%PEXE%\%EXE%"|
DEL "%~f0"
```

**Fig. 4** Contents of "startinfo.bat". The binary was named "SamSam2.exe" in the analysis environment

*endeavour2.exe* reads each file in 10.5 KB chunks during the encryption process with a 3072 byte AES key value inserted into the file header, having been encrypted with the RSA public key. An extract of an encrypted file is shown in Fig. 5.

Earlier versions of the ransomware component SamSam.exe (32a2d1a9d91ce7d9c130a9b0616c40ac4003355d) included two PE files embedded in the resource section of the ransomware binary (Tables 2 and 3):

- Samsam.del.exe:
- Samsam.selfdel.exe

After the ransomware has finished encrypting files, the tool selfdel.exe deletes the ransomware binary.

| Test.txt | Test.txt.mention9823 | | | | | | | | | | | | | | | |
|---|---|---|---|---|---|---|---|---|---|---|---|---|---|---|---|---|
| Offset(h) | 00 | 01 | 02 | 03 | 04 | 05 | 06 | 07 | 08 | 09 | 0A | 0B | 0C | 0D | 0E | 0F |
| 00000000 | 3C | 00 | 41 | 00 | 41 | 00 | 41 | 00 | 41 | 00 | 41 | 00 | 41 | 00 | 41 | 00 | <.A.A.A.A.A.A. |
| 00000010 | 41 | 00 | 41 | 00 | 41 | 00 | 41 | 00 | 41 | 00 | 41 | 00 | 41 | 00 | 41 | 00 | A.A.A.A.A.A.A. |
| 00000020 | 41 | 00 | 41 | 00 | 41 | 00 | 41 | 00 | 41 | 00 | 41 | 00 | 3E | 00 | 0D | 00 | A.A.A.A.A.>... |
| 00000030 | 0A | 00 | 3C | 00 | 41 | 00 | 41 | 00 | 41 | 00 | 3E | 00 | 4C | 00 | 77 | 00 | ..<.A.A.>.L.w. |
| 00000040 | 56 | 00 | 44 | 00 | 30 | 00 | 43 | 00 | 6C | 00 | 30 | 00 | 75 | 00 | 32 | 00 | V.D.0.C.1.0.u.2. |
| 00000050 | 39 | 00 | 61 | 00 | 6B | 00 | 43 | 00 | 43 | 00 | 68 | 00 | 73 | 00 | 6D | 00 | 9.a.k.C.C.h.s.m. |
| 00000060 | 32 | 00 | 76 | 00 | 6D | 00 | 61 | 00 | 2F | 00 | 35 | 00 | 49 | 00 | 6E | 00 | 2.v.m.a./.5.I.n. |
| 00000070 | 47 | 00 | 4D | 00 | 2B | 00 | 4A | 00 | 56 | 00 | 31 | 00 | 75 | 00 | 37 | 00 | G.M.+.J.V.1.u.7. |
| 00000080 | 67 | 00 | 63 | 00 | 61 | 00 | 6A | 00 | 55 | 00 | 4E | 00 | 75 | 00 | 6F | 00 | g.c.a.j.U.N.u.o. |
| 00000090 | 62 | 00 | 62 | 00 | 38 | 00 | 78 | 00 | 78 | 00 | 57 | 00 | 79 | 00 | 52 | 00 | b.b.8.x.x.W.y.R. |
| 000000A0 | 4B | 00 | 77 | 00 | 2F | 00 | 32 | 00 | 31 | 00 | 57 | 00 | 6A | 00 | 31 | 00 | K.w./.2.1.W.j.1. |
| 000000B0 | 6C | 00 | 62 | 00 | 73 | 00 | 73 | 00 | 49 | 00 | 33 | 00 | 46 | 00 | 63 | 00 | l.b.s.s.I.3.F.c. |
| 000000C0 | 73 | 00 | 76 | 00 | 4E | 00 | 31 | 00 | 6E | 00 | 55 | 00 | 32 | 00 | 34 | 00 | s.v.N.1.n.U.2.4. |

**Fig. 5** Extract of header for an encrypted file

**Table 2** Samsam.del.exe

| Filename | Samsam.del.exe/sdelete.exe |
|---|---|
| Internal name | SDelete |
| Description | Sysinternals Sdelete |
| Size | 148 KB |
| SHA1 | 964f7144780aff59d48da184daa56b1704a86968 |
| Compile time stamp | 14/01/2012 23:06:53 UTC |

**Table 3** Samsam.selfdel.exe

| Filename | Samsam.selfdel.exe/selfdel.exe |
|---|---|
| Internal name | selfdel.exe |
| Description | Custom tool to delete SamSam.exe |
| Size | 5.6 KB |
| SHA1 | 5e70502689f6bf87eb367354268923e6a7e875c6 |
| Compile time stamp | 02/12/2015 22:24:42 UTC |

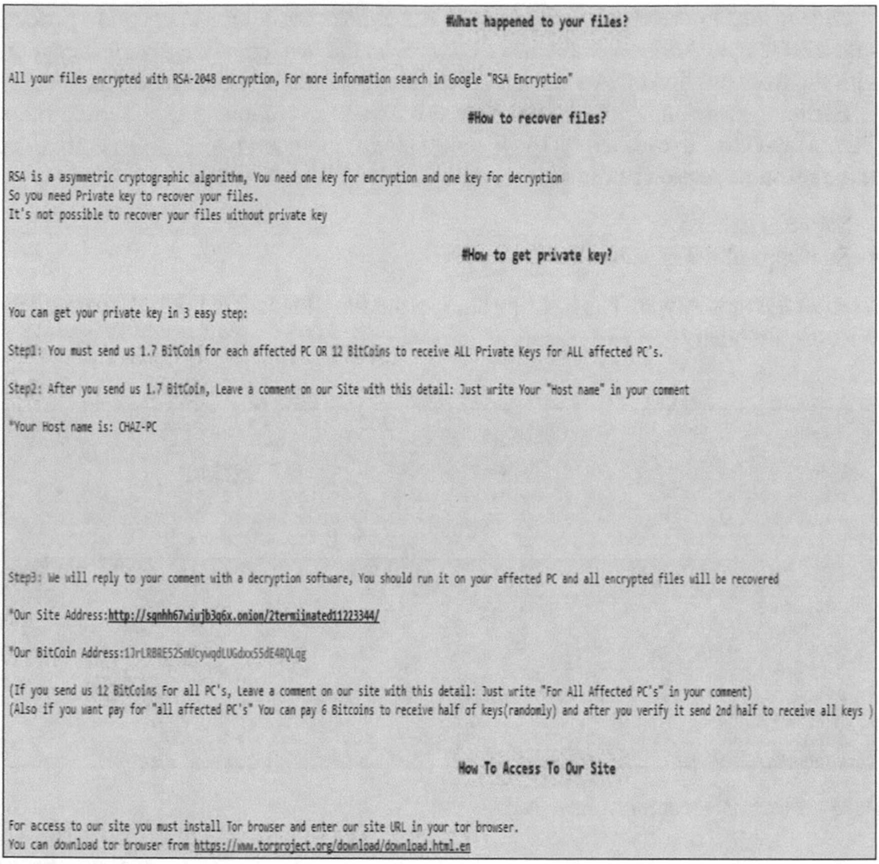

```
                                    #What happened to your files?

All your files encrypted with RSA-2048 encryption, For more information search in Google "RSA Encryption"

                                    #How to recover files?

RSA is a asymmetric cryptographic algorithm, You need one key for encryption and one key for decryption
So you need Private key to recover your files.
It's not possible to recover your files without private key

                                    #How to get private key?

You can get your private key in 3 easy step:

Step1: You must send us 1.7 BitCoin for each affected PC OR 12 BitCoins to receive ALL Private Keys for ALL affected PC's.

Step2: After you send us 1.7 BitCoin, Leave a comment on our Site with this detail: Just write Your "Host name" in your comment

*Your Host name is: CHAZ-PC




Step3: We will reply to your comment with a decryption software, You should run it on your affected PC and all encrypted files will be recovered

*Our Site Address:http://sqnhh67wiujb3q6x.onion/2termiinated11223344/

*Our BitCoin Address:13rLRBRE525mUcywqdLUGdxx55dE4RQLqg

(If you send us 12 BitCoins For all PC's, Leave a comment on our site with this detail: Just write "For All Affected PC's" in your comment)
(Also if you want pay for "all affected PC's" You can pay 6 Bitcoins to receive half of keys(randomly) and after you verify it send 2nd half to receive all keys )

                                    How To Access To Our Site

For access to our site you must install Tor browser and enter our site URL in your tor browser.
You can download tor browser from https://www.torproject.org/download/download.html.en
```

**Fig. 6** Extract of HTML file explaining the decryption process

## 3.3    Payment

The payment help URL is a TOR onion site: hxxp://sqnhh67wiujb3q6x[dot]onion/2termiinated11223344/.

The payment HTML code is encrypted in the binary. Dumping the expression "e19" reveals the HTML code (Fig. 6).

## 3.4    Summary

The structure of the SamSam attack is akin to a modern-day intrusion and required a knowledgeable and skilled adversary to execute successfully. The multi-phase attack, coupled with the use of batch scripts and legitimate tools such as Sysinternals PsExec lowers the likelihood of detection. Once infected, a host initiates a HTTP connection to the payment help site but apart from that the ransomware component is self-contained. The targeted nature of the campaign is a new venture in the history of ransomware attacks; a user did not click a link to become infected, rather the attacker found a vulnerability, identified systems of interest and deployed the ransomware. The lack of post-infection call-back is also notable.

# 4   WannaCry Technical Analysis

## 4.1   *WannaCry Introduction*

WannaCry ransomware began spreading worldwide on May 12, 2017. WannaCry consists of two components: a core ransomware component, and an SMB worm component, which is capable of spreading over both local-area networks (LAN) and the public Internet. The worm component exploits a vulnerability in Server Message Block (SMB) protocol: CVE-2017-0143, CVE-2017-0144, CVE-2017-0145, CVE-2017-0146, CVE-2017-0147, CVE-2017-0148 [29]. The Shadow Brokers, a group widely believed to have stolen a large amount of NSA exploits, released an exploit for the vulnerability, called EternalBlue on April 14, 2017. On March 14, 2017, Microsoft released an emergency patch for the vulnerability: MS17-010 [30]. A search of Shodan during the troubled days revealed close to two million open ports for SMB/445 [31].

WannaCry components:

1.  The dropper: mssecsvc.exe

    - Contains the worm component that attempts to exploit the SMB vulnerability on remote hosts.
    - Also contains the ransomware component as an embedded resource.

2.  The ransomware component: taskche.exe

    - Contains an embedded encrypted DLL used for the file encryption functionality.
    - Also contains the decryption program "WanaDecrypt0r 2.0" and a password protected zip containing various configurations files.

## 4.2   *WannaCry Dropper Details*

Upon execution, Mssecsvc.exe (Table 4), first attempts to connect the kill-switch domain: hxxp://www[dot]iuqerfsodp9ifjaposdfjhgosurijfaewrwergwea[dot]com

**Table 4** Mssecsvc.exe

| Filename | Mssecsvc.exe |
| --- | --- |
| Internal name | Ihdfrgui.exe |
| Description | Attack SMB vulnerability, create tasksche.exe |
| Size | 3.65 MB |
| SHA1 | e889544aff85ffaf8b0d0da705105dee7c97fe26 |
| Compile time stamp | 20/11/2010 Sat 09:03:08 UTC |

If the connection attempt is successful, the dropper will stop executing and terminate. If the connection attempt is unsuccessful, the return value of the function call
to *InternetOpenUrl* will be NULL, and the dropper will continue to execute. Within
hours of the outbreak, the kill-switch domain was sinkholed by a security researcher
[29] which resulted in the connection attempt succeeding, thus causing the dropper
to terminate.

The kill-switch conditional check, with the payload function call labelled
InfectWithWannaCry, is shown in Fig. 7. The value of parameter *dwAccessType* ("IN-
TERNET_OPEN_TYPE_DIRECT") to the function call *InternetOpen* indicates the
worm is not proxy aware, and hence could call the payload on machines situated
behind a proxy.

The function labelled *InfectWithWannaCry* first checks the number of command
line arguments provided with a conditional jump taken depending on the result.

If run with no command line arguments, the dropper creates and starts a service
named *mssecsvc2.0* with the parameters "-m security" and runs under services.exe.
The service masquerades as a legit Microsoft service and is responsible for running
the worm functionality that attempts to exploit the SMB vulnerability on remote
machines over port 445.

**Fig. 7** Conditional check in IDA machines situated behind a proxy device

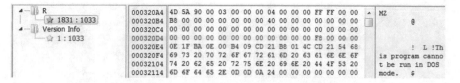

**Fig. 8** Embedded PE file from resource section R/1831:1033

Next, the dropper loads a PE file from its resource section named "R/1831:1033" and saves it as a file on disk: c:\windows\tasksche.exe (Fig. 8)

The dropped executable is moved from C:\Windows\tasksche.exe to C:\Windows\qeriuwjhrf. Finally, the dropper executes the dropped executable with the argument/i by calling *kernel32.CreateProcessA*, discussed further in Sect. 4.4.

## 4.3 The Worm Component

*mssecsvc2.0* running as a service with two command line arguments executes the worm functionality. First, the dropper calls the function *OpenSCManager* to gain access to the service *mssecsvc2.0*. The dropper then calls *StartServiceCtrlDispatcherA*, which initiates the SMB worm thread (Fig. 9).

For networking functionality, the worm calls the Windows Sockets function: *WSAStartup*. Next, the worm extracts two DLLs, a 32-bit and a 64-bit DLL, from its data section and subsequently calls the Windows function *GlobalAlloc* to store both DLLs in the process memory. Both extracted DLLs are functionally identical and act as the worm payloads to initiate an infection on a target host.

Next, the worm creates two threads. One thread to scan the local network of the victim's machine, and one thread to scan public internet IP addresses.

The LAN propagation thread initially calls the function *GetAdaptersInfo* to perform three tasks:

```
█ N ωↄ
00408101
00408101 loc_408101:              ; CODE XREF: InfectWithWannaCry+3D↑j
00408101 lea      eax, [esp+14h+ServiceStartTable] ; Load Effective Address
00408105 mov      [esp+14h+ServiceStartTable.lpServiceName], offset ServiceName ; "mssecsvc2.0"
0040810D push     eax              ; lpServiceStartTable = 0018FE84
0040810E mov      [esp+18h+ServiceStartTable.lpServiceProc], offset EternalBlue_Worm ; THIS IS THE WORM PROPAGATION
00408116 mov      [esp+18h+var_8], 0
0040811E mov      [esp+18h+var_4], 0
00408126 call     ds:StartServiceCtrlDispatcherA ; Connects the main thread of a service process to the service control manager,
00408126                           ; which causes the thread to be the service control dispatcher thread for the calling process
00408126                           ;
00408126                           ; Start Service Crtl Dispatcher (Run SMB Exploit)
0040812C pop      edi
0040812D add      esp, 10h         ; Add
00408130 retn                      ; Return Near from Procedure
00408130 InfectWithWannaCry endp
00408130
```

**Fig. 9** Run SMB exploit

| Source | Destination | Protocol | Lengt | Info |
|--------|-------------|----------|-------|------|
| 172.16.0.26 | 172.16.0.37 | TCP | 54 | 49838 → 445 [ACK] Seq=1 Ack=1 Win=65536 Len=0 |
| 172.16.0.26 | 172.16.0.37 | TCP | 54 | 49838 → 445 [FIN, ACK] Seq=1 Ack=1 Win=65536 Len=0 |
| 172.16.0.37 | 172.16.0.26 | TCP | 60 | 445 → 49838 [ACK] Seq=1 Ack=2 Win=65536 Len=0 |
| 172.16.0.37 | 172.16.0.26 | TCP | 60 | 445 → 49838 [RST, ACK] Seq=1 Ack=2 Win=0 Len=0 |
| 172.16.0.26 | 172.16.0.37 | TCP | 66 | 49839 → 445 [SYN] Seq=0 Win=8192 Len=0 MSS=1460 WS=256 SACK_PERM=1 |
| 172.16.0.37 | 172.16.0.26 | TCP | 66 | 445 → 49839 [SYN, ACK] Seq=0 Ack=1 Win=8192 Len=0 MSS=1460 WS=256 SACK_PERM=1 |
| 172.16.0.26 | 172.16.0.37 | TCP | 54 | 49839 → 445 [ACK] Seq=1 Ack=1 Win=65536 Len=0 |
| 172.16.0.26 | 172.16.0.37 | SMB | 142 | Negotiate Protocol Request |
| 172.16.0.37 | 172.16.0.26 | SMB | 185 | Negotiate Protocol Response |
| 172.16.0.26 | 172.16.0.37 | SMB | 157 | Session Setup AndX Request, User: .\ |
| 172.16.0.37 | 172.16.0.26 | SMB | 179 | Session Setup AndX Response |
| 172.16.0.26 | 172.16.0.37 | SMB | 149 | Tree Connect AndX Request, Path: \\172.16.0.26\IPC$ |
| 172.16.0.37 | 172.16.0.26 | SMB | 104 | Tree Connect AndX Response |
| 172.16.0.26 | 172.16.0.37 | SMB Pipe | 132 | PeekNamedPipe Request, FID: 0x0000 |
| 172.16.0.37 | 172.16.0.26 | SMB | 93 | Trans Response, Error: STATUS_INSUFF_SERVER_RESOURCES |
| 172.16.0.26 | 172.16.0.37 | TCP | 54 | 49839 → 445 [FIN, ACK] Seq=365 Ack=346 Win=65280 Len=0 |
| 172.16.0.37 | 172.16.0.26 | TCP | 60 | 445 → 49839 [ACK] Seq=346 Ack=366 Win=65280 Len=0 |
| 172.16.0.37 | 172.16.0.26 | TCP | 60 | 445 → 49839 [RST, ACK] Seq=346 Ack=366 Win=0 Len=0 |
| 172.16.0.26 | 172.16.0.35 | TCP | 66 | [TCP Spurious Retransmission] 49833 → 445 [SYN] Seq=0 Win=8192 Len=0 MSS=1460 WS=256 SACK_PERM=1 |
| 172.16.0.35 | 172.16.0.26 | TCP | 60 | 445 → 49833 [RST, ACK] Seq=1 Ack=1 Win=0 Len=0 |

**Fig. 10** WannaCry infected machine probing a potential target. Output from Wireshark

1. Obtain the IP address of the host machine.
2. Generate a list of IP addresses on the host local area network.
3. Finally, attempt to connect to each IP address on port 445.

The WAN thread first generates a random IP addresses using *CryptGenRandom()* modulo 255 for each octet. To avoid non-routable IP addresses, the first octet cannot be equal to 225 or 127. The thread then attempts to connect to each generated IP. If the worm successfully connects it enters a loop and iterates through all possible addresses in the Class C subnet, creating an exploit threat for each successful connection.

SMB, which operates over port 139 and port 445, is a protocol for file sharing and general-purpose remote transactions. The exploit thread weaponizes the NSA developed exploit, EternalBlue, which affects SMB version 1 in Windows: XP, 7, Server 2003 and Server 2008.

SMB transaction request and responses enable read and write operations between a SMB client and an SMB server. When an SMB message request is greater than the SMB *MaxBufferSize*, the remainder of the message is sent as a Trans2 secondary request [32]. The vulnerability is exploited by sending a specially crafted packet in the Trans2 secondary request packet (Fig. 10).

To exploit a remote host, the infected host (hereafter 'worm') takes the following steps:

1. The worm attempts to establish a remote SMB session by submitting two SMB requests.

   a. SMB_COM_NEGOTIATE (0x72) used to start an SMB session between a server and client.
   b. SMB_COM_SESSION_SETUP_ANDX (0x73) used to configure an SMB session.

2. The SMB command SMB_COM_TREE_CONNECT_ANDX (0x75) packet is sent to the Inter-Process Communication (IPC$) tree of the remote host to establish a client connection to the server share. Valid credentials are not required in default server configurations, logging on as user "\" and connecting to IPC$ is permitted [32].

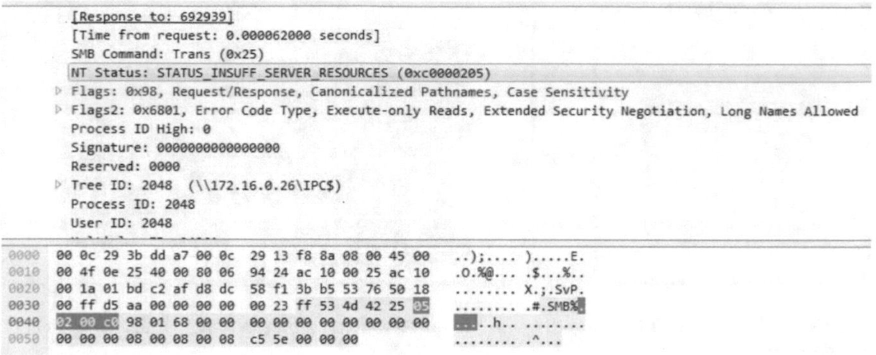

```
[Response to: 692939]
[Time from request: 0.000062000 seconds]
SMB Command: Trans (0x25)
NT Status: STATUS_INSUFF_SERVER_RESOURCES (0xc0000205)
▷ Flags: 0x98, Request/Response, Canonicalized Pathnames, Case Sensitivity
▷ Flags2: 0x6801, Error Code Type, Execute-only Reads, Extended Security Negotiation, Long Names Allowed
  Process ID High: 0
  Signature: 0000000000000000
  Reserved: 0000
▷ Tree ID: 2048   (\\172.16.0.26\IPC$)
  Process ID: 2048
  User ID: 2048
```

```
0000  00 0c 29 3b dd a7 00 0c  29 13 f8 8a 08 00 45 00   ..);.... ).....E.
0010  00 4f 0e 25 40 00 80 06  94 24 ac 10 00 25 ac 10   .O.%@... .$...%..
0020  00 1a 01 bd c2 af d8 dc  58 f1 3b b5 53 76 50 18   ........ X.;.SvP.
0030  00 ff d5 aa 00 00 00 00  00 23 ff 53 4d 42 25 05   ........ .#.SMB%.
0040  02 00 c0 98 01 68 00 00  00 00 00 00 00 00 00 00   .....h.. ........
0050  00 00 00 08 00 08 00 08  c5 5e 00 00 00            ........ .^...
```

**Fig. 11** Response STATUS_INSUFF_SERVER_RESOURCES

    a. TreeID = \\${ip_address_of_remote_host}\IPC$

    b. UserID = User ID returned from session setup command in step 1.

3. The Worm sends SMB packet SMB_COM_TRANSACTION (0x25) on FID 0x0, testing for MS17-010 vulnerability.
4. Next, the Worm checks of value of "NT Status". A return status of "STATUS_INSUFF_SERVER_RESOURCES (0xC0000205)" on FID 0x0 against IPC$ means the machine does not have the MS17-010 patch and is therefore exploitable. Microsoft added additional authentication in MS17-010 which results in an "INVALID" response (Fig. 11).
5. Next, the worm sleeps for 3 seconds before calling a function to check for the presence of DoublePulsar [33] on the remote host.

If the remote host does not have DoublePulsar installed, the worm will take the following steps in an attempt to exploit the MS17-010 vulnerability:

1. Connect to remote host using SMB_COM_NEGOTIATE (0x72), SMB_COM_SESSION_SETUP_ANDX (0x73) and SMB_COM_TREE_CONNECT_ANDX (0x75).
2. Submit specifically crafted SMB_COM_TRANSACTION2 (0x33) packet.
   The encrypted payload is contained in the SMB data request portion.

3. Srv.sys, the kernel-mode SMB driver, is compromised and injects launcher.dll into the user-mode process lsass.exe (Fig. 12).
4. *PlayGame*, the export function of launcher.dll, is called and drops mssecsvc.exe.
5. lssass.exe executes mssecsvc.exe by calling *CreateProcessA*, thus repeating the infection cycle.

The new infection begins by extracting and running mssecsvc.exe from the resource contained within the DLL. Repeating the infection cycle—the worm module running as a service and scanning for vulnerable hosts, and the execution of the ransomware binary.

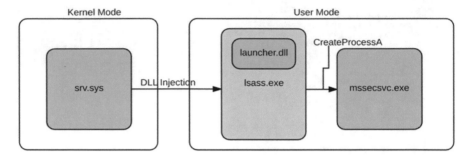

**Fig. 12** SrvOs2FeaToNt function of srv.sys is vulnerable resulting in launcher.dll being injected into lsass.exe

## 4.4 Ransomware Component

The initial dropper creates and starts a service that probes and attempts to spread the ransomware, whereas the dropped resource executes the ransomware component of the infection. The initial dropper executes the extracted resource with the /i parameter.

Initially, a conditional statement checks the number of command line arguments. If tasksche.exe (Table 5) is executed without the/i switch (not executed by the dropper), it first creates and starts a service. PE file tasksche.exe contains an embedded ZIP file named "XIA/2058:1033" in its resource section (Fig. 13).

Upon execution, tasksche.exe obtains the name of the victim's host and generates a random a string using the Microsoft *srand* function. Using the host name as a checksum, the function generates a pseudo random string of 15 characters plus three numbers in length.

The pseudo random string is used as the service name and as part of a new installation directory where tasksche.exe is copied. Next, the registry key HKEY_LOCAL_MACHINE\SOFTWARE\WanaCrypt0r is created. The registry value *wd* points to the installation path of tasksche.exe.

Next, the embedded ZIP is extracted to the installation directory using the hardcoded password "WNcry@2oI7". The ZIP folder contains a total of nine items.

**Table 5** tasksche.exe

| Filename | tasksche.exe |
|---|---|
| Internal name | diskpart.exe |
| Description | Ransomware component |
| Size | 3.5 MB |
| SHA1 | 5ff465afaabcbf0150d1a3ab2c2e74f3a4426467 |
| Compile time stamp | 20/11/2010 Sat 09:05:05 TUC |

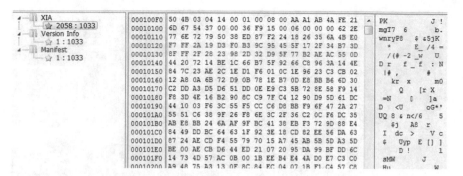

**Fig. 13** Embedded ZIP file with the 'PK' magic bytes

- taskdl.exe—application to delete files.
- taskse.exe—tool to launch decryption tool.
- b.wnry—image of the ransom note.
- c.wnry—configuration file.
- r.wnry—ransom note.
- s.wnry—ZIP archive containing TOR with required libraries.
- t.wnry—Encrypted ransomware DLL.
- u.wnry—GUI application.
- msg—directory containing ransom payment notes in 30 languages.

Next, tasksche.exe parses c.wnry to load the onion sites required for Command and Control (C2) (Fig. 14).

After file extraction, tasksche.exe calls a function to load the hardcoded bitcoin wallets and update the configuration file *c.wnry*. Afterward, the following two commands are executed:

- *attrib +h*: Hiding the directory containing the unzipped files.
- *icacls ./grant Everyone:F/T/C/Q*: give full access to directory and all sub-directories.

Next a function is called to load the following exports: *CreateFileW, WriteFile, ReadFile, MoveFileW, MoveFileExW, DeleteFileW, CloseHandle*

Following this, a 2048-bit RSA public key is extracted from the PE file using the function *CryptImportKey*. The key is 1172-bytes in length and is stored within the data section of the executable, as shown in Fig. 15.

The RSA public key is used to decrypt an AES key at the beginning of the resource file *t.wnry*. The recovered AES key is subsequently used to decrypt the rest of *t.wnry*. Dumping the decrypted file in OllyDbg reveals it to be a PE file, specifically a DLL.

```
000000b0:  0000 3132 7439 5944 5067 7775 655a 394e  ..12t9YDPgwueZ9N
000000c0:  794d 6777 3531 3970 3741 4138 6973 6a72  yMgw519p7AA8isjr
000000d0:  3653 4d77 0000 0000 0000 0000 0000 0000  6SMw............
000000e0:  0000 0000 6778 3765 6b62 656e 7632 7269  ....gx7ekbenv2ri
000000f0:  7563 6d66 2e6f 6e69 6f6e 3b35 3767 3773  ucmf.onion;57g7s
00000100:  7067 727a 6c6f 6a69 6e61 732e 6f6e 696f  pgrzlojinas.onio
00000110:  6e3b 7878 6c76 6272 6c6f 6f78 7269 7932  n;xxlvbrloxvriy2
00000120:  6335 2e6f 6e69 6f6e 3b37 366a 6464 3269  c5.onion;76jdd2i
00000130:  7232 656d 6279 7634 372e 6f6e 696f 6e3b  r2embyv47.onion;
00000140:  6377 776e 6877 686c 7a35 326d 6171 6d37  cwwnhwhlz52maqm7
00000150:  2e6f 6e69 6f6e 3b00 0000 0000 0000 0000  .onion;.........
00000160:  0000 0000 0000 0000 0000 0000 0000 0000  ................
00000170:  0000 0000 0000 0000 0000 0000 0000 0000  ................
00000180:  0000 0000 0000 0000 0000 0000 0000 0000  ................
00000190:  0000 0000 0000 0000 0000 0000 0000 0000  ................
000001a0:  0000 0000 0000 0000 0000 0000 0000 0000  ................
000001b0:  0000 0000 0000 0000 0000 0000 0000 0000  ................
000001c0:  0000 0000 0000 0000 0000 0000 0000 0000  ...............
000001d0:  0000 0000 0000 0000 0000 0000 0000 6874  ..............ht
000001e0:  7470 733a 2f2f 6469 7374 2e74 6f72 7072  tps://dist.torpr
000001f0:  6f6a 6563 742e 6f72 672f 746f 7262 726f  oject.org/torbro
00000200:  7773 6572 2f36 2e35 2e31 2f74 6f72 2d77  wser/6.5.1/tor-w
00000210:  696e 3332 2d30 2e32 2e39 2e31 302e 7a69  in32-0.2.9.10.zi
00000220:  7000 0000 0000 0000 0000 0000 0000 0000  p...............
```

**Fig. 14** ASCII representation of the Onion sites used for C2, righmost column

```
00401875 lea      ecx, [esi+8]    ; Load Effective Address
00401878 push     ecx
00401879 push     eax
0040187A push     eax
0040187B push     1172
00401880 push     offset unk_40EBF8 ; Hardcoded RSA key
00401885 push     dword ptr [esi+4]
00401888 call     dword_40F898    ; ADVAPI32.CryptImportKey
00401888                          ; CryptImportKey function transfers a cryptographic key
00401888                          ; from a key BLOB into a cryptographic service provider(CSP).
0040188E jmp      short loc_4018A3 ; Jump
```

**Fig. 15** Hardcoded Public RSA Key

## 4.5 Encrypted DLL

The file encryption process begins begins with a call to the DLL *TaskStart*
function. The decrypted DLL (Table 6) first creates a mutex named
"Global\MsWinZonesCacheCounterMutex" by calling the function *CreateMutex*.
Following this, it obtains the name of current directory and reads the configuration
file c.wnry which contains TOR addresses for C2 and Bitcoin wallets previously
loaded by tasksche.exe. The DLL then creates four configurations files required for
the file encryption process.

**File Encryption Process**

Figure 16 illustrates the WannaCry encryption workflow. Each new infection
generates a RSA-2048 keypair. The public RSA key is exported to 00000000.pky.

**Table 6** Decrypted DLL (Kbdlv.dll) from t.wnry

| Filename | Kbdlv.dll |
|---|---|
| Internal name | Kbdlv.dll |
| Description | DLL decrypted from t.wnry |
| Size | 3.7 MB |
| SHA1 | 19384a2c51329c80e8b13d8bc2324fcf152a263b |
| Compile time stamp | 14/07/2009 01:12:55 UTC |

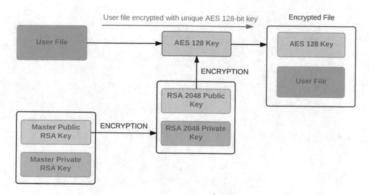

**Fig. 16** WannaCry file encryption

1. Each file on the filesystem is encrypted with a unique 128-bit AES key. Each AES key is generated using the Windows function *CryptGenRandom*.
2. The AES key used to encrypt each file is then encrypted with the RSA public key stored in 0000000.pky. Decrypting each file requires the private RSA key of the public key used for encrypting the AES key.
3. Each encrypted file header is prepended with a custom signature "WANACRY!" and each encrypted file will be appended with the ".WNCRY" extension.
4. The generated victim's RSA private key is encrypted with the Master RSA public key.
5. The Master RSA public key used to encrypt the generated RSA private key is embedded inside the DLL.
6. WannaCry authors have possession of the Master RSA private key.

WannaCry encrypts files with the following extensions: .doc, .docx, .xls, .xlsx, .ppt. The format of an encrypted file is displayed in Table 7.

Figure 17 shows the hex format of an encrypted user file. An encrypted file has the file marker "WANACRY!" and the file extension ".WNCRY". Next, a function at address 0x10032C0 is called to whitelist the following list of directories from the encryption process:

**Table 7** Format of encrypted file

| Offset | Value | Description |
| --- | --- | --- |
| 0x0000 | WANACRY! | Encrypted file header |
| 0x000C | AES Key | RSA encrypted AES key |
| 0x0110 | Size in bytes | Original file size |
| 0x0118 | Beginning of file contents | Encrypted file |

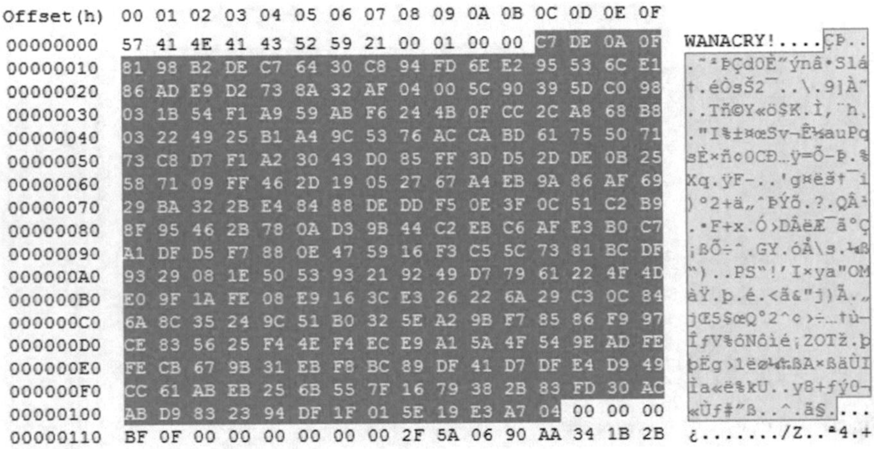

```
Offset(h)  00 01 02 03 04 05 06 07 08 09 0A 0B 0C 0D 0E 0F
00000000   57 41 4E 41 43 52 59 21 00 01 00 00 C7 DE 0A 0F   WANACRY!....ÇÞ..
00000010   81 98 B2 DE C7 64 30 C8 94 FD 6E E2 95 53 6C E1   .˜²ÞÇd0È"ýnâ•Slá
00000020   86 AD E9 D2 73 8A 32 AF 04 00 5C 90 39 5D C0 98   †.éÒsŠ2¯..\.9]À˜
00000030   03 1B 54 F1 A9 59 AB F6 24 4B 0F CC 2C A8 68 B8   ..Tñ©Y«ö$K.Ì,¨h.
00000040   03 22 49 25 B1 A4 9C 53 76 AC CA BD 61 75 50 71   ."I%±¤œSv¬Ê½auPq
00000050   73 C8 D7 F1 A2 30 43 D0 85 FF 3D D5 2D DE 0B 25   sÈ×ñ¢0CÐ…ÿ=Õ-Þ.%
00000060   58 71 09 FF 46 2D 19 05 27 67 A4 EB 9A 86 AF 69   Xq.ÿF-..'g¤ëš†¯i
00000070   29 BA 32 2B E4 84 88 DE DD F5 0E 3F 0C 51 C2 B9   )º2+ä„ˆÞÝõ.?.QÂ¹
00000080   8F 95 46 2B 78 0A D3 9B 44 C2 EB C6 AF E3 B0 C7   .•F+x.Ó›DÂëÆ¯ã°Ç
00000090   A1 DF D5 F7 88 0E 47 59 16 F3 C5 5C 73 81 BC DF   ¡ßÕ÷ˆ.GY.óÅ\s.¼ß
000000A0   93 29 08 1E 50 53 93 21 92 49 D7 79 61 22 4F 4D   ").. PS"!'I×ya"OM
000000B0   E0 9F 1A FE 08 E9 16 3C E3 26 22 6A 29 C3 0C 84   àŸ.þ.é.<ã&"j)Ã.„
000000C0   6A 8C 35 24 9C 51 B0 32 5E A2 9B F7 85 86 F9 97   jŒ5$œQ°2^¢›÷…†ù—
000000D0   CE 83 56 25 F4 4E F4 EC E9 A1 5A 4F 54 9E AD FE   Îƒ V%ôNôìé¡ZOTž.þ
000000E0   FE CB 67 9B 31 EB F8 BC 89 DF 41 D7 DF E4 D9 49   þËg›1ëø¼‰ßA×ßäÙI
000000F0   CC 61 AB EB 25 6B 55 7F 16 79 38 2B 83 FD 30 AC   Ìa«ë%kU..y8+ƒý0¬
00000100   AB D9 83 23 94 DF 1F 01 5E 19 E3 A7 04 00 00 00   «Ùƒ#"ß..^.ã§.....
00000110   BF 0F 00 00 00 00 00 00 2F 5A 06 90 AA 34 1B 2B   ¿......./Z..ª4.+
```

**Fig. 17** Encrypted AES key stored at the beginning of a user file. Offset 0x0C to 0x10C

- \\
- $\
- Intel
- ProgramData
- WINDOWS
- Program Files
- Program Files (x86)
- AppData\\Local\\Temp
- Local Settings\\Temp
- Temporary Internet Files
- Content.IE5

Next the following commands are executed to terminate the processes listed:

- Taskkill.exe/f/im Microsoft.Exchange.*
- Taskkill.exe/f/im MSExchange*
- Taskkill.exe/f/im sqlserver.exe
- Taskkill.exe/f/im sqlwriter.exe
- Taskkill.exe/f/im mysqld.exe

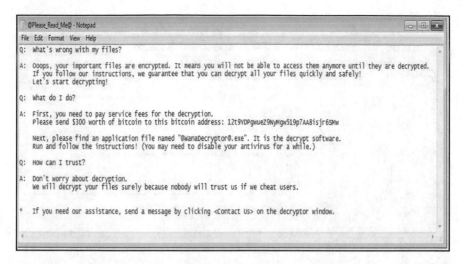

**Fig. 18** Displays an extract of '@Please_Read_Me@.txt'

At this stage each folder on the filesystem contains a copy of the ransom note '@Please_Read_Me@.txt' (Fig. 18), and a copy of the application @WanaDecryptor@.exe.

Next the DLL launches the decryptor GUI application (Fig. 19) by copying u.wnry and then calling it. *@WanaDecryptor@.exe* runs the TOR client to connect to the .onion sites, with the TOR server running on localhost:9050. It then creates @WanaDecryptor@.exe.lnk for persistence. Finally, the Volume Shadow Copies are deleted.

## 4.6 Summary

WannaCry was developed to abort execution if it successfully connected to the hard-coded kill-switch domain. As a result, the early discovery of the kill-switch domain severely lessened the potential impact. The SMB component is essentially a copy and paste from a stolen NSA exploit. There have been several theories as to why include the kill-switch domain, one being the worm escaped prematurely. Whatever the reason, the attackers and other cybercriminals learned a valuable lesson in malware resiliency.

**Fig. 19** @WanaDecryptor@.exe GUI Application

## 5 Ransomworms in the Future

Ransomware attacks of the future will continue to evolve in sophistication and become increasingly targeted at enterprise networks. The Cryptolocker campaign introduced a successful business model affording cybercriminals anonymity and scalability. The SamSam campaign introduced a targeted attack methodology akin to a modern day intrusion, with several distinct phases and tools. The WannaCry campaign introduced a truly self-contained ransomworm with devastating impact. Cybercriminals are increasingly moving away from the "cast a net" approach to a more directed approach against enterprise networks, spurred on by the potential for large financial gain.

It is reasonable to infer targeted ransomworm attacks of the future will continue to be modular in design, using tools and exploits at each stage of the kill-chain. A potential ransomware framework may include the following modules:

1. Access module: A module to gain an initial foothold to an organisations network through compromising externally facing assets. An example would be an adversary scanning an organisations DMZ, finding vulnerable applications, and leveraging an arsenal of exploits to compromise and gain access to the internal network.

2. Internal reconnaissance module: The modules purpose would to be to map the internal network to identify targets of interest. Targets may include: (a) Critical systems; (b) DR sites and backups.
3. Exploit module: This module would transmit a self-contained ransomware payload to the targeted hosts.
4. Detonation module: Detonate the ransomware payload across all targets at once, which may be tracked via a unique serial number, such as a GUID.
5. C2 Infrastructure: Optional. Module may be omitted to reduce the likelihood of detection.

## 5.1 Epidemiology

Epidemiology is defined as the study and analysis of the patterns, causes, and effects of health and disease conditions in defined populations [34]. The definition of epidemiology denotes a data centric approach in finding patterns and causes of disease in populations. Historic breakthroughs in epidemiology emphasise this, such as that of John Snow. A famous pioneer of epidemiology, John Snow, lead investigations into Cholera outbreaks during the nineteenth Century. Using a data-centric approach, Snow created a map representation of cholera cases occurring in London, and found them to be clustered in an area supplied by water pump known as the Broad Street pump [35]. Snow used Chlorine to clean the water and remove the pump, thus ending an outbreak of Cholera. The approach and rational adopted by Snow has a relevance in fighting the ransonworm outbreaks of the future. An endpoint detection equals an endpoint infection. Runbooks prescribe cleaning the infected endpoint and resumption of normal business operations. It is reasonable to suggest that detection of an infection occurs at the later stages of the kill-chain, suggesting a high likelihood of additional host-infections within the environment [36]. In a ransomworm outbreak, discovering the root cause is an essential step in helping stop an infection spreading further. A data driven approach, where defenders categorize infections and their relationships to one another, may prove vital in stopping the ransomworm outbreaks of the future.

Similarly, patching can be thought of an antidote or remedy. Fortunately in the case of WannaCry, the registration of the kill-switch curtailed the attack before serious damage occurred, and before an antidote was made available by Microsoft. Consider the infection rate had similar attack occurred where no kill-switch existed and no patch was made available.

In terms of attacker intent, it is reasonable to presume the WannaCry attackers sought financial gain, perhaps with a desire to cause some destruction along the way. Imagine a situation where a hacktivist group take possession of a high grade, zero-day exploit similar to EternalBlue. Groups such as Anonymous pursue their goals with little regard for collateral damage. A high grade, contagious exploit in the possession of a Hacktivist group such as Anonymous, a group motivated by beliefs rather than financial gain, could be catastrophic. Recent events prove this a possibility. The Snowden [37] leaks, and more recently the Hal Martin [38] leaks, prove unequivocally that the NSA and CIA have stockpiled a cyber arsenal of high grade exploits for popular operating systems in use today.

## 5.2   A Note on IoT

The number of Internet of Things (IoT) devices continues to increase exponentially, everything is becoming a computer. A smoothie-maker can now be a computer, cars are computers on wheels, and when these devices are connected to the internet, they become vulnerable to ransomware and other threats [39, 40]. Attacks on these devices have occurred already [23]. What happens when a ransomworm campaign targets an IoT device or platform? If a particular device is widely used, and the potential for financial gain exists, then it may become an attractive target for cybercriminals.

WannaCry targeted Windows, the most popular desktop operating system in the world. Microsoft even released patches for unsupported versions of Windows, such as Windows XP.

For the most part, IoT manufacturers do not provide such a service.

Inherent IoT problems include:

1. Severe lack of protection.
2. Lack of regular updates.
3. Not supported after release—what happens if a manufacturer goes out of business?

## 6   Conclusion and Future Work

Ransomware has evolved into a sophisticated and potent weapon to wield for cyber-criminals to wield. Investigation and analysis of new threats provide an insight to new and emerging techniques being used by cybercriminals. The research performed here is geared towards investigating the techniques employed by the SamSam and WannaCry campaigns. The information obtained can be used to improve existing knowledge on both campaigns.

Targeted ransomware will continue to rise due to the reliable business model and the significant potential for financial gain. Future attacks may not be as wide reaching as WannaCry, but the destructive effect on a target organisation may result in severe business impact and potential insolvency.

Future research could follow an epidemiology-type approach and attempt to discover relationships or patterns existing between similar campaigns, from a behavioural or functionality perspective. This may help identify future campaigns and aid the development of detection signatures to locate an attack before its too late.

# References

1. cert-ist: The Reveton ransomware (2013). Available: https://www.cert-ist.com/public/en/SO_detail?code=201301_article. Last accessed 04 Dec 2019
2. Symantec: ISTR (2016). Available: https://www.symantec.com/content/dam/symantec/docs/reports/istr-21-2016-en.pdf. Last accessed 01 Dec 2019
3. Schaefer, E., Le-Khac, N-A., Scanlon M.: Integration of ether unpacker into ragpicker for plugin-based malware analysis and identification. In: 16th European Conference on Cyber Warfare and Security, Dublin, Ireland, June 2017
4. Linke, A., Le-Khac, N-A.: Control flow change in assembly as a classifier in malware analysis. In: 4th IEEE International Symposium on Digital Forensics and Security, Arkansas, USA, April 2016. https://doi.org/10.1109/ISDFS.2016.7473514
5. Pilkey, A.: Ransomware likely to continue exponential growth unless governments act, says f-secure labs (2017). Available: https://www.f-secure.com/en_GB/web/press_gb/news/news-archive/-/journal_content/56/1075444/1992103?p_p_auth=TAbI9XWV&refererPlid=1769223. Last accessed 25 June 2019
6. Wikipedia: AIDS (Trojan horse) (2006). Available: https://en.wikipedia.org/wiki/AIDS_(Trojan_horse). Last accessed 29 May 2019
7. Young, A., et al.: Cryptovirology: extortion-based security threats and countermeasures (1996). Available: http://ieeexplore.ieee.org/document/502676/?reload=true. Last accessed 22 May 2019
8. Ducklin, P.: Reveton/FBI ransomware—exposed, explained and eliminated (2012). Available: https://nakedsecurity.sophos.com/2012/08/29/reveton-ransomware-exposed-explained-and-eliminated/. Last accessed 18 Jan 2019
9. SONICWALL: 2017 annual threat report (2017). SONICWALL, UK. Last accessed 01 May 2019
10. US Government: How To Protect Your Networks From Ransomware (2017). Available: https://www.justice.gov/criminal-ccips/file/872771/download. Last accessed 03 Aug 2017
11. Blake, A.: Cybercriminals rake in $325M from CryptoWall ransomware: report (2015). Available: http://www.washingtontimes.com/news/2015/nov/2/cybercriminals-rake-in-325m-cryptowall-ransomware/. Last accessed 09 Aug 2019
12. Bursztein, E., McRoberts, K., Invernizzi, L.: Tracking desktop ransomware payments (2017). Available: https://www.blackhat.com/docs/us-17/wednesday/us-17-Invernizzi-Tracking-Ransomware-End-To-End.pdf. Last accessed 31 July 2019
13. Schrott, U.: Social engineering and ransomware (2017). Available: https://blog.eset.ie/2017/07/28/social-engineering-and-ransomware/. Last accessed 29 July 2019
14. quttera: From compromised website to ransomware infection (2016). Available: https://blog.quttera.com/post/from-compromised-website-to-ransomware-infection/. Last accessed 17 Jun 2019
15. Zamora, W.: Malvertising and ransomware: the Bonnie and Clyde of advanced threats (2016). Available: https://blog.malwarebytes.com/101/2016/06/malvertising-and-ransomware-the-bonnie-and-clyde-of-advanced-threats/. Last accessed 21 Feb 2019
16. Ducklin, P.: Destructive malware "CryptoLocker" on the loose—here's what to do. Available: https://nakedsecurity.sophos.com/2013/10/12/destructive-malware-cryptolocker-on-the-loose/. Last accessed 12 Feb 2019
17. Wikipedia: List of cryptocurrencies (2017). Available: https://en.wikipedia.org/wiki/List_of_cryptocurrencies. Last accessed 19 May 2019
18. Wikipedia: Conficker (2016). Available: https://en.wikipedia.org/wiki/Conficker. Last accessed 22 Dec 2019
19. Microsoft: Microsoft Security Bulletin MS02-039—Critical (2003). Available: https://technet.microsoft.com/library/security/ms02-039. Last accessed 24 Dec 2019
20. Marvin the Robot: ZCryptor: The conqueror worm (2016). Available: https://www.kaspersky.com/blog/zcryptor-ransomware/12268/. Last accessed 16 Jan 2019

21. Symantec Security Response: Samsam may signal a new trend of targeted ransomware (2016). Available: https://www.symantec.com/connect/blogs/samsam-may-signal-new-trend-targeted-ransomware. Last accessed 30 June 2019
22. Lockheed Martin: The Cyber Kill Chain (2016). Available: http://www.lockheedmartin.com/us/what-we-do/aerospace-defense/cyber/cyber-kill-chain.html. Last accessed 29 Jan 2019
23. Mitre: CVE-2010-0738 (2010). Available: http://cve.mitre.org/cgi-bin/cvename.cgi?name=cve-2010-0738. Last accessed 10 Aug 2019
24. Joaomatosf: jexboss (2017). Available: https://github.com/joaomatosf/jexboss. Last accessed 13 Apr 2019
25. Beek, C., Furtak, A.: Targeted ransomware no longer a future threat (2016). Available: http://www.intelsecurity.com/advanced-threat-research/content/Analysis_SamSa_Ransomware.pdf. Last accessed 06 Aug 2019
26. gentilkiwi: mimikatz (2015). Available: https://github.com/gentilkiwi/mimikatz. Last accessed 07 Aug 2019
27. xd4d: de4dot (2017). Available: https://github.com/0xd4d/de4dot. Last accessed 16 May 2019
28. ILSpy: ILSpy.NET Decompiler (2017). Available: http://ilspy.net/. Last accessed 17 May 2019
29. cve.mitre.org: CVE-2017-0146 (2017). Available: https://www.cve.mitre.org/cgi-bin/cvename.cgi?name=CVE-2017-0146. Last accessed 08 July 2019
30. Microsoft: Microsoft Security Bulletin MS17-010—Critical (2017).Available: https://technet.microsoft.com/en-us/library/security/ms17-010.aspx. Last accessed 04 Aug 2019
31. Shodan: Port 445 exposed externally (2017). Available: https://www.shodan.io/search?query=port%3A445&language=en. Last accessed 02 Feb 2019
32. MSDN: SMB requests (2017). Available: https://msdn.microsoft.com/en-us/library/ee441730.aspx. Last accessed 24 May 2019
33. Trend Micro: Malware using exploits from shadow brokers leak reportedly in the wild (2017). Available: https://www.trendmicro.com/vinfo/us/security/news/cybercrime-and-digital-threats/malware-using-exploits-from-shadow-brokers-in-the-wild. Last accessed 07 Aug 2019
34. Wikipedia: Epidemiology (2017). Available: https://en.wikipedia.org/wiki/Epidemiology. Last accessed 09 Aug 2019
35. Tuthill, K.: John Snow and the broad street pump (2003). Available: http://www.ph.ucla.edu/epi/snow/snowcricketarticle.html. Last accessed 10 Aug 2017
36. Mandiant: RPT-M-Trends (2017). Mandiant US 47
37. BBC: Edward Snowden: leaks that exposed US spy programme. Available: http://www.bbc.com/news/world-us-canada-23123964. Last accessed 10 Aug 2017
38. Chirgwin, R.: Ex-NSA contractor Harold Martin indicted: he spent 'up to 20 years stealing top-secret files' (2017). Available: https://www.theregister.co.uk/2017/02/08/us_grand_jury_indicts_harold_martin_nsa/. Last accessed 10 Aug 2017
39. Goudbeek, A., Choo, K.-K.R., Le-Khac, N.-A.: A Forensic investigation framework for smart home environment. In: 17th IEEE international conference on trust, security and privacy in computing and communications (IEEE TrustCom-18), New York, USA, Aug 2018. https://doi.org/10.1109/TrustCom/BigDataSE.2018.00201
40. Le-Khac, N.-A., Jacobs, D., Nijhoff, J., Bertens, K., Choo, K.-K.R.: Smart vehicle forensics: challenges and case study. Future generation of computer systems. Elsevier, New York, July 2018. https://doi.org/10.1016/j.future.2018.05.081

**Cian Young** is the Global Security Operation Centre Manager at Workday Inc. He holds a MSc in Forensics Computing and Cybercrime Investigation from University College Dublin. He has been working in Security Operations for global or-ganisations since 2015 following a 10 year career in the Irish Defence Forces. He currently resides in San Francisco, California.

**Robert McArdle** is a Director in Trend Micro's Forward Looking Threat Research team, where he is involved in analysing the latest Cybercrime threats, specializing in researching the future threat landscape, Open Source Intelligence (OSINT) and coordinating investigations with international law enforcement. Robert is a regular presenter for the press and at security conferences, and has contributed to many papers on Cybercrime and future crime research. He also lectures in Malware Analysis and Cybercrime Investigations on MSc modules at Cork IT and University College Dublin (UCD).

**Nhien-An Le-Khac** is a lecturer at the School of Computer Science (CS) University College Dublin (UCD), Ireland. He is currently the program director of MSc program in Forensic Computing and Cybercrime Investigation (FCCI)—an international program for the law enforcement officers specializing in cybercrime investigations. He is also the co-founder of UCD-GNECB Postgraduate Certificate in fraud and e-crime investigation. Since 2008, he is a research fellow in Citibank, Ireland (Citi). He obtained his Ph.D. in Computer Science in 2006 at the Institut National Polytechnique de Grenoble (INPG), France. His research interest spans the area of Cybersecurity and Digital Forensics, Data Mining/Distributed Data Mining for Security, and Fraud and Criminal Detection. Since 2013, he has collaborated on many research projects as a principal/co-PI/funded investigator. He has published more than 150 scientific papers in peer-reviewed journal and conferences in related research fields, and his recent edited book has been listed the Best New Digital Forensics Book according to the Book Authority.

**Kim-Kwang Raymond Choo** received the Ph.D. in Information Security in 2006 from the Queensland University of Technology, Australia. He currently holds the Cloud Technology Endowed Professorship at the University of Texas at San Antonio (UTSA). In 2015, he and his team won the Digital Forensics Research Challenge organized by the Germany's University of Erlangen-Nuremberg. He is the recipient of the 2019 IEEE Technical Committee on Scalable Computing (TCSC) Award for Excellence in Scalable Computing (Middle Career Researcher), 2018 UTSA College of Business Col. Jean Piccione and Lt. Col. Philip Piccione Endowed Research Award for Tenured Faculty, British Computer Society's 2019 Wilkes Award Runner-up, 2019 EURASIP Journal on Wireless Communications and Networking (JWCN) Best Paper Award, Korea Information Processing Society's Journal of Information Processing Systems (JIPS) Survey Paper Award (Gold) 2019, IEEE Blockchain 2019 Outstanding Paper Award, Inscrypt 2019 Best Student Paper Award, IEEE TrustCom 2018 Best Paper Award, ESORICS 2015 Best Research Paper Award, 2014 Highly Commended Award by the Australia New Zealand Policing Advisory Agency, Fulbright Scholarship in 2009, 2008 Australia Day Achievement Medallion, and British Computer Society's Wilkes Award in 2008. He is also a fellow of the Australian Computer Society, an IEEE senior member, and co-chair of IEEE Multimedia Communications Technical Committee's Digital Rights Management for Multimedia Interest Group.

# Forensic Investigation of Ransomware Activities—Part 2

Christopher Boyton, Nhien-An Le-Khac, Kim-Kwang Raymond Choo, and Anca Jurcut

**Abstract** Ransomware is a particularly predatory form of Cybercrime which feeds on a person's sentimental value for personal data such as family photos, videos and sometimes a lifetime's collection of data. In general, a banking Trojan causes a temporary monetary loss, Ransomware however, has the potential to have irreversible, catastrophic loss of data for the victim. Ransomware has grown exponentially since 2015, and it is this staggering growth that poses a problem for future detection. Signature-based detection alone cannot cope with the number of signatures that will continue to be created. Distribution of databases becomes a greater task with the growth of signatures. Anomaly-based detection is another option to consider as it looks at behaviour traits rather than signatures alone. However, many traits found in malware are just as easily found in legitimate software. This fact leads to the possibility of false positives which in turn leads to a lack of confidence from the user. This chapter proposes an approach that uses a hybrid detection system of both signature based detection and anomaly based detection. Analysis was carried out on the crypto-worm variant known as zCrypt, with the goal of analysing attack vectors to counter them effectively. The main aim of this work is to maximize detection rates, minimise false-positives, and protect the best defence against Ransomware—online backups.

**Keywords** Ransomware · Signature-based detection · Anomaly-based detection

C. Boyton (✉)
Irish Defence Forces, Newbridge, Ireland
e-mail: chrisboyton84@gmail.com

N.-A. Le-Khac · A. Jurcut
University College Dublin, Dublin, Ireland

K.-K. R. Choo
University of Texas at San Antonio, San Antonio, TX, USA

© The Editor(s) (if applicable) and The Author(s), under exclusive license to Springer Nature Switzerland AG 2020
N.-A. Le-Khac and K.-K. R. Choo (eds.), *Cyber and Digital Forensic Investigations*, Studies in Big Data 74, https://doi.org/10.1007/978-3-030-47131-6_5

# 1  Introduction

Being in the eras of Internet of Things (IoT) [1] and Internet of Everything (IoE), it is a fashion to use computers connected via the Internet. As a result, cybercrimes have become a major concern for the users in the digital world. Ransomware can be considered as one major type of malwares that makes users to prey upon when accessing the computer systems either by locking the system's screen or by locking the victim's data and files unless an extort payment (ransom) is paid [2]. Hence, the evolution of ransomware has created a lucrative environment for cybercriminals to generate more and more income from potential victims, consequently occurred at an alarming rate day by day.

Since ransomware is often highly automated, the ubiquitous nature inherent with modern computing systems enables criminals to target a wide range of victims not only as individuals, but also like industries and governmental agencies, causing significant financial losses for them. Hence, the need for a comprehensive method for both detecting ransomware activity and preventing infection is highlighted in modern computer systems, thus substantially limiting the loss of data within an organisation.

## 1.1  Evolution of Ransomware

Ransomware acts in the same way as other viruses such as Trojan, Worms and Spyware in the victim's computer. Normally, a Trojan horse attacks the system by hiding the software such as email attachments and downloads, whereas Worms spread itself from file to file, computer to computer, via emails and other network traffic. On the other hand, Spyware is a client side software, mostly comes hidden with free downloaded software, that it uses the computer victim's CPU for monitoring the client activities and send the client's storage or data to a remote machine. By comparison, ransomware is second generation malicious software which targets a slew of files, manly seeks system vulnerabilities potentially caused by its precedents. It makes the victim inaccessible to the system by encrypting the files and locking the system [3].

Ransomware attacks can differ depending on its methodology used for corrupting data files of the victims. The first ransomware PC Cyborg/AIDS was delivered using floppy disks. It principally replaces autoexec.bat file of the computer and counts the number of times that the system reboots. When the counted value exceeds 90, it hides directories and encrypts all the file names in the system root directory [4].

After developing this first ever ransomware PC Cyborg, different facets and features were added for ransomware in order to exploit additional vulnerabilities with cybercrimes in computer systems as shown in Table 1.

**Table 1** Timeline of representative ransomware [4]

| Name | Year | Notable features |
| --- | --- | --- |
| PC Cyborg | 1989 | Spreads using floppy disks |
| GPCoder | 2005–2008 | Spreads via emails; encrypts a large set of files |
| Archiveus | 2006 | First ransomware to use RSA encryption |
| WinLock | 2010 | Blocks PCs by displaying a ransom message |
| Reveton | 2012 | Warning purportedly from a law enforcement |
| DirtyDecrypt | Summ. 2013 | Encrypts eight different file formats |
| CryptLocker | Sept. 2013 | Fetches a public key from the C&C |
| CryptoWall | Nov. 2013 | Requires TOR browser to make payments |
| Android Defender | 2013 | First Android locker-ransomware |
| TorDroid | 2014 | First Android crypto-ransomware |
| Critroni | July 2014 | Similar to CryptoWall |
| TorrentLocker | Aug. 2014 | Stealthiness: indistinguishable from SSH connections |
| CTB-Locker | Dec. 2014 | Uses Elliptic Curve Cryptography, TOR and Bitcoins |
| CryptoWall 3.0 | Jan. 2015 | Uses exclusively TOR for payment |
| TeslaCrypt | Feb. 2015 | Adds the option to pay with PayPal My Cash Cards |
| Hidden Tear | Aug. 2015 | Open source ransomware released for educational purposes |
| Chimera | Nov. 2015 | Threatens to publish users' personal files |
| CryptoWall 4.0 | Nov. 2015 | Encrypts also filenames |
| Linux.Enoder.1 | Nov. 2015 | Encrypts Linux's home and website directories |
| DMA-Locer | Jan. 2016 | Comes with a decrypting feature built-in |
| PadCrypt | Feb. 2016 | Live chat support |
| Locky Ransomware | Feb. 2016 | Installed using malicious macro in a word document |
| CTB-Locker for WebSites | Feb. 2016 | Targets Wordpress |
| KeRanger | Mar. 2016 | First ransomware for Apple's Mac computers |
| Cerber | Mar. 2016 | Offered as RaaS (and quote in Latin) |
| Samas | Mar. 2016 | Pentesting on JBOSS servers |
| Petya | Apr. 2016 | Overwrites MBT with its own loader and encrypts MFT |
| Rokku | Apr. 2016 | Use of QR code to facilitate payment |
| Jigsaw | Apr. 2016 | Press victims into paying ransom |
| CryptXXX | May 2016 | Monitors mouse activities and evades sandboxed environment |

(continued)

**Table 1**  (continued)

| Name | Year | Notable features |
|------|------|------------------|
| Mischa | May 2016 | Installed when PETYA fails to gain administrative privileges |
| RAA | June 2016 | Entirely written in Javascript |
| Satana | June 2016 | Combines the features of PETYA and MISCHA |
| Stampado | July 2016 | Promoted through aggressive advertising campaigns on the Dark web |
| Fantom | Aug. 2016 | Uses a rogue Windows update screen |
| Cerber3 | Aug. 2016 | Third iteration of the Cerber ransomware |

## *1.2   Types of Ransomware*

Based on the variants used for preventing the victims' interaction for computers and the actions it performs on the devices, ransomware can be classified into several groups as follows [5]:

- **File Encryption Ransomware**: They encrypt data and files in the victim's computer and ask to pay the ransom when the user attempts to open the file. This involves symmetric 256-bit AES (Advanced Encryption Standard) key for encryption and asymmetric RSA (Rivest–Shamir–Adleman) private key for decryption in Ransomware activity. This requires a server and an access to the Internet but not always the case, as an example, CBT locker do not require the access to the Internet.
- **Screen Lock Ransomware**: A Trojan with constantly generated messages using the APIs (Application Program Interfaces) from the OS (Operating System) execute a continuous loop and lock the computer screen. Although it asks for a ransom, but the data and files in the computer are not encrypted.
- **Master Boot Record (MBR) Ransomware**: They attack to the portion of hard disk where the operating system boot is located and changes the boot state by displaying another type of message.
- **Web Server or Browser Lock Ransomware**: They attack web servers and encrypt their files. However, sometimes, the ransomware is actually executable, thus only pop up ransom message page which consists of images and HTML codes running on JavaScripts that controls the background threads and applications.
- **Mobile Phone Ransomware**: Recently, ransomware has evolved not only to target computer systems but also for less secured areas like mobiles and M2M (Machine-to-Machine Communication). They are usually embedded in the downloaded applications. However, mobile phone forensics [6] and malware analysis are not in the scope of this chapter.

Some of the major crypto ransomware considered to be very dangerous [7, 8] include: Crypto Wall, CTB Locker, Cryptolocker, Locky, WannaCry, Petya and zCrypt.

- **Crypto Wall**—Crypto Wall mainly encrypts the files and filenames on the victim's computer. This is spread via emails with an attachment of zip file which consists a script file and an exploit kit. These exploit kits have rogue advertisements (ADs) which can direct the victims to attacker's website. Then, it starts injecting itself into explorer.exe, creates a new instance of the process and copies it to %APPDATA%. Thus, it create a registry value run key in the local user registry root path.
- **CTB Locker**—CTB stands as Curve Tor Bitcoin where curve refers to Elliptic Curve Cryptography. This creates an environment for a TOR network which conceals its users' identities and their online activities from network traffic analysis. It can initially process of the victim's without internet connection also. Along with the encryption, it also disables the volume Shadow copies feature in Windows.
- **Cryptolocker**—Now this is also threatening and this was consumed by operation Tovar in the year 2014. Its cost is also breathtaking nearly $300 million over several systems.
- **Locky**—This type of ransomware is distributed via the spam emails. These emails contain Microsoft Office document as an attachment. When the file is downloaded, the macro written in it generates the Locky ransomware. It can also delete the shadow volume copies and can also encrypt external hard drives and database files.
- **WannaCry**—This (also known as WCry or WanaCryptor) is one of the most dangerous types of ransomware. It was all over UK, and spread over 200,000 systems around the world. It is a ransomware program that first utilizes CVE-2017-0199, then adds vulnerabilities for Microsoft Office in Widows machines and finally spreads via EternalBlue.
- **Petya**—Petya ransomware has the lower detection rate by anti-virus search engines since it has lower payload. It also uses the same vulnerability created by the WannaCry ransomware.
- **zCrypt**—This type of ransomware behaves like a virus. It focuses on an uncommon method of spreading and it doesn't depend on malicious emails to find victims. Further, it can spread on USB sticks. Like other ransomware zCrypt does not simply attack all files but rather it finds the important directories that can be altered and then damage these.

The most virulent and aggressive types of ransomware are the *crypto ransomware*, which encrypts the victim's data and the *locker ransomware*, which completely locks the victim's computer by preventing users from accessing the system or input devices to the system. Moreover, crypto ransomware is capable of encrypting any file located on both mapped and unmapped network drives in addition to the files encrypted in the victim's computer; thus can create an instance that brings a department or the entire organization to be halted, if one system is infected within a short period time interval. Very often crypto ransomware does not encrypt the whole hard-disk, but searches for specific extensions only. For examples, files containing text documents, presentations and images (having extensions such as .doc, .jpg, .pdf) that usually contain valuable information of a user [4].

## *1.3   Contribution*

The detection of Ransomware can be a very challenging process [9], whereby failure to detect and identify Ransomware activity can lead to a catastrophic loss of data. To the corporate and enterprise victims, this can mean a considerable loss in earnings and a substantial loss in customer confidence. To the personal user this can mean the loss of years of personal memories and personal data. Current methods of detection include signature based detection and anomaly based detection. Signature based detection is reliant on previously detected samples. Anomaly based detection focuses on known malware infection processes which can lead to false positives [10, 11].

This chapter presents an in-depth look at how a hybrid solution of both signature based and anomaly based detection system can be used in a lab environment to both *identify known samples* and *to detect unknown samples*. The analysis of how this can be achieved will be done through an investigative and technical analysis of the "zCrypt" samples. This analysis will be used to investigate methods used by crypto-worms to propagate and encrypt files. The information gathered based on this investigation is implemented into an automated tool to detect and identify Ransomware.

## 2   Analysis and Detection of Ransomware

This section presents the current techniques used for the analysis and the detection of ransomware. Firstly, the behaviour and the activity of ransomware are briefly introduced. Secondly, the existing techniques used for the detection of ransomware are classified, reviewed and briefly discussed.

## *2.1   Crypto-Ransomware Behaviour and Activity*

The most notable behaviour of Crypto-ransomware is encrypting the original data and files on the victim's device using robust cryptography in a silent manner. After the data and files are encrypted, this malware notifies the user that his data and files have been encrypted and demands a ransom in order to decrypt and release the original data and files. Normally, it is required to pay the ransom in Bit-Coins or in other crypto-currency or digital currency. Note that the irreversible nature and only the receiver is allowed to make refunds for a paid activity, Bit-Coins has become the standard choice for crypto-ransomware cybercrimes.

Since the attacker keeps the decryption key only stored on his server to prevent victim from recovering the encrypted data and files, the victim has no other choice and are forced to pay the ransom. However, if a timer or deadline warning is posted and the user cannot not pay the ransom within the given interval, the data and files are no longer accessible for recovery by the recovery key. Therefore, any victim can only regain his data and through the use of anonymous payment (e.g., Bit-coin [12, 13]). This financial success of ransomware activities lead to introduce several families of ransomware and it is difficult to track these attackers using conventional detection and prevention algorithms.

In the case that the ransomware payloads are not using encryption, then it attempts to lock the screen or modify the MBR (master boot record and/or partition table) of the victim's computer by using a simple application which restricts the interaction with victim's computer system. Hence, this type of ransomware is considered relatively weak and even the damage can be reversed without paying the ransom.

Conventionally, Cryptographic Ransomware stores the decrypt key on the victim's computer, thus it is possible to grab the encrypted key using reverse engineering procedures. However, the modern crypto-ransomware uses a command and control (C&C) server, where the encryption and decrypt keys reside on server and not on the victim's computer. Sometimes, this process enables security teams to identify and audit the traffic of ransomware activity beforehand and discontinue the connection prior to complete the attack. But, attackers has already corrected this flaw by encrypting files before communicating to the C&C [14].

Hence, the crypto-ransomware activity is divided into three types, based on the processes employed by cryptosystem as Symmetrical Cryptosystem Ransomware, Asymmetrical Cryptosystem Ransomware and Hybrid Cryptosystem Ransomware.

1. **Symmetrical Cryptosystem Ransomware**—The same encryption key generated on the infected computer is used as the decryption key to encrypt the victim's data and files by employing a symmetrical encryption algorithm. Typically, file encryption ransomware uses 256-bit length AES (Advanced Encryption Standard) key or DES (Data Encryption Standard) key. Hence, in this method, even the victim can recover the secret key by applying reverse engineering or memory scanning techniques.
2. **Asymmetrical Cryptosystem Ransomware**—In this method, a public key, embedded in the ransomware file or downloaded during communication established with C&C server, is used to encrypt the victim's file. The RSA (Rivest–Shamir–Adleman) encryption generates two user-specific keys, a public key and a private key. Since the private key is never shared by the attacker, the victim is not aware about the private key. Therefore, the victim must pay the ransom to decrypt the files. However, this technique consumes more resources while encrypting the files.
3. **Hybrid Cryptosystem Ransomware**—This uses a dynamically generated symmetric key to encrypt the victim's data and the files and a pre-loaded public key to encrypt the symmetric key itself.

## 2.2  Analysis and Detection of Ransomware

Ransomware is generally analysed using two main methods: (i) static analysis and (ii) dynamic analysis.

(i)  *Static analysis*—is the study of malicious files without executing the files. Packers and obfuscation can make this type of analysis difficult.
(ii) *Dynamic analysis*—is the study of malicious files by executing the file in a controlled environment such as a sandbox or virtual machine. Anti-virtual machine and anti-debugging techniques can make dynamic analysis more difficult to complete.

Ransomware detection can be completed using *host-based signatures* and *network-based signatures*. Host-based signatures detect malicious codes on the victim computer as new files creations, modifications to existing files or specific changes made to the registry. The network-based signatures techniques are used to detect malicious code by monitoring the network traffic.

Basically, the techniques used to classify ransomware activities can be discussed under three categories [15].

1. *Local Static Information*—The information is extracted from the malware before the program is executed.
2. *Dynamic Information*—The information is extracted based on the action of ransomware that takes on the infected computer while the program is running.
3. *Information Extracted from Network Traffic*—This is the information obtained from the network traffic created by the running ransomware.

Some of the most well-known methods that are currently used for the Ransomware detection are briefly discussed in the following subsections.

### 2.2.1  Fast and Efficient Linguistic-Based Ransomware Detection

Signature checking is very effective against known threats. However, the main disadvantage lies in the inability to detect unknown threats. This can be resolved by using distinguishing features and recognising Ransomware behaviour. With Ransomware, there are three key features to be found: the users are threatened, the devices that are locked and the data encrypted. In most cases a ransom demand is made for reward. Linguistic-based Ransomware Detection uses this feature to detect an attack. By analysing strings extracted from the application resources, threatening text can be extrapolated and used to identify a malicious application. This system is used in collaboration with encryption detection, whereby APIs known for use in encryption are used as a filter [16].

### 2.2.2 Early Warning System for Ransomware Detection

CryptoDrop is a system that uses file monitoring to act as an early indication of Ransomware activity. This system proves effective against detecting Ransomware while reducing the occurrence of false positives. CryptoDrop uses an incremental counter to monitor suspicious file activity. Once the threshold is surpassed the flagged process is suspended and further investigation of a protected directory is done to determine whether encryption is being used [17].

### 2.2.3 Malware Detection

A combination of signature-based IDS and anomaly-based IDS can be used to mitigate against malware. It is possible to generate rules for anomaly based IDS for the detection of both known and unknown detection systems. Anomaly based IDS can employ statistical techniques, data mining, artificial neural networks, and genetic algorithms and have the advantage of being able to detect zero-day malware. Markov chains use a random variable which changes over time. The response to detection can include controlling user activity, terminating the host, restarting services and delaying suspicious system calls using hooks. This system allows IDS to discover attacks consistently with minimum false positives [18].

### 2.2.4 Website Protection Schemes Based on Behavior Analysis of Malware Attackers

AV vendors collect malware from the Internet using honeypots. To achieve this, known blacklisted URLs are used to detect diverse collections of samples. These are then analysed and signatures are created and added to databases. A client honeypot mimics vulnerable clients to maximise the number of samples collected [19].

### 2.2.5 Improving the Effectiveness of Behavior Based Malware Detection

Sequences of API calls were grouped into n-grams and compared with benign and malware profiles. Large n-grams result in the requirement for a large database while small n-grams increase the risk of false positives. The research investigates whether anomaly-based and signature-based detectors perform better when used together. NSL-KDD is a benchmark dataset used to compare detection methods. Search file, copy file, delete file, get file information, move file, read file, a write file are APIs are commonly called by malware. API hooking to capture API calls does not have to cause a noticeable delay in common application software although Microsoft Developer Network reports that performance can be hampered. Tests show that API hooking does not affect performance any worse than if using AV software. API monitor and Detours are software programs which can assist with intercepting API calls for Windows OS. Memory related API calls occur at a very high frequency which could impact the operation of the program if hooked [20].

### 2.2.6 Malware Detection Through Call Graphs

A call graph is a graphical representation of a binary executable, whose functions are shown as vertices and calls between functions are shown as edges. IDA Pro call graphs show where local functions start and sub and external calls are given by name. This can be used to investigate the sequence of calls used by specific malware [21].

### 2.2.7 Large-Scale Malware Analysis, Detection, and Signature Generation

A system can be created to speed up the process of identifying suspicious samples by determining whether a new sample is like currently known malware using function-call graphs. This can be used to achieve malware clustering using static features. A SMIT (Scalable Malware Indexing Tree) optimises the number of nodes used in the Hungarian algorithm, which uses a cost analysis approach when analysing function-call graphs. This is achieved by computing a common function set using a combination of each function's mnemonic sequence, CRC of this sequence, and symbolic name. For malware variants within the same family, the common function leads to a match of over 90%. This contrasts with a match of less than 20% for malware from different families. MutantX is also used to create clusters corresponding to malware which can label new threats while reducing the time spent on manual analysis.

Typical hash signatures are derived by hashing the malware binary. This method has a low false positive rate but leads to a problem with signature distribution due to the exponential growth in malware. String signatures which use a byte sequence present in malware can reduce the problem with signature distribution. Hancock uses a number of patterns to determine a good string signature. Unusual constant values can indicate an IP address or port number. Unusual address offsets can indicate a large class or structure used by malware. Local or non-library function calls dependant on how its parameters are prepared. Math instructions can indicate obfuscation, such as XOR operations [22].

### 2.2.8 Adaptive Rule-Based Detection Employing Learning Classifier Systems

Specification-based detection makes use of manually created security specifications for correct behaviour of critical objects. Deviations from these rules indicate possible malware activity. Static analysis of system calls using addresses, names, and return addresses compared to the same system calls monitored during dynamic analysis can show deviations due to injection by a malicious process [23].

### 2.2.9 Dealing with Next-Generation Malware

System calls are hooked using a standard user-space technique. All processes are started in a suspended state and system calls are hooked by hooking KiIntSystem-Call and KiFastSystemCall. User-space applications create handles to access data structures representing a resource of the system [24]. Using a similar approach in NoCrypt would allow each process to be examined upon starting. A global hook can be used to insert functions into each running process to check if that process is malicious.

### 2.2.10 Behavior Based Worm Detection

A worm running on an infected host will attempt to propagate to additional victims by actively scanning the network. Elevated privileges are usually required in order to execute on a remote host. These privileges can be acquired by utilizing an exploit in software or by exploiting a poor configuration. SQL-Slammer was found to use UDP packets to infect additional hosts. UDP was used because it is a stateless protocol which does not need to establish a connection or receive a response from the target address. To identify new targets on a network a worm can perform a scan in a number of ways. These include Random, Local Preference, Sequential, Permutation, Topological and Hit-List scanning.

Worm infection mechanisms include buffer overflow attacks and code injection. Buffer overflow attacks write excess data which overwrites adjacent memory addresses which can then be used to execute code. Code injection exploits web applications whereby data is not properly checked for code markers allowing extra code to be executed [25].

To prevent infection a combination of reactive antibodies (signature based defences and system patching) and protective measures such as address space randomization can hinder worm behaviour. Examining known patterns of system calls can be used to detect worm activity on an individual host. Snapshots of multiple hosts with an intersection rate of greater than 50% can indicate worm infection. However, using snapshots in this manner can lead to false positives.

## 3   Proposed Approach

### 3.1   Analysis of zCrypt

zCrypt was analysed mainly to find a method for identifying Crypto-Worm behaviour. A 32-bit Windows 7 Malware Lab VM was used for the analysis. The typical infection vector for the analysed samples was distributed as an infected email attachment. zCrypt makes an effort to hide itself and spread as much as possible. It applies hidden attributes to dropped files. It calls on the *GetDriveTypeA* function to enumerate all drives with the specific goal of seeking out removable drives. This function is the key to its propagation. zCrypt uses typical methods to achieve persistence by making registry changes and adding itself to the start-up programs. It uses PKI (Public Key Infrastructure) which indicates that decryption would be extremely difficult. Unlike some other Ransomware variants which have the key hardcoded into the program, this creates a unique key for each victim. A test is carried out to ensure files can be encrypted, where failure results in termination of the program.

#### 3.1.1   Infection Vector for zCrypt

In general, Ransomware may possibly infect a victim using several different methods such as vulnerability exploitation, social engineering, infected email attachments and infected external media. However, zCrypt is interesting since it can be downloaded from the Internet via spam emails, macro malware, Fake Flash Player installs or propagated using removable drives. Then, zCrypt attempts to confuse the user by displaying a popup when files are being encrypted. This macro-based malware acts as a distraction to the user.

#### 3.1.2   Installation Process of zCrypt

Usually, zCrypt drops a copy of itself into the following location with hidden attributes.

- C:\Users\Lab\AppData\Roaming\zcrypt.exe
- C:\Users\Lab\AppData\Roaming\btc.addr
- C:\Users\Lab\AppData\Roaming\cid.ztxt
- C:\Users\Lab\AppData\Roaming\public.key
- C:\Users\Lab\AppData\Roaming\private.key.

The computer's personal identifier, or Computer ID (CID) in the "cid.ztxt" file is used to uniquely identify the machine to the cyber criminals. This CID is found embedded in the URL that is contacted. This CID is retrieved from the client machine.

During the Public/Private key encryption process, the strings were examined, basically four references were found as shown in Fig. 1. They are RSA routines which

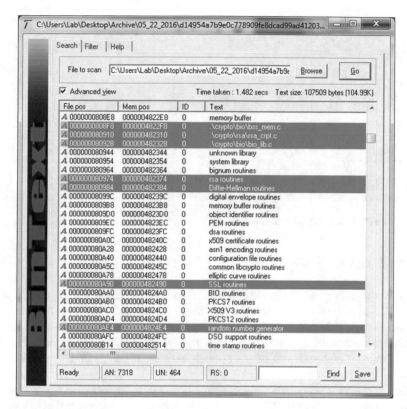

**Fig. 1** Indicators of key exchange and encryption

are used in encryption, Diffie-Hellman routines which are used for key exchanges, SSL routines which form the Secure Sockets Layer and Random Number Generator which is often used as a seed for key generator.

### 3.1.3 Achieving the Persistence

Once zCrypt is successfully installed to the victim's computer, it takes precautionary action to ensure a complete and full infection. zCrypt adds a registry entry to ensure that it is automatically started when the system is rebooted. Along with this entry, numerous registry entries are created for smooth propagation of zCrypt.

Additionally, an *autorun.inf* file is dropped into the removable drive which allows for wormlike propagation. On examination of this file, it was found not to contain anything discernible. This could have been designed with future development in mind. A file called system is also dropped with hidden attributes. This was found to contain the same strings as zCrypt. Further, the start-up folder also contains a shortcut link to *zcrypt.exe* to ensure execution after a reboot of the system.

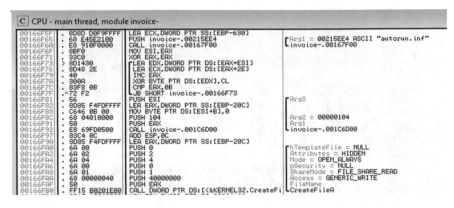

**Fig. 2** Setting hidden attributes to "*autorun.inf*"

### 3.1.4 Propagation of zCrypt

After enumerating all connected drives using the Windows API *GetDriveTypeA*, zCrypt copies itself into all removable drives with hidden attributes, i.e. F:\system.exe. This demonstrates that zCrypt enumerates the drives on execution to map out other drives which are currently available. By hooking an API, *GetDriveTypeA*, it is possible to intercept such activity, if there is a Crypto-Worm in action.

The executable was disassembled using *IDA Pro* to further investigate how the Ransomware is propagated. When the *GetDriveTypeA* function is called, a value of 2 is pushed onto the stack and then it is compared with the return value. The value of 2 signifies the presence of a removable drive. If a removable drive is present, the "*autorun.inf*" file is dropped. However, when this *autorun.inf* file is created, the attributes are changed to hidden thus it will not be visible to the user. This process was further confirmed using *OllyDbg* as shown in Fig. 2.

*BinText*, as shown in Fig. 3, was used to analyse the strings found in the "*system*" application dropped to the removable media. The strings showed the same instructions for payment found in the original zCrypt sample. This confirms the use of worm-like propagation to spread through the medium of connected drives. The strings also showed the URL containing the ransom demand at *http*[*.*]//qwertyuiop[*.*]gp. The domain name is the reverse of the domain contacted for the key exchange.

### 3.1.5 Encryption

zCrypt attempts to download a RSA public key from the C2 Server. If it fails, there is a contingency in plan to allow for off-line execution. zCrypt attempts to use the hidden public key in the *AppData* folder if it cannot connect to the C2 Server. This Ransomware uses statically linked with *OpenSSL* which can be verified by creating a public/private key pair in *OpenSSL* and replacing the dropped public key in the

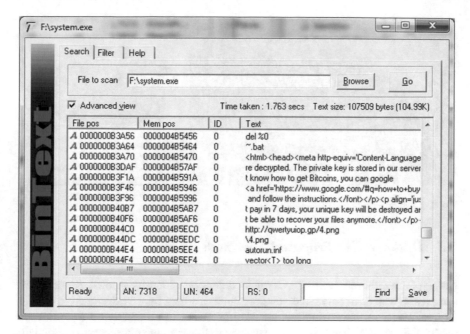

**Fig. 3** Strings showing instructions identical to zCrypt sample

*AppData* folder. Offline encryption of the target system's files can be reversed using the private key. zCrypt carries out a test to see if it can decrypt the target system's files. If this is successful, the remainder of the encrypted files are decrypted. zCrypt creates a file to store encrypted content as *Imager_Lite_3.1.1.zip*.

zCrypt terminates, if it is unable to encrypt the user's files. Files are encrypted slowly as a result of RSA being used. Once encryption is complete, zCrypt checks to see if the private key has been released, indicating that payment has been made. zCrypt downloads the private key and decrypts the target system's files. When the decryption process is completed, then deletes itself and the keys associated with it.

### 3.1.6 Static Analysis

DIE is an IDA python plugin designed to enrich IDA's static analysis with dynamic data. DIE was used to test the samples to see if they were packed. Although the sample was found not to be packed there was evident compression/encryption (Fig. 4).

The entropy of the executable indicated that there was a possibility of packing. However, when the executable was examined in the disassembler, this was found not to be the case.

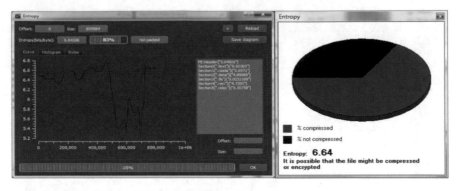

**Fig. 4** Check for packing

## 3.2 Design of the Proposed Approach

NoCrypt uses the analysis of zCrypt to detect and identify known Ransomware samples as well as to detect unknown variants. It was primarily designed to counter Crypto-Worm variants using enumeration of drives to propagate up to removable media and network shares. NoCrypt was created as a modular, top down design, with each individual function tested before being imported into one complete program. The program was written in C/C++ using Microsoft Visual Studio 2015. It was designed for use in a Windows 7 32-bit VM lab environment, with 4 GB RAM, 4 processors, 60 GB pre-allocated memory and NAT. The flowchart in Fig. 5 illustrates the execution of the program.

NoCrypt begins by enumerating drives to determine which drives require protection. To demonstrate the core features of this program, the currently logged user directory is used as the protected directory. This could also be expanded to protect specified directories in specified drives. Files are implanted in the protected directory to be used as a future reference for NoCrypt when baseline being surpassed.

Once the implantation of files has been completed, all currently running processes are enumerated to establish if there are any malicious processes running. This is accomplished using a whitelist (containing SHA256 hashes of known windows processes), a blacklist (containing SHA256 hashes of known ransomware samples), and finally, the transmission of unknown hashes to *VirusTotal*. Although, the hash has still not being recognised at this stage, the implanted files can monitor any change in the system.

A global hook is used to monitor API calls to *GetDriveType* using *NoCrypt.dll*. Upon termination of the program for a user's request, all implanted files are deleted from the protected directory to prevent any accumulation of unused files with time.

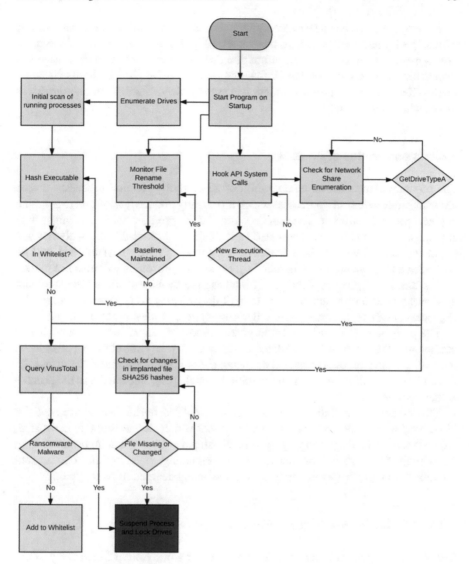

**Fig. 5** Flow control of NoCrypt program

### 3.2.1 Enumeration of Drives

The first task carried out by the NoCrypt program is, enumeration of drives on the client machine. This creates a complete list of associated drives, which is used to show the program such that which drives will have files implanted on them. These implanted files will be used to check for the Ransomware activity in the event of an unusual file change.

In here, the *GetLogicalDriveStrings* function is used to fill a buffer with strings which specify any valid drives in the system [26]. The returned value of the function is passed to the *GetDriveType* function that identifies whether a disk drive is a removable, fixed, CD-ROM, RAM disk or a network drive [27]. It should be noted that the *GetDriveTypeA* (ANSII) function is used by zCrypt itself to enumerate drives to assist in propagation.

### 3.2.2    Enumeration of Processes

The program carries out an enumeration of all currently running processes on the client machine to identify processes which are already in the whitelist and to examine any new processes further. The first execution of the program will take longer than any subsequent executions. The reason for this is that normally the whitelist is a populated one. However, since processes vary from system to system and will be created at an exponential rate, hardcoding for known processes will also not work.

The amount of process identifiers (PIDs) as used by Microsoft, is calculated and used as the maximum value in a '*for*' loop [28]. Each process is passed to *pathfinder* () process and *hash_process* () while ignoring system processes at this stage.

The *pathfinder* () function takes a given process ID as an input parameter and extrapolates the file pathname. A handle is created to take control of this process. A query of the open process is carried out using *GetModuleFileNameEx* to retrieve the full path name of file containing the specified module [29]. The handle to the process is then returned.

This process enumeration is used to create *SHA256* hashes which are used for identification of trusted processes. Any processes which are not found to be trusted, are further examined by querying *VirusTotal* using a Public API code or monitoring file activity for a suspicious increase in file renaming. If any of the enumerated processes alters the implanted files, it is immediately deemed to be malicious.

### 3.2.3    Monitoring the System for an Abnormal File Activity

The *WatchDirectory* () function is used to monitor a baseline of file activity. If the assigned threshold is surpassed, it will indicate a possible threat. This in itself is not a definitive declaration of threat, as normal file activity could surpass this threshold. It is just another method of indicating that activity needs to be investigated further. A directory name is given as an argument and in the case of this program, it is the root of each drive which has been previously enumerated.

To measure whether or not the baseline is being surpassed, the program uses *SYSTEMTIME*. Anti-debugging techniques use the system time to query the amount of time between a given number of machine code instructions. This provided the inspiration for checking if there was in increase in file activity.

As mentioned, this function is used to monitor the amount of activity occurring with respect to file changes. This monitoring occurs at two levels, the monitoring of directories and the monitoring of files within the directories. The *Find-FirstChangeNotification* function sets up a handle with initial change notification filters, if a change matching the filter occurs the action is processed [30]. In the case of this function *lpDir* is the variable containing the directory to be watched. For the purpose of this program *lpDir* will be the root of each selected drive. This handle is configured not to watch the subtree to reduce resources required by the system. The *FILE_NOTIFY_CHANGE_FILE_NAME* change notification filter is used to monitor changes to files. Although the program is only interested in file renaming, this parameter returns any file name change within the directory including renaming, creating or deleting. Error handling is presented to mitigate against failures to create a handle.

The same process is carried out for monitoring each sub-directory. The parameters are the same as the previous section with the exception of two changes. This process uses a second separate handle and subtree directories are monitored.

### 3.2.4   Implanting Files to Monitor Changes

One of the key features of the execution of this program is the implanting of files throughout the system. These files are used to detect Ransomware activity by identifying filename changes.

The *ListDirectoryContents* () function was used to recursively list directory contents of a directory path argument. For this program, the directory is the root directory of the drive. The path names of all files within the directory are retrieved using '*' as a file mask for all contents. The *FindFirstFile* function searches for a file or subdirectory using a specified name and attributes [31]. The function uses the fact that an existing directory will always have "." and ".." as the first directories in it. If these directories are present, the file path is built up using the parent directories in the name.

A check is carried out to see if the entity is a file or folder. The *FILE_ATTRIBUTE-_DIRECTORY* is used to determine if an entity is a folder. This ensures that implanted files are only written to directory folders. The use of recursion ensures that files are dispersed down along the subtree to child folders. This prevents Ransomware skipping directories to evade detection. The *gen_random* () function solves the issue of files being created with the same name as user files. The *create_file* function creates the implanted file using a randomly generated file extension. The purpose of this is to create files with extensions known to be encrypted by Ransomware.

The final stage of this function resolves the issue of how to minimise the impact of implanted files on user activity. The *GetFileAttributes* function is used to retrieve the file's system attributes. The *SetFileAttributes* function is used to set the file's attributes to hidden [32]. If the user has set the Windows folder options to be able to view hidden items, these files will be visible. If the user attempts to open any of these files, there will be no impact to the system as the files do not contain any information.

### 3.2.5    Transmitting Files to *VirusTotal*

This section of the program uses a modified version of a program called *"check_first"* written by Adam Kramer. The program is licenced under the *GNU GENERAL PUBLIC LICENCE* Version 3, 29 June 2007. The source code has been modified for use in this program [33].

*VirusTotal.com* provides a public API which allows the uploading and scanning of files to access information generated by *VirusTotal*. The API uses *HTTP POST* requests in conjunction with JSON responses. The public API limits requests to 4 per minute. The API key used in this program is a personal API key used for testing purposes. Misuse of this key will result in a ban from using the *VirusTotal* public API [34].

The reason the file size is important is that *VirusTotal* has imposed a file size limit for its public API. A check is done to see if the file to be uploaded is greater than 32 MB. A warning will be displayed if the file is too large. This should not be an issue as most malware executables have a size in the order of kilobytes.

### 3.2.6    Global API Hook

A global or system-wide API hook is used to intercept specified function calls from running processes [35]. First, it is required to change registry key values of *AppInit_DLLs* and *LoadAppInit_DLLs*. The *RegSetValueEx* function is called to set the data of the specified *LoadAppInit_DLLs* value using the on variable [36]. The same process is used to add the pathname of the custom DLL to the *AppInit_DLLs* value. An extra character is added to the *strlen* parameter to account for the NULL terminator.

When the program is executed, the *AppInit_DLLs* infrastructure is used to load the custom DLL containing the hook in all user-mode processes that are linked to *User32.dll*. The DLL itself uses a header file containing the API specific functions. There are four distinct phases identified when using a hook [37] as initialize the hook, hook the API, unhook the API and remove the hook.

A data structure is used in the header file to manage API hooks. When a hook is initialized it monitors the entire system for a specific function call. In the event of the function call being detected, the first six bytes of the function are overwritten with a jump instruction to an alternate function. Once the hook function is carried out, execution resumes from the original function address in what is known as trampolining. The data structure stores the function address of the original function and the hook function as well as an array for the first six bytes containing op codes for execution.

Once the alternative function has completed its execution the *UnhookAPIHook* function is used to return the flow of execution to the first 5 bytes of the original function. This leads to normal execution of the previously hooked function. The final step in the hooking process is to remove the hook using the *RemoveAPIHook* function. In this stage, the allocated virtual memory is released back to the system for use.

**Fig. 6** *DebugView* showing hooked processes

# 4 Evaluation of Proposed Approach

## 4.1 Evaluation Methodology

### 4.1.1 Testing the Global API Hook

The global API Hook was tested to determine if all processes were being hooked by the *NoCrypt DLL* file. The notifications of successful injection into processes are the outputs to *DebugView* which helps to monitor the hooking process. Current processes were hooked and the *ExitProcess* and *GetDriveTypeA* functions were monitored. *DebugView* was also used to confirm that new processes were also hooked as shown in Fig. 6. When a process was terminated, the *ExitProcess* hook was executed and the output to *DebugView* documented the process as finishing and the hook being released.

### 4.1.2 Detection of Known Sample Hashes

NoCrypt contains a blacklist of 706 known Ransomware sample hashes. Each of these samples was checked for detection rates from 60 AV (Anti-Virus) vendors which use *VirusTotal*. NoCrypt detects these samples and identifies the family which they are belonged to. The pervious literature shows that signature-based detection alone is not sufficient to detect Ransomware activity.

Testing the NoCrypt to confirm detection of known Ransomware hashes results a warning to the user in each instance. Along with identifying the Ransomware

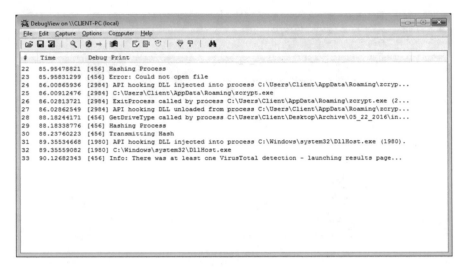

**Fig. 7** *DebugView* output of hash transmission to *VirusTotal*

variant, a link to a decryptor was also provided if exists. When a sample is found to be malicious, the process is immediately terminated and a warning is given to the user.

### 4.1.3   Detection of Sample Hashes by Querying *VirusTotal*

As each process is hooked, NoCrypt checks for a match in the whitelist or in the blacklist and if still unknown, the hash is transmitted to *VirusTotal*. Testing was carried out to confirm NoCrypt's performance in querying *VirusTotal*. This was done using known malicious files which were not in the blacklist and known trusted applications to confirm that NoCrypt would not impact the operation of legitimate programmes. The Task Manager was monitored to ensure that the sample was running. The *NoCrypt DLL* was injected into the sample "*invoice-order.exe*". The output of *DebugView* shows in line 28 that the *GetDriveType* function was called by the sample as shown in Fig. 7. The hash was transmitted and it was found on *VirusTotal*.

If there is a *VirusTotal* match found, the results page is launched in Google Chrome. This allows the user to identify the malicious file as shown in Fig. 8 and there is also an indication in the top right of the page whether other user have flagged the process as malicious or legitimate.

### 4.1.4   Detection of Worm Based Propagation

Crypto Worms use the same system calls as legitimate functions since detection of *GetDriveType* function by itself is not an indicator for Ransomware activity. The

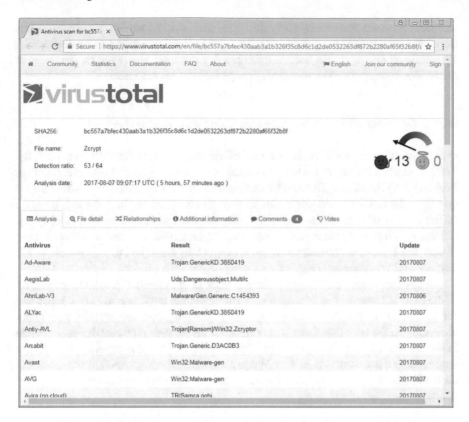

**Fig. 8** Results page of *VirusTotal*

global hook detects this function call. Three functions were used to gauge the effectiveness of NoCrypt's propagation detection as introducing a new removable media to the system, introducing an external network adapter and running a sample of zCrypt. In all three cases, *GetDriveType* function call was detected and a warning was given to the user. To ensure no false-positives, the function call detection results in further monitoring of Ransomware file activity.

### 4.1.5 Performance of Baseline Monitoring

Windows Resource Monitor was used to assess the impact of NoCrypt on system resources when monitoring baseline file activity. First, CPU usage can be seen before the execution of NoCrypt. When NoCrypt is not running the system is using on average 4% of the CPU capability. There is a significant increase when NoCrypt first runs. This is a result of the enumeration of all running processes and an increase in read/write activity as files are implanted to the protected directories. The CPU usage returns to normal once the checking of implanted files is completed. The baseline monitoring results in a check, every time the baseline is surpassed. The baseline has been set as 4 file changes per second. The CPU usage was found to be acceptable, with little or no impact on normal operation by the user. However, in the event of many files being deleted or renamed, the implanted files are checked.

### 4.1.6 Detection of Unknown Samples Using File Monitoring

File monitoring is the third in the line of defence against Ransomware activity. Testing was carried out to ensure that changes made to implanted files are detected. While testing the *Watch_Directory* function, one of the implanted files was changed and then the baseline was surpassed to force a check of implanted files. Because of the change of data in the implanted file, the hash of the file was different and did not match the reference hash. Then, the implanted files were checked using the *Check_Implants* function and the verbose output was displayed to the console as file number 352 was changed (Fig. 9).

An unknown sample of "*CryptoHasYou*" was used to further test the functionality of the baseline monitoring. NoCrypt was running while the Ransomware sample was run and because *VirusTotal* had not seen it. A warning was successfully generated, indicating to the user that there was possible Ransomware Activity due to a change in the implanted files. Therefore, this test confirmed that the baseline monitoring successfully detects encrypted files. When a change in the implanted files is detected, all non-essential processes are terminated and the file which has been changed is displayed as shown in Fig. 10.

## 4.2   Evaluation Results

See Table 2.

**Fig. 9**  Demonstration of hash change of implanted file

**Fig. 10** Termination of non-essential processes

**Table 2** Comparison of different ransomware detection methods

| Sample | Detection method | | | Result | Remarks |
|---|---|---|---|---|---|
| | Hash detection | VirusTotal detection | File monitoring | Encrypted files | |
| zCrypt | | ✓ | | 0 | |
| Z2 | ✓ | | | 0 | |
| BTCWare | | ✓ | | 0 | |
| Cloudsword | | ✓ | | | Over 5 min |
| CryptoHasYou | ✓ | | | 0 | |
| Dcry | | | ✓ | 15 | |
| GlobeImposter | | ✓ | | 0 | |
| GoldForYou | | | ✓ | 40 | |
| Nemesis | | ✓ | | 1 | Admin |

## 4.3   Discussion

### 4.3.1   Practicality of the Propose Approach

NoCrypt can be used as an Analysis Tool to immediately identify known samples and monitor unknown samples. It can be used to confirm whether the sample is Ransomware and if the sample is in the blacklist, then it provides useful information to the user. NoCrypt would not be a substitution for AV software as it specifically deals with Ransomware only.

### 4.3.2   Assessment of *NoCrypt* Against Other Approaches

Heldroid uses machine learning to detect Ransomware using a threatening text detector, an encryption detector and a locking detector [38]. In testing, there were 19 undetected Ransomware samples out of 375 because of the system's reliance on linguistics. The system could be fooled by using another language whereas *NoCrypt* uses file activity as an indicator of malicious behaviour.

CryptoDrop is a program that uses a similar method of file monitoring to check for file encryption. In a test of 492 samples, on average, there was a file loss of 10 files per sample executed [39]. This performance is better than *NoCrypt*, which has a greater file loss due to the *VirusTotal* public API time delay. However, this system does not use known hashes to detect or identify Ransomware.

While these systems are very effective at detecting Ransomware using a single detection technique, there is a clear benefit to using a combination of signature- and anomaly-based detection. While signature-based detection is quick at identifying known threats, it is ineffective against new ones. On the other hand, anomaly-based detection has the disadvantage of resulting in false-positives. *NoCrypt* attempts to capitalize on the benefits of using both these techniques, while mitigating against the disadvantages associated with them.

## 5 Conclusions

The design and evaluation of NoCrypt has highlighted the benefits associated with using a hybrid system of both signature and anomaly based detection. The results shown in the comparison of AV vendor's detection rate of Ransomware samples demonstrate the shortcomings of using signature based detection alone.

NoCrypt uses three lines of defence, known hashes, hash query and file monitoring. Monitoring implanted files to detect file changes eliminates any false positives associated with anomaly-based detection. While the proof of concept works, file monitoring is only effective if files are encrypted. This leads to some files being encrypted before the process is terminated. While it does not prevent encryption, it does prevent a complete loss of data. The number of files encrypted is low and the path name of each file that has been changed is shown in the message box.

Known samples are immediately identified and the user is shown the name of the sample, the location of the malicious file, and where applicable, a link to a decryptor. Unknown Ransomware variants are detected and terminated to reduce loss of data. All non-essential processes are terminated as a precaution.

Worm-like Propagation is detected and prevented using the global API hook. This method of preventing propagation is successful as once the process calling the API to enumerate drives is deemed malicious, it is terminated and the user is warned of an attempt to enumerate drives.

There are several ways that NoCrypt could be improved as better solution:

- Mitigate against users accidently removing implanted files causing the file monitoring function to produce errors while comparing implanted file hashes with reference file hashes.

- Reducing the number of files encrypted before detection from the file monitoring function could be done by creating a honeypot directory at the root of each protected directory.
- The public API key provided by *VirusTotal* restricts the user to 4 queries per minute, causing a considerable delay in the enumeration of processes. Use of a private API key would improve NoCrypt's performance and the speed at which it detects the threats.
- The option to add a process to the whitelist if *VirusTotal* users deem it a false-positive would eliminate any false-positives. The option to add trusted applications would also improve NoCrypts functionality and efficiency while reducing the number of queries to *VirusTotal* required.
- Having an update service like that used by AV vendors, updating the blacklist with sample hashes would improve NoCrypt's ability to identify specific Ransomware variants.

# References

1. Alabdulsalam, S., Schaefer, K., Kechadi, M.-T., Le-Khac, N.-A.: Internet of things forensics: challenges and case study. In: Gilbert, P., Sujeet, S. (eds.) Advances in Digital Forensics XIV. Springer Berlin Heidelberg, New York (2018). https://doi.org/10.1007/978-3-319-99277-8_3
2. Gonzalez, D., Hayajneh, T.: Detection and prevention of crypto-ransomware. In: Proceedings of IEEE 8th Annual Ubiquitous Computing, Electronics and Mobile Communication Conference (UEMCON), Oct 2017, pp. 472–478
3. Aidan, J.S., Garg, Z.U.: Advanced Petya ransomware and mitigation strategies. In: 2018 First International Conference on Secure Cyber Computing and Communication (ICSCCC), 2018, pp. 23–28
4. Sgandurra, D., Muñoz-González, L., Mohsen, R., Lupu, E.C.: Automated dynamic analysis of ransomware: benefits, limitations and use for detection (2016). arXiv:1609.03020. [Online]. Available: https://arxiv.org/abs/1609.03020
5. Shinde, R., Van der Veeken, P., Van Schooten, S., van den Berg, J.: Ransomware: studying transfer and mitigation. In: Computing Analytics and Security Trends (CAST) International Conference, 2016, pp. 90–95
6. Faheem, M., Le-Khac, N.-A., Kechadi, M.-T.: Smartphone forensics analysis: a case study for obtaining root access of an android Samsung S3 device and analyse the image without an expensive commercial tool. J. Inf. Secur. 5(3), 83–90 (2014). https://doi.org/10.4236/jis.2014.53009
7. Kok, S.H., Abdullah, A., Jhanjhi, N.Z., Supramaniam, M.: Ransomware, threat and detection techniques: a review. Int. J. Comput. Sci. Netw. Secur. 19, 136–146 (2019)
8. Dunn, J., Macaulay, T., Magee, T.: The worst types of ransomware attacks. Computerworld, 12 June 2018. https://www.computerworlduk.com/galleries/security/worstransomware-attacks-we-name-internets-nastiest-extortion-malware3641916/
9. Connolly, L.Y., Wall, D.S.: The rise of crypto-ransomware in a changing cybercrime landscape: taxonomising countermeasures. Comput. Secur. 87 (2019). https://doi.org/10.1016/j.cose.2019.101568
10. Schaefer, E., Le-Khac, N.-A., Scanlon, M.: Integration of ether unpacker into ragpicker for plugin-based malware analysis and identification. In: 16th European Conference on Cyber Warfare and Security, Dublin, Ireland, June 2017

11. Linke, A., Le-Khac, N.-A.: Control flow change in assembly as a classifier in malware analysis. In: 4th IEEE International Symposium on Digital Forensics and Security, Arkansas, Apr 2016. https://doi.org/10.1109/isdfs.2016.7473514
12. Zollner, S., Choo, K.-K.R., Le-Khac, N.-A.: An automated live forensic and postmortem analysis tool for bitcoin on windows systems. IEEE Access **7** (2019). https://doi.org/10.1109/access.2019.2948774
13. Van der Horst, L., Choo, K.-K.R., Le-Khac, N.-A.: Process memory investigation of the bitcoin clients electrum and bitcoin core. IEEE Access **5** (2017). https://doi.org/10.1109/access.2017.2759766
14. Almashhadani, A., Kaiiali, M., Sezer, S., O'Kane, P.: A multi-classifier network-based crypto ransomware detection system: a case study of locky ransomware. IEEE Access **7**, 47053–47067 (2019)
15. Berrueta, E., Morato, D., Magaña, E., Izal, M.: A survey on detection techniques for cryptographic ransomware. IEEE Access **7**, 44925–44944 (2019)
16. Andronio, N.: Heldroid: fast and efficient linguistic-based ransomware detection. Master Thesis. [Online]. Indigo.uic.edu. Available at: http://indigo.uic.edu/bitstream/handle/10027/19676/Andronio_Nicolo.pdf?sequence=1 (2012)
17. Scaife, N., Carter, H., Traynor, P., Butler, K.R.B.: Cryptolock (and drop it): stopping ransomware attacks on user data. In: Proceedings of IEEE 36th International Conference on Distributed Computing Systems (ICDCS), June 2016, pp. 303–312. https://doi.org/10.1109/icdcs.2016.46
18. Alazab, A., Hobbs, M., Abawajy, J., Khraisat, A.: Malware detection and prevention system based on multi-stage rules. Int. J. Inf. Secur. Privacy (IJISP) **7**(2), 29–43 (2013)
19. Yagi, T.: Website protection schemes based on behavior analysis of malware attackers. Master Thesis. [Online]. /ir.library.osaka-u.ac.jp. Available at: http://ir.library.osaka-u.ac.jp/dspace/bitstream/11094/51137/1/25863_%e8%ab%96%e6%96%87.pdf (2013)
20. Fadsli Marhusin, M.: Improving the effectiveness of behaviour-based malware detection. Master of Information Technology (Computer Science), UKM, Malaysia. [Online]. Unsworks.unsw.edu.au. Available at: http://unsworks.unsw.edu.au/fapi/datastream/unsworks:10868/SOURCE02?view=true (2012)
21. Kinable, J.: Malware Detection Through Call Graphs. [Online]. Brage.bibsys.no. Available at: https://brage.bibsys.no/xmlui/bitstream/handle/11250/262290/353049_FULLTEXT01.pdf?sequence=1&isAllowed=y (2010)
22. Hu, X.: Large-scale malware analysis, detection, and signature generation. Doctor of Philosophy (Computer Science and Engineering), The University of Michigan. [Online]. Deepblue.lib.umich.edu. Available at: https://deepblue.lib.umich.edu/bitstream/handle/2027.42/89760/huxin_1.pdf?sequence=1&isAllowed=y (2011)
23. Blount, J.: Adaptive rule-based malware detection employing learning classifier systems. Masters Theses. 5008. https://scholarsmine.mst.edu/masters_theses/5008 (2011)
24. Paleari, R.: Dealing with next-generation malware. PhD Thesis. [Online]. air.unimi.it. Available at: https://air.unimi.it/retrieve/handle/2434/155496/138529/phd_unimi_R07627.pdf (2010)
25. Stafford, J.: Behaviour-based worm detection. PhD Thesis. [Online]. Scholarsbank.uoregon.edu. Available at: https://scholarsbank.uoregon.edu/xmlui/bitstream/handle/1794/12341/Stafford_oregon_0171A_10322.pdf?sequence=1&isAllowed=y (2012)
26. Msdn.microsoft.com: GetLogicalDriveStrings Function (Windows). [Online]. Available at: https://msdn.microsoft.com/en-us/library/windows/desktop/aa364975(v=vs.85).aspx (2017)
27. Msdn.microsoft.com: GetDriveType Function (Windows). [Online]. Available at: https://msdn.microsoft.com/en-us/library/windows/desktop/aa364939(v=vs.85).aspx (2017)
28. Msdn.microsoft.com: Enumerating All Processes (Windows). [Online]. Available at: https://msdn.microsoft.com/en-us/library/windows/desktop/ms682623(v=vs.85).aspx (2017)
29. Msdn.microsoft.com: GetModuleFileNameEx Function (Windows). [Online]. Available at: https://msdn.microsoft.com/en-us/library/windows/desktop/ms683198(v=vs.85).aspx (2017)
30. Msdn.microsoft.com: FindFirstChangeNotification Function (Windows). [Online]. Available at: https://msdn.microsoft.com/en-us/library/windows/desktop/aa364417(v=vs.85).aspx (2017)

31. Msdn.microsoft.com: FindFirstFileEx Function (Windows). [Online]. Available at: https://msdn.microsoft.com/en-us/library/aa364419(VS.85).aspx (2017)
32. Msdn.microsoft.com: Retrieving and Changing File Attributes (Windows). [Online]. Available at: https://msdn.microsoft.com/en-us/library/windows/desktop/aa365522(v=vs.85).aspx (2017)
33. Kramer, A.: adamkramer/check_first. [Online]. GitHub. Available at: https://github.com/adamkramer/check_first/blob/master/check_first.cpp (2015)
34. Virustotal.com: Public API Version 2.0—VirusTotal. [Online]. Available at: https://www.virustotal.com/en/documentation/public-api/#getting-url-scans (2017)
35. Podobry, S.: Easy way to set up global API hooks—CodeProject. [Online]. Codeproject.com. Available at: https://www.codeproject.com/Articles/49319/Easy-way-to-set-up-global-API-hooks (2012)
36. Msdn.microsoft.com: RegSetValueEx Function (Windows). [Online]. Available at: https://msdn.microsoft.com/en-us/library/windows/desktop/ms724923(v=vs.85).aspx (2017)
37. rohitab.com—Forums: Header file for API hooking—Source Codes. [Online]. Available at: http://www.rohitab.com/discuss/topic/40192-header-file-for-api-hooking/#entry10106168 (2013)
38. Andronio, N.: Heldroid: Fast and Efficient Linguistic-Based Ransomware Detection. [Online]. Indigo.uic.edu. Available at: http://indigo.uic.edu/bitstream/handle/10027/19676/Andronio_Nicolo.pdf?sequence=1 (2012)
39. Scaife, N., Carter, H., Traynor, P., Butler, K.: Stopping Ransomware Attacks on User Data. [Online]. https://www.cise.ufl.edu/. Available at: https://www.cise.ufl.edu/~traynor/papers/scaife-icdcs16.pdf (2016)

**Christopher Boyton** received a bachelor of Engineering in Electronic Engineering and Military Communications Systems from the Institute of Technology Carlow (2014), and a Master of Science in Forensic Computing and Cybercrime Investigation from University College Dublin (2016). He was a member of the Defence Forces from 2004 to 2017. While in the Defence Forces he worked in the Communications and Information Services Corp. Since 2017 he has been working for Trend Micro in his current role as a Security Incident Analyst for Trend Micro's Managed Detection and Response service. His research interests focus on network and endpoint security, malware analysis, proactive threat hunting, automating investigation and incident response and remediation.

**Nhien-An Le-Khac** is a lecturer at the School of Computer Science (CS), University College Dublin (UCD), Ireland. He is currently the program director of MSc program in Forensic Computing and Cybercrime Investigation (FCCI)—an international program for the law enforcement officers specializing in cybercrime investigations. He is also the co-founder of UCD-GNECB Postgraduate Certificate in fraud and e-crime investigation. Since 2008, he is a research fellow in Citibank, Ireland (Citi). He obtained his Ph.D. in Computer Science in 2006 at the Institut National Polytechnique de Grenoble (INPG), France. His research interest spans the area of Cybersecurity and Digital Forensics, Data Mining/Distributed Data Mining for Security, and Fraud and Criminal Detection. Since 2013, he has collaborated on many research projects as a principal/co-PI/funded investigator. He has published more than 150 scientific papers in peer-reviewed journal and conferences in related research fields, and his recent edited book has been listed the Best New Digital Forensics Book according to the Book Authority.

**Kim-Kwang Raymond Choo** received the Ph.D. in Information Security in 2006 from the Queensland University of Technology, Australia. He currently holds the Cloud Technology Endowed Professorship at the University of Texas at San Antonio (UTSA). In 2015, he and his team won the Digital Forensics Research Challenge organized by the Germany's University of

Erlangen-Nuremberg. He is the recipient of the 2019 IEEE Technical Committee on Scalable Computing (TCSC) Award for Excellence in Scalable Computing (Middle Career Researcher), 2018 UTSA College of Business Col. Jean Piccione and Lt. Col. Philip Piccione Endowed Research Award for Tenured Faculty, British Computer Society's 2019 Wilkes Award Runner-up, 2019 EURASIP Journal on Wireless Communications and Networking (JWCN) Best Paper Award, Korea Information Processing Society's Journal of Information Processing Systems (JIPS) Survey Paper Award (Gold) 2019, IEEE Blockchain 2019 Outstanding Paper Award, Inscrypt 2019 Best Student Paper Award, IEEE TrustCom 2018 Best Paper Award, ESORICS 2015 Best Research Paper Award, 2014 Highly Commended Award by the Australia New Zealand Policing Advisory Agency, Fulbright Scholarship in 2009, 2008 Australia Day Achievement Medallion, and British Computer Society's Wilkes Award in 2008. He is also a fellow of the Australian Computer Society, an IEEE senior member, and co-chair of IEEE Multimedia Communications Technical Committee's Digital Rights Management for Multimedia Interest Group.

**Anca Delia Jurcut** received a bachelor of Mathematics and Computer Science from West University of Timisoara, Romania (2007) and a Ph.D from University of Limerick, Ireland (2013). From 2008 to 2013, she was a research assistant with the Data Communication SecurityLaboratory at University of Limerick, and from 2013 to 2015, she was working as a postdoctoral researcher in the Department of Electronic and Computer Engineering at the University of Limerick and as a software engineer at IBM, Ireland. Since 2015, she has been an assistant professor with the School of Computer Science, University College Dublin, Ireland. Her research interests focus on network and data security, security for internet of things (IoT), security protocols, formal verification techniques and attack detection.

# CCTV Forensics in the Big Data Era: Challenges and Approaches

**Richard Gomm, Ryan Brooks, Kim-Kwang Raymond Choo, Nhien-An Le-Khac, and Kien Wooi Hew**

**Abstract** The video security market has seen rapid expansion over the last few years, expanding from the old analogue into new and emerging IP systems whilst running on multiple platform digital video recorders (DVR) of both Open and Proprietary format. Accordingly to the 2017 video surveillance report from IFSEC Global (https://www. ifsecglobal.com/video-surveillance-report-2017), for example, proprietary formats accounted for 56% of their surveyed responses. It is primarily within these proprietary formats that numerous challenges arise with the forensic acquisition and analysis of the closed circuit television camera (CCTV) data. Such challenges are expanded when consideration is given to the volume of data and the complexity of data due to the different manufacturers, data formats, file systems, volume storage methods, etc. In this paper, we review current approach on CCTV forensics in literature, and identify challenges of CCTV forensics in the context of big data forensics. Then, we describe a new forensic process/workflow for acquiring and analysing artefacts from large amounts data from CCTV devices of different models/manufacturers, prior to presenting and analysing three real world case studies.

**Keywords** Digital forensics · CCTV forensics · DVR · File system · Big data forensics

R. Gomm (✉)
Garda Síochana Ombudsman Commission, Dublin, Ireland
e-mail: Richard.Gomm@gsoc.ie

R. Brooks
Avon and Somerset Constabulary, Bristol, UK

K.-K. R. Choo
University of Texas at San Antonio, San Antonio, TX, USA

R. Gomm · N.-A. Le-Khac
University College Dublin, Dublin, Ireland

K. W. Hew
Royal Canadian Mounted Police, Vancouver, BC, Canada

© The Editor(s) (if applicable) and The Author(s), under exclusive license to Springer Nature Switzerland AG 2020
N.-A. Le-Khac and K.-K. R. Choo (eds.), *Cyber and Digital Forensic Investigations*, Studies in Big Data 74, https://doi.org/10.1007/978-3-030-47131-6_6

# 1 Introduction

According to The British Security Industry Authority (BSIA), there is one CCTV for every 14 people in the UK, totalling 4.9 million cameras. In 2009, the Metropolitan Police, London, England used CCTV evidence in 95% of its murder cases [1]. IMS Research [2], an independent supplier of market research, estimated that the video surveillance equipment market was worth over $9 billion in 2010. Yet only around 40% of the market was accounted for by the 15 top video surveillance equipment supplier. The remaining market share was split amongst the numerous other smaller companies who provide CCTV solutions, usually at lower prices resulting from components with lower specification.

The variety of the components used within the DVRs has resulted in numerous proprietary systems being developed ranging from simple modifications of standard file systems, such as FAT, NTFS, and ext2, through to the creation of proprietary file systems such as those seen in some AVTECH, HIKVISON, and GANZ systems. Each DVR manufacturer tends to provide their own program and interface from which the user can set recording options, select live view options and review previously recorded video. During an investigation, in the majority of cases, the investigator may simply utilize the DVR built-in program to review, select and export any relevant video evidence, while adhering to established best practices.

This, however, is not always possible or preferred for a variety of reasons for example the DVR may be damaged, unknown errors may be encountered, the time-frame of the video data of interest may not be available, the video may have been intentionally or unintentionally deleted, the crime being investigated is of a very serious nature, or any combination of the above.

Despite the high market shares of companies such as AVTECH, GANZ, and HIKVISION, there is very little in the way of official manufacturer documented digital forensic analysis due to the closed nature of their proprietary systems. Hence, researchers and forensic examiners have come up with various and novel approaches to the analysis, whether it is reverse engineering the file system, interpreting the custom logging methods, or examining for lost or deleted video data.

The examination of DVR and CCTV footage raises significant challenges with regards to big data. Firstly, the volume of data concerned is huge, both in data size (e.g. GB/TBs) and in time format of many days/weeks/months recorded, all of which require examination. Secondly, with so many different systems, manufacturers, formats etc. the variety of the data is constantly changing.

This paper and the proposed new approach aims to unify the fundamental and core approaches of existing research along with the principles of digital evidence into a cohesive general methodology of DVR forensic analysis. The methodology is designed to apply to the forensic analysis of any DVR, known or yet unknown, and achieve success.

## 2 CCTV Forensics: A Literature Review

Dongen in [3] detailed the examination of a Samsung DVR for video evidence that could not be found through the video recorder's own menus. Dongen's methodology include a preliminary investigation to identify make, model and characteristics of the video recorder; create a forensic image of the data and a clone to be placed back into the device for testing; examination of the data using forensic software; and lastly, utilizing the information to extract video data. Dongen's experiment showed that it could be possible to obtain results through first recovering bookkeeping files that contain pointers to areas of the hard drive. This paper focused on examination of bookkeeping data and not examination of the video data. It also did not examine for video data outside the date range indicated by the bookkeeping data.

In [4], authors reverse-engineered a typical 4-camera DVR system by conducting controlled tests with the system. The methodology employed include initial device investigation to familiarize with its capabilities and operating modes, conducting a controlled experiment by deploying a wiped hard drive into the video recorder to record test video, creating a forensic image of the hard drive, and the examination of the file system, and video data. This paper's experiment was successful in interpreting much of the data storage and MPEG4 video structure. However, a controlled experiment is not always possible or feasible due to equipment failure, both DVR and hard drive, and time constraints.

Ariffin et al. [5] proposed a technique to carve video files with timestamps without referring to a file system. They propose a more detailed digital forensics framework, which comprised of (1) identification, (2) preservation, (3) analysis and (4) presentation. The research centred on a DVD based Sony Camcorder recorded DVD that was unfinalised, and hence was not recognized by the operating system, Windows XP, as there was no file system. The forensic analysis focused on the carving of DVD based VOB file video data based on known headers, and analysis of video timestamp via the associated IFO data. This paper, however, only focused on known headers of full video files already written on the disc.

Tobin et al. [6] conducted a reverse engineering of a hard drive from an initially unknown DVR system. Their methodology revolves around the use of the process monitoring tool Procmon and the AVTECH software Disktools.exe that successfully recognized the disk image. While meaningful, the work did not go into the interpretation of video data and is disadvantaged with its reliance on an unknown third-party software along with its limitations.

In [7], authors presented a forensic analysis of the data recorded by a Hikvision DVR system. In this research, the structure and mechanics of the unnamed Hikvision file system was analysed and identified. With the detailed understanding of the file system, the proposed procedure can be useful to counter anti-forensic activities, such as system initialization or data overwriting. The research shares a good understanding of this particular proprietary file system. While it is a large corporation, Hikvision has only recently entered the North American market and is not currently very widely used.

Authors in [8] provided some details in the workings of their independently developed video recovery software, which was capable of recovering video data from

Dahua and Hikvision DVRs. This project solely focused on the analysis of DVR systems which are not common outside of Asia.

Gomm et al. [9] detailed a reverse engineering of a Ganz DVR file system with no prior information. Bookkeeping type data was located and interpreted which included information such as camera number, date, pointer to sector containing video data and the number of data sectors. The researchers then analysed the carved video data, which was determined to be of a proprietary format. In their scenario, the date stamp was acquired from the bookkeeping data while the date stamp of the video data could not be successfully decoded. The developed forensic process was then compared to the manufacturer provided tool. It was shown that the forensic process was able to recover more video data than the official tool.

The FBI Regional Computer Forensics Laboratory's "Best Practices for the Retrieval of Video Evidence for Digital CCTV Systems" [10] provides a useful guide and step-by-step consideration for investigators in video surveillance evidence acquisition. The detailed considerations include information gathering upon arrival on scene, assessing recording system for output (which includes most advisable to least advisable media format and data storage to use), proprietary data retrieval, evidence handling procedures, legal issues, and some checks prior to leaving the scene. While many of the key considerations remain relevant, such as the legal issues, quality deterioration of converted video versus native video, note-taking and proper evidence handling, much of the technical side of the CCTV systems are out of date and are virtually non-relevant today (i.e. S-video vs. Composite video output, Zip and Jaz disks). Moreover, this guide is aimed at first responders and does not discuss the forensic analysis of the seized data.

In [11] authors detailed an in-depth analysis of the 'Hikvision' proprietary file system. The authors developed a reliable method for forensic recovery of footage from these systems. 'Hikvision' is a very common system type and this paper will be relevant for many years and a useful reference for technicians. The downside of this paper is that it only covers one filesystem.

Authors in [12] discussed techniques and tools for authenticating video though spotting differences in videos' naturally occurring properties (differences in picture noise), through techniques such as interlacing and field duplication to provide tools for authenticating video.

## 3    CCTV Forensic Challenges Within the Context of Big Data

Moore's Law was the observation made in 1965 that the number of transistors per square inch on integrated circuits had doubled every year since the integrated circuit was invented, and that this can assist in predicting the development of technology. Kryder's Law [13] observed that since the introduction of hard disk drives in 1956 the density of information stored in a square inch has seen a 50 million fold increase. This has resulted in massive data storage being available with the largest commercially available single hard drive currently storing 60 TB of data, with that expected to grow as technology continues to progress. This increase in storage capacity has allowed for DVRs to be installed with a previously unseen storage capacity, even the most basic models on the market now are being equipped with 500 GB/1 TB drives which, depending on the compression, codecs, number of cameras etc. can allow for months of footage to be retained by the unit. It is within this volume of data available that we encounter this first challenge of CCTV forensics, namely: specifically how to handle, examine and report on the extensive volume of data.

Secondly, we must consider the variety challenge encountered by the vast number of different suppliers and their proprietary approaches to data storage. This issue is compounded by the lack of legislation or working groups (like for example the IEEE or ISO) in determining a standard in which CCTV should be recorded and the means for which to play it back. Reportedly, there may be a variety of 2500–3000 different types of CCTV recording formats currently in use, anyone of which could be handed to an investigator anywhere in the world on a USB stick or optical media and causing forensic challenges in its identification and playback, potentially causing delay in time critical cases. This particular issue has arisen historically—in 1992, Microsoft released the 'AVI' format as part of Windows 3.1. AVI has a standard file header (signature 0x52 49 46 46/ASCII = RIFF), but from the header onwards the media can be encoded in any codec (codec meaning 'compress/decompress') developed by who or whatever encoded the video. The AVI header cannot tell the computer where to find this codec and if the end user does not have the codec installed on their computer, the file will simply not play or play a distorted or incorrectly decoded image. This is not just on AVI files but on many other common looking containers such as .mpeg, .mpg, .h264, .vob, .flv, .swf, .asf, .mov, .wmv, where the same issue applies.

This variety issue poses the second challenge addressed in this paper.

## 4 Proposed CCTV Forensic Approach

The proposed approach builds on the digital camcorder forensics framework proposed by Ariffin et al. [5]. The methodology employed here has 4 broad steps, (1) identification, (2) preservation, (3) analysis, and (4) presentation. As the name implies, the focus of the development of this general methodology of CCTV forensic analysis will be the analysis portion. That said, all of the other components are vital and contribute to an efficient and successful analysis. This methodology presumes that the DVR records locally, such as to a connected hard drive, and not to the cloud or variations of it. The objective of this framework is to overcome the challenges of CCTV forensics today as mentioned above.

Figure 1 describes the proposed general methodology for DVR forensic analysis.

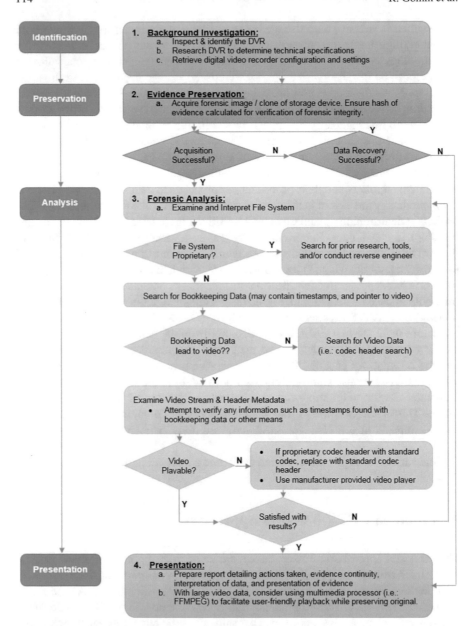

**Fig. 1** Proposed DVR forensic analysis general methodology

This methodology is designed to apply to the forensic analysis of any DVR, known or yet unknown.

### Step 1: Background Investigation

In the general digital forensic process, the first step is Preparation where the examiners encompass research on the digital devices to ensure that the relevant requisite technical knowledge was known to them. It is also in the case of CCTV forensics. However, to cope with the recent challenges in big data forensics, our proposed first step is Identification. Eventually, digital evidence today can be found in not only traditional storage devices (hard drives, SD cards, memory, etc.) but also in smart portable devices (phones, tablets, and GPS devices) [14], smart wearable devices [15], smart TV [16], connected vehicle [17, 18], drones [19], remote sensing devices, etc. This is also true for CCTV forensic cases when there is no standard in which CCTV should be recorded as mentioned above.

Hence, our first step is to identify relevant data source for the investigation to ensure that the data is collected in a manner that maximizes its integrity. Specially, our new proposed CCTV forensic analysis methodology looks to build on this and establish a Background Investigation to equip the forensic examiner with as much available information as possible prior to the actual examination of data. A thorough background investigation provides for a well-informed starting point in the subsequent forensic analysis and will greatly reduce the effort, complexities and time required for a successful forensic analysis. In locating this information, the investigator has reduced the Variety seen within big data by now having a specific start location, further it may allow previous works to 'lead the way' in the reducing the volume of data to be examined.

For example, knowing that the DVR was initialized and setup two days prior, gives us a good starting date to look for bookkeeping data, assuming the system timestamp is sufficiently accurate. Another example may be that knowing that the DVR records to the MPEG4 codec tells us that the video data may use the standard MPEG header "0x 00 00 00 01 ??", where "??" is the stream ID.

Background investigation involves the following, when applicable:

1. Inspection and identification of the DVR (i.e. make, model, series, number and type of external port, type of internal connections). Also note any damage or anomalies.
2. Research the identified DVR to determine its technical specifications (i.e. number of channels, file format, video codec, internet capabilities and features).
3. Retrieve DVR configuration and settings (i.e. DVR internal date/time, quality settings, length of recording before overwriting loop, cloud credentials, video format). Ideally, this step is to be done without the evidence drive in the DVR.

The big data volume can be reduced by conducting a thorough Background investigation the examiner can help set search terms and criteria for examination, which in turn may help to reduce the amount of the data to be examined.

### Step 2: Evidence Preservation

The second step of our proposed methodology includes two principal tasks, namely:

acquisition and data recovery. With regards to both volume and veracity of big data, the forensic acquisition and the data obtained must be proven to be trustworthy. Following from Step 1, the examiner maybe in a position to reduce the volume data acquisition to one subset of the evidence acquired.

**Acquisition**

Original evidence preservation is a vital principle in digital forensics. This involves creating forensic image files, forensic clone of the evidence, or both. A hash of the evidence will also be created for verification of the forensic integrity of the image or clone to ensure that they are identical to the original evidence. Any subsequent examination will then be conducted on the forensic image or clone; thus, preserving the original evidence. The preservation of the original evidence enables future verification or re-examination of the evidence. Further, this process mitigates the potential for hardware failures of the evidential hardware caused by the continuous use and possible strain of the forensic analysis.

If an examination of the video data through the DVR's own interface is warranted and possible, a forensic clone of the original evidence is highly recommended. Not only does this comply with the vital principle of preservation of evidence, some DVR's will automatically commence video recording upon boot, even when no cameras are connected and before the user is able to interact with the DVR. The examiner, however, should be aware that this may not always work as some DVR will not recognize cloned drives and will only work with the original.

**Data Recovery**

Should the acquisition of a forensic image not complete successfully, considerations have to be made for data recovery attempt depending on the amount of data that was acquired, if any. If an unstable drive with a large portion of bad sectors is suspected, consider utilizing data recovery systems such as the DeepSpar Disk Imager. In some circumstances, a hard drive repair can be undertaken which may include circuit board replacement and head adjustment. This step can be onerous as a clean room is often necessary and a circuit board replacement not only requires a board of the same make and model, it often requires a board of the same firmware, chipset, site code and batch. A combination of these methods can also be employed.

***Step 3: Forensic Analysis***

Forensic analysis is a principal step in any digital forensic process. In current approaches of CCTV forensics such as in [9], the focus of their model is determining the video file format from the file signature. This would allow the examiner to carve, extract and exam the data present. Unfortunately with the passage of time and the significant increase in manufactures such file signatures are now becoming less common due to proprietary codecs and formats being developed. Hence, in the Forensic Analysis step of our new approach, we presents a new method of examining 'bookkeeping data' for cases where file signatures are not present or are unrecognised. Again when dealing with the Volume of Big Data the examiner will need every assistance in reducing the subset of data being examined.

The proposed Step 3: Forensic Analysis takes into account 3 core groups: file system, bookkeeping data and video data. While this sequence of analysis is preferred, it is by no means a requirement, nor is the successful analysis of one core group a pre-requisite for the successful analysis of the next. The groups are presented in this preferred order as the analysis and understanding of the system as a whole is more effectively built in this order. This will thus lead to a greater chance of a successful analysis.

## File System

The first step in the forensic analysis phase is to determine the file system used. If a standard file system is used, such as FAT16/32, EXT2/3, or NTFS, then the identification of files and folders, and their metadata such as timestamps will be straightforward.

If a non-standard or proprietary file system is encountered, then a thorough research into any prior successful reverse engineer should be conducted. There may a tool that may assist in the interpretation of the proprietary file system and the retrieval of video evidence [8]. When prior research or 3rd party software is found, it is important to understand its mechanics and limitations. Independent verification and additional forensic examination may be required.

If prior successful research is not found, then a reverse engineer of the file system may be required. This process could be performed through a detailed analysis of the raw data to explain data structures and data offsets [4, 7] or monitoring processes while performing tasks utilizing available manufacturer software [6].

## Bookkeeping Data

The operating system and integrated software of DVR keeps track of recorded video with a bookkeeping data in one form or another. This bookkeeping data can be in the form of log, database or similar files, or perhaps not in discrete files at all. Some vital information in this data include timestamp and the physical location of the corresponding video. With this data at hand in an accessible form, the DVR is able to quickly provide the user with the time frame of available video data. However, in a forensic analysis, if the examiner relies solely on this data to determine the availability of video data, crucial evidence could be missed. Some previous research investigated some methods of examining for and possibly retrieving video data not within the timeframe of the bookkeeping data by searching for discrete file signature [5], and by comparing video extracted via locators in a proprietary file system to video presented by a manufacturer provided computer program.

In some cases, bookkeeping data such as DVD IFO files do not actually contain information to the location of the corresponding video data, other than in terms of a referenced filename. In the case of DVR's however, the analysis and interpretation of bookkeeping data can in some cases reveal the location of corresponding video data with the timestamp of interest [9].

Careful analysis and interpretation would be needed to locate and decipher meaningful information from this data. Success of which could lead directly to the discovery of video evidence and/or a deeper understanding of the mechanics of the DVR.

## Video Data

It is not uncommon for the storage of video data in DVR's to be in a large unallocated area which reside in its own partition of the storage device. This is especially true in a DVR that utilizes standard or traditional file systems such as FAT, NTFS or EXT. These file systems use multi-level index to create and retrieve files, and can have difficulty supporting oversize storage file capacity. Moreover, the core function of a DVR which involves the continuous and repeated creation and/or overwriting of video files can cause significant file fragments and non-contiguous storage of files which can greatly reduce the performance and stability of the file system. This is believed to be the primary reason for the use of simple, non-sophisticated proprietary file systems.

Following the identification of the general location of the video data, be it in its own series of files, in one large video file, or in large portion of raw disk or unallocated area, we now attempt to extract playable video data. In some cases, simply inputting the video data into open format players such as Media Player Classic would suffice in viewing video. However, the number of and complexity of forensic research into this area suggests that this is often not the case.

Video data can be distinguished between the video data stream, and the video data header or metadata. Prior research suggest that while in most cases, the video data stream may be relatively standard, perhaps with some minor differences, the video data header is anything but. Unless video data is stored in a standard video file format such as MP4, the video data header mostly likely requires careful analysis and interpretation.

Some elements of the header metadata to examine for are timestamps and channel or camera number. When available, this data can then be compared to the corresponding bookkeeping data for verification.

One technique to locate video stream data is to look for known video codec headers. This is made simpler armed with the knowledge gained through Step 1: Background Investigation. Video stream data is to be distinguished from video file headers. A common codec used in DVR's is MPEG-4. Some DVR's are advertised as utilizing the H.264 codec. It is important to note that H.264 is also known as MPEG-4 AVC (advanced video codec) and is jointly maintained with MPEG so that they have identical technical content. The standard MPEG header is 0x 00 00 00 01 ??, where ?? is the stream ID.

When standard video codec headers cannot be found, it is possible that the video stream could still utilize a standard codec but with the header modified. It is also possible that the video stream uses a non-standard or proprietary codec. Attempts can be made to replace the suspected custom header with a standard header and then attempt to play it with an open format player. Alternatively, or in conjunction, the carved video can be inputted as is into the manufacturer's provided video player.

This is explored and demonstrated further in Case Studies 1 and 2 below, wherein two separate examinations are conduct by way of reverse engineering proprietary file systems and bookkeeping data.

### Step 4: Presentation

In a general digital forensic process, this step is aim to have the data/information interpreted into forms that is acceptable as evidence to the courts. Today, with the advent of Big Data this can mean trying to explain the forensic examination of multiple devices that contained multiple sources of data ranging from just a few bytes in terms of jpg images through to multi gigabytes in terms of videos. Any difficulty or failure to present that evidence in a manner understandable by a Jury could led the case to collapse.

Therefore, in our new proposed approach, the presentation phase consist of a report detailing actions taken, evidence continuity, interpretation of data, and presentation of relevant evidence found. It is a good practice to attempt to present this material in the simplest way possible as it has to be communicated to investigators, lawyers (prosecutors and defence), judges, juries and other stakeholders.

When a large amount of video evidence is located, consider processing the video data with a comprehensive multimedia processor such as FFMPEG. The goal of this processing is for the creation of appropriate standard headers, footers, and index of the video stream and key frames to support video seeking and potentially to provide smoother playback. This facilitates the user-friendly playback of the video evidence by non-technical stakeholders.

When opting for this process, care has to be taken not to deteriorate the video quality with a re-encoding of the video data, especially to an inferior codec. Choosing the option to maintain the video codec stream with no re-encoding should be chosen. The original carved video data should of course be preserved as evidence.

This is explored and demonstrated further in Case Study 3 below, wherein the presentation of evidence obtained by the forensic examination of multiple CCTV sources was made to a UK Court. The evidence led to multiple convictions.

## 5 Case Studies

In this section, different case studies are described to demonstrate how to apply our proposed model in tackle the large and variety of CCTV forensic data.

### 5.1 Case Study 1—GANZ DVR C-MPDVR-16

Scenario: a criminal investigation was undertaken in relation to a serious criminal offence of which significant video footage had been captured on a Ganz CCTV DVR model C-MPDVR-16. The unit was setup to save the last 31 days of footage to an internal hard drive, however despite the referenced footage being within this timeframe it could not be located using the manufacturers DVR playback software. The Ganz DVR unit was submitted for forensic examination for the retrieval of specified video footage which, according to the proprietary video backup application, was not retrievable.

The Ganz C-MPDVR-16 was equipped with a 1 Tb Hard Drive and contained footage covering a period of approximately 30 days. The investigation was concerned with a period of approximately three minutes from one specific date. The Volume of the data to be explored compared to the size of the required evidence, when compound with the proprietary software being unable to find the exact data, created a challenge of searching Big Data.

Additionally the Ganz C-MPDVR-16 uses a proprietary storage system and video encoder, of which no details or reverse engineering had been completed at the time of the examination. As previously highlight the Variety of data is an issue within Big Data, in this one scenario we have identified at least two proprietary unknown formats, with additional aspects such as the book keeping data still to be examined.

Using the General Methodology of DVR Forensic Analysis model proposed above the case study starts with Step 1: Background Investigation.

**Step 1: Background Investigation**
Open Source research confirmed that the Ganz DVR recorded in a proprietary MPEG codec produced by AV-Tech. No Open Source forensic examination tools were located, however it was established that that the device used the AV-Tech file system which again had not been decoded or had its details made public. This meant the only known avenue for recovery of the relevant CCTV was via the manufacturer's software, which had already proven ineffective. All relevant dates and times of footage required were established from the investigator. Further the unit was identified as being able to control a maximum of 16 cameras.

**Step 2: Evidence Preservation**
Following the Background Investigation we progressed to a visual examination of the unit, including the internal functioning and components.

In the visual inspection phase the Ganz C-MPDVR-16 DVR was photographed (Fig. 2).

The next stage is preservation. Within the law enforcement environment it is imperative that any data is recovered in a forensically-sound method. The general accepted criterion is that no changes are made to the original data source, and that any copies made are identical to the original data source. In our experiment, the Ganz DVR unit was seized from the working environment by law enforcement agents on the 4th of April 2013 shortly after 08:00 h. A strict chain of custody from seizure to examination was in place and no access to the Ganz DVR Unit was permitted. For this examination the DVRs internal hard disk drive was removed and connected to a forensic device known as a write blocker. Next a bit copy was made of the hard disk drive. This is a method to ensure that an exact copy of the entire hard disk drive is produced; including all used and used space. The resulting image was MD5 hashed

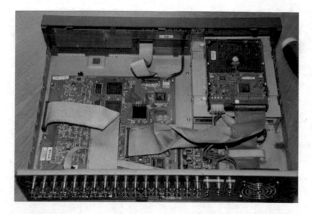

**Fig. 2** Internal view of GANZ DVR

and compared to the original to ensure a full and correct copy. The source (original hard disk drive) was then securely retained under locked room and access controlled logged environment.

As a successful acquisition had occurred, the case proceeded to Step 3: Forensic Analysis.

**Step 3: Forensic Analysis**
In line with the General Methodology of DVR Forensic Analysis model the Forensic Analysis consisted of firstly examining the cloned hard disk drive at a sector/byte level, specifically the Master Boot Record and Sector 0 in order to establish whether any of the industry standard file systems were present. These can be easily identified by the presence of unique hexadecimal values, commonly referred to as a 'Magic Marker' or 'Magic Byte', being present in Sector 0 of the hard disk drive (Fig. 3).

No Magic Marker was present on the Ganz DVR hard disk drive, although the ASCII text of AVTECH and FSS16A was located, this led to feedback into Step 1 and resulted in further Open Source research as previously described.

Continuing with the General Methodology of DVR Forensic Analysis model we look for any bookkeeping data, be it in the form of log files, timestamps, or other regular patterns of data present on the hard disk drive. Sector 0 of the hard disk drive

```
  Offset  | 0  1  2  3  4  5  6  7   8  9  A  B  C  D  E  F |
0000000000| 00 00 44 A9 4E 39 00 00  FB FF FF 39 00 00 00 FF |  D©N9    ûÿÿ9    ÿ
0000000010| 00 B0 4B BA 01 00 00 00  00 00 58 6D 01 00 00 00 |  °K²       Xm
0000000020| 30 7E F7 00 00 00 00 00  60 6F F7 00 00 00 00 00 | 0~÷       `o÷
0000000030| 55 41 56 54 45 43 48 AA  46 53 53 31 36 41 00 55 | UAVTECHªFSS16A U
0000000040| 00 3C AD BA 02 00 00 00  00 30 5E 38 3A 00 00 00 | <-²      0^8:
0000000050| 00 00 72 74 00 00 00 00  00 FF AC BA 02 00 00 00 |  rt      ÿ-²
0000000060| 00 02 00 00 00 00 00 00  00 FF 71 74 00 00 00 00 |         ÿqt
0000000070| 00 00 AD BA 02 00 00 00  00 FF AC BA 02 00 00 00 |  -²      ÿ-²
0000000080| 00 00 00 00 00 00 00 00  00 00 00 00 00 00 00 00 |
```

**Fig. 3** Ganz internal hard disk drive—MBR (Sector 0)

| Offset | 0 | 1 | 2 | 3 | 4 | 5 | 6 | 7 | 8 | 9 | A | B | C | D | E | F | | |
|---|---|---|---|---|---|---|---|---|---|---|---|---|---|---|---|---|---|---|
| 0000000980 | 20 | DB | 51 | 77 | 00 | 00 | 00 | F0 | 06 | 08 | 01 | 00 | 00 | 00 | 00 | 00 | ÛQv | ð |
| 0000000990 | 20 | C8 | 5A | 77 | 00 | 00 | 00 | F0 | 06 | 08 | 01 | 01 | 00 | 00 | 00 | 00 | ÈZv | ð |
| 00000009A0 | 80 | B5 | 63 | 77 | 00 | 00 | 00 | F0 | 06 | 08 | 01 | 02 | 00 | 00 | 00 | 00 | µcv | ð |
| 00000009B0 | 80 | A2 | 6C | 77 | 00 | 00 | 00 | F0 | 06 | 08 | 01 | 03 | 00 | 00 | 00 | 00 | ¢lv | ð |
| 00000009C0 | C0 | 8F | 75 | 77 | 00 | 00 | 00 | F0 | 06 | 08 | 01 | 04 | 00 | 00 | 00 | 00 | À uv | ð |
| 00000009D0 | C0 | 7C | 7E | 77 | 00 | 00 | 00 | F0 | 06 | 08 | 01 | 05 | 00 | 00 | 00 | 00 | À|~v | ð |
| 00000009E0 | 00 | 6A | 87 | 77 | 00 | 00 | 00 | F0 | 06 | 08 | 01 | 06 | 00 | 00 | 00 | 00 | jv | ð |
| 00000009F0 | 40 | 57 | 90 | 77 | 00 | 00 | 00 | F0 | 06 | 08 | 01 | 07 | 00 | 00 | 00 | 00 | @W v | ð |
| 0000000A00 | 60 | 44 | 99 | 77 | 00 | 00 | 00 | F0 | 06 | 08 | 01 | 08 | 00 | 00 | 00 | 00 | `Dv | ð |
| 0000000A10 | C0 | 31 | A2 | 77 | 00 | 00 | 00 | F0 | 06 | 08 | 01 | 09 | 00 | 00 | 00 | 00 | À1¢v | ð |
| 0000000A20 | C0 | 1E | AB | 77 | 00 | 00 | 00 | F0 | 06 | 08 | 01 | 0A | 00 | 00 | 00 | 00 | À «v | ð |
| 0000000A30 | E0 | 0B | B4 | 77 | 00 | 00 | 00 | F0 | 06 | 08 | 01 | 0B | 00 | 00 | 00 | 00 | à ´v | ð |
| 0000000A40 | 20 | F9 | BC | 77 | 00 | 00 | 00 | F0 | 06 | 08 | 01 | 0C | 00 | 00 | 00 | 00 | ù¼v | ð |
| 0000000A50 | 60 | E6 | C5 | 77 | 00 | 00 | 00 | F0 | 06 | 08 | 01 | 0D | 00 | 00 | 00 | 00 | `æÅv | ð |
| 0000000A60 | A0 | D3 | CE | 77 | 00 | 00 | 00 | F0 | 06 | 08 | 01 | 0E | 00 | 00 | 00 | 00 | ÓÎv | ð |
| 0000000A70 | E0 | C0 | D7 | 77 | 00 | 00 | 00 | F0 | 06 | 08 | 01 | 0F | 00 | 00 | 00 | 00 | àÀ×v | ð |
| 0000000A80 | 20 | AE | E0 | 77 | 00 | 00 | 00 | F0 | 06 | 08 | 01 | 10 | 00 | 00 | 00 | 00 | ®àv | ð |
| 0000000A90 | 40 | 9B | E9 | 77 | 00 | 00 | 00 | F0 | 06 | 08 | 01 | 11 | 00 | 00 | 00 | 00 | @év | ð |
| 0000000AA0 | 60 | 88 | F2 | 77 | 00 | 00 | 00 | F0 | 06 | 08 | 01 | 12 | 00 | 00 | 00 | 00 | `òv | ð |
| 0000000AB0 | 80 | 75 | FB | 77 | 00 | 00 | 00 | F0 | 06 | 08 | 01 | 13 | 00 | 00 | 00 | 00 | uûv | ð |
| 0000000AC0 | C0 | 62 | 04 | 78 | 00 | 00 | 00 | F0 | 06 | 08 | 01 | 14 | 00 | 00 | 00 | 00 | Àb x | ð |
| 0000000AD0 | E0 | 4F | 0D | 78 | 00 | 00 | 00 | F0 | 06 | 08 | 01 | 15 | 00 | 00 | 00 | 00 | àO x | ð |
| 0000000AE0 | 00 | 3D | 16 | 78 | 00 | 00 | 00 | F0 | 06 | 08 | 01 | 16 | 00 | 00 | 00 | 00 | = x | ð |
| 0000000AF0 | 20 | 2A | 1F | 78 | 00 | 00 | 00 | F0 | 06 | 08 | 01 | 17 | 00 | 00 | 00 | 00 | * x | ð |
| 0000000B00 | 60 | 17 | 28 | 78 | 00 | 00 | 00 | F0 | 06 | 08 | 02 | 00 | 00 | 00 | 00 | 00 | ` (x | ð |
| 0000000B10 | 80 | 04 | 31 | 78 | 00 | 00 | 00 | F0 | 06 | 08 | 02 | 01 | 00 | 00 | 00 | 00 | 1x | ð |

**Fig. 4** Ganz internal hard disk drive—Sector 2

had already been identified as containing a proprietary master boot record, or filing record so we proceeded to Sector 1 which was empty of any data, and finally onto Sector 2.

It was within Sector 2 that a repeating pattern 24 entries incrementing by 1, starting at hex 06 08 01 00 through to 06 08 01 17 was identified. At offset 0B00 the 06 08 01 increases to 06 08 02 and the 24 entry process repeats (Fig. 4).

Considering that file system was to record CCTV encoded with Date/Time stamps it was proposed likely that 24 entries related to the 24 h of a day. Assuming that the 24 entries related to the 24 h of the day then it was logical to assume the date was also recorded. Using the top offset:

20 DB 51 77 00 00 00 F0 06 08 01 00 00 00 00 00

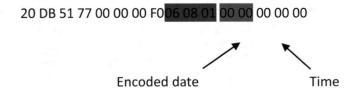

Encoded date             Time

Considering the values present it was likely that the encoded date was in reverse format, in this case it would equal 06/08/01 or the 1st of August 2006. Based on Step 1 we knew that the DVR unit was seized on the 4th of April 2013 shortly after 8 a.m., in order to explore the above reasoning a hex search was conducted for the following values:

| Offset | 0 | 1 | 2 | 3 | 4 | 5 | 6 | 7 | 8 | 9 | A | B | C | D | E | F | | | |
|--------|---|---|---|---|---|---|---|---|---|---|---|---|---|---|---|---|---|---|---|
| 0001EEFB50 | F8 | D1 | 94 | B7 | 01 | 00 | 00 | F0 | 0D | 04 | 03 | 12 | 00 | 00 | 00 | 00 | øÑ∎· | ŏ | |
| 0001EEFB60 | 78 | 32 | C4 | B7 | 01 | 00 | 00 | F0 | 0D | 04 | 03 | 13 | 00 | 00 | 00 | 00 | x2Ä· | ŏ | |
| 0001EEFB70 | 78 | 93 | F3 | B7 | 01 | 00 | 00 | F0 | 0D | 04 | 03 | 14 | 00 | 00 | 00 | 00 | x∎ó· | ŏ | |
| 0001EEFB80 | D8 | F3 | 22 | B6 | 01 | 00 | 00 | F0 | 0D | 04 | 03 | 15 | 00 | 00 | 00 | 00 | Øó", | ŏ | |
| 0001EEFB90 | 58 | 55 | 52 | B8 | 01 | 00 | 00 | F0 | 0D | 04 | 03 | 16 | 00 | 00 | 00 | 00 | XUR, | ŏ | |
| 0001EEFBA0 | E8 | 48 | 7B | B8 | 01 | 00 | 00 | F0 | 0D | 04 | 03 | 16 | 33 | 33 | 00 | 00 | èH{, | ŏ | 33 |
| 0001EEFBB0 | 98 | B6 | 81 | B8 | 01 | 00 | 00 | F0 | 0D | 04 | 03 | 17 | 00 | 00 | 00 | 00 | ∎¶ , | ŏ | |
| 0001EEFBC0 | 78 | 17 | B1 | B8 | 01 | 00 | 00 | F0 | 0D | 04 | 04 | 00 | 00 | 00 | 00 | 00 | x ±, | ŏ | |
| 0001EEFBD0 | D8 | 78 | E0 | B8 | 01 | 00 | 00 | F0 | 0D | 04 | 04 | 01 | 00 | 00 | 00 | 00 | Øxà, | ŏ | |
| 0001EEFBE0 | 38 | DA | 0F | B9 | 01 | 00 | 00 | F0 | 0D | 04 | 04 | 02 | 00 | 00 | 00 | 00 | 8Ú ¹ | ŏ | |
| 0001EEFBF0 | 98 | 3A | 3F | B9 | 01 | 00 | 00 | F0 | 0D | 04 | 04 | 03 | 00 | 00 | 00 | 00 | ∎:?¹ | ŏ | |
| 0001EEFC00 | D8 | 9B | 6E | B9 | 01 | 00 | 00 | F0 | 0D | 04 | 04 | 04 | 00 | 00 | 00 | 00 | Ø∎n¹ | ŏ | |
| 0001EEFC10 | 18 | FD | 9D | B9 | 01 | 00 | 00 | F0 | 0D | 04 | 04 | 05 | 00 | 00 | 00 | 00 | ý ¹ | ŏ | |
| 0001EEFC20 | 98 | 5D | CD | B9 | 01 | 00 | 00 | F0 | 0D | 04 | 04 | 06 | 00 | 00 | 00 | 00 | ∎]Í¹ | ŏ | |
| 0001EEFC30 | B8 | 85 | DB | B9 | 01 | 00 | 00 | F0 | 0D | 04 | 04 | 06 | 11 | 37 | 00 | 00 | ,∎Û¹ | ŏ | 7 |
| 0001EEFC40 | F8 | BE | FC | B9 | 01 | 00 | 00 | F0 | 0D | 04 | 04 | 07 | 00 | 00 | 00 | 00 | ø¾ü¹ | ŏ | |
| 0001EEFC50 | 38 | 20 | 2C | BA | 01 | 00 | 00 | F0 | 0D | 04 | 04 | 08 | 00 | 00 | 00 | 00 | 8 ,º | ŏ | |

**Fig. 5** Ganz hard drive—Sector 63,358

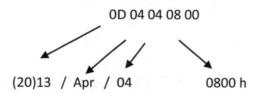

The specific code was located in sector 63,358.

Significantly the next few sectors following 63,358 had no data present; this would coincide with the device not being active (as having been seized) and therefore no data being recorded. Further the data recorded in sector 63,358 was minimal in size, there was insufficient size for video footage (Fig. 5).

A further search for the hex code 0D 04 04 08 00 revealed a second find in sector 28,978,208, offset 0374584070 (Fig. 6).

Despite the second appearance of the code there was still was no actual video data present. However a repeating pattern was noticed, in that the first byte of each offset ranged from 00 to 0F hexadecimal, which was theorized to equate to one of the available sixteen cameras on the DVR system, as identified in Step 1: Background Information.

It was further theorised that the data seen at the first finding (offset 1EEFC50) contained a locator/pointer to the data for each individual camera as recorded in the second finding area (offset 0374584070).

```
0001EEFC50  38 20 2C BA 01 00 00 F0   0D 04 04 08 00 00 00 00 | 8 ,º   ŏ
```

Going back to the data at offset 1EEFC50 the only data which remained after removing the date/time stamp was the first 8 bytes:

| Offset | 0 | 1 | 2 | 3 | 4 | 5 | 6 | 7 | 8 | 9 | A | B | C | D | E | F | |
|---|---|---|---|---|---|---|---|---|---|---|---|---|---|---|---|---|---|
| 0374584070 | 00 | 42 | A8 | 50 | 38 | 39 | 00 | 00 | 0D | 04 | 04 | 08 | 00 | 00 | 00 | 03 | B¨¨P89 |
| 0374584080 | 04 | 42 | AB | 50 | 38 | 39 | 00 | 00 | 0D | 04 | 04 | 08 | 00 | 00 | 00 | 06 | B«P89 |
| 0374584090 | 08 | 42 | B1 | 50 | 38 | 39 | 00 | 00 | 0D | 04 | 04 | 08 | 00 | 00 | 00 | 04 | B±P89 |
| 03745840A0 | 0C | 42 | B5 | 50 | 38 | 39 | 00 | 00 | 0D | 04 | 04 | 08 | 00 | 00 | 00 | 07 | BµP89 |
| 03745840B0 | 11 | D0 | BC | 50 | 38 | 39 | 00 | 00 | 0D | 04 | 04 | 08 | 00 | 00 | 00 | 13 | Đ¼P89 |
| 03745840C0 | 05 | D0 | CF | 50 | 38 | 39 | 00 | 00 | 0D | 04 | 04 | 08 | 00 | 00 | 00 | 0F | ĐÏP89 |
| 03745840D0 | 09 | D0 | DE | 50 | 38 | 39 | 00 | 00 | 0D | 04 | 04 | 08 | 00 | 00 | 00 | 16 | ĐÞP89 |
| 03745840E0 | 0D | D0 | F4 | 50 | 38 | 39 | 00 | 00 | 0D | 04 | 04 | 08 | 00 | 00 | 00 | 12 | ĐôP89 |
| 03745840F0 | 02 | B0 | 06 | 51 | 38 | 39 | 00 | 00 | 0D | 04 | 04 | 08 | 00 | 00 | 00 | 14 | ° Q89 |
| 0374584100 | 06 | B0 | 1A | 51 | 38 | 39 | 00 | 00 | 0D | 04 | 04 | 08 | 00 | 00 | 00 | 1E | ° Q89 |
| 0374584110 | 0A | B0 | 38 | 51 | 38 | 39 | 00 | 00 | 0D | 04 | 04 | 08 | 00 | 00 | 00 | 18 | °8Q89 |
| 0374584120 | 0E | B0 | 50 | 51 | 38 | 39 | 00 | 00 | 0D | 04 | 04 | 08 | 00 | 00 | 00 | 1C | °PQ89 |
| 0374584130 | 03 | E0 | 6C | 51 | 38 | 39 | 00 | 00 | 0D | 04 | 04 | 08 | 00 | 00 | 00 | 26 | àlQ89 &|
| 0374584140 | 07 | E0 | 92 | 51 | 38 | 39 | 00 | 00 | 0D | 04 | 04 | 08 | 00 | 00 | 00 | 17 | à´Q89 |
| 0374584150 | 0B | E0 | A9 | 51 | 38 | 39 | 00 | 00 | 0D | 04 | 04 | 08 | 00 | 00 | 00 | 1B | à©Q89 |
| 0374584160 | 0F | E0 | C4 | 51 | 38 | 39 | 00 | 00 | 0D | 04 | 04 | 08 | 00 | 00 | 00 | 11 | àÄQ89 |

**Fig. 6** Ganz hard drive—Sector 28,978,208

38 20 2C BA 01 00 00 F0

offsets 5, 6, and 7 were the same throughout the sector which indicated they were not likely to point towards a specific location so they could be discarded, leaving:

38 20 2C BA 01

If the above theory was to be correct then 38 20 2C BA 01 hex had to relate to sector 28,978,208.

It was evident that the data was stored in little endian format, in that hex 01 BA 2C 20 equalled decimal 28,978,208.

With the above method established the process was repeated with the remaining data in sector 28,978,208:

0374584070  00 42 A8 50 38 39 00 00   0D 04 04 08 00 00 00 03   B¨¨P89

Again the date was represented in offsets 8, 9 and 10, leaving 2–5 for the 'pointer' to the video data. Also offsets 6 and 7 remained the same throughout the sector, as seen previously above. To that end they were deemed irrelevant and ignored.

This left the following hexadecimal code: 00 42 A8 50 38 39.

Previously only the last 4 offsets were used and they were in stored in little endian format, using the same criteria again provided: 39 38 50 A8 which in decimal = 959,991,976.

An examination of sector 959,991,976 revealed a sector full of an unknown data, likely to be the video footage, interesting the data continued for 3 sectors which was evident at offset 15 of the data in sector 28,978,208.

In summary there were 3 distinct book keeping areas on the hard drive, the first contains entries which provide a general date and time for recorded footage and a pointer to the second area. The second area contained separate entries holding a refined date and time stamps for each of the individual 16 cameras video footage recorded, and a pointer to the video footage. Whilst the third area contained the actual video footage which was carved using standard industry tools and playable by the Ganz DVR video player software.

**Step 4: Reporting**

A proprietary tool called DiskTools.exe was obtained from AV-Tech which allowed access to the Ganz DVR unit internal hard disk drive. The DiskTools.exe application stated that the earliest available video footage recoverable from the Ganz hard disk drive was from the 18th of March 2013 at 00:48:09 h. Having established the filing structure employed on the Ganz hard disk drive the investigator was able to establish that footage from 20:00 h on the 17th of March 2013 was still present and recoverable. Therefore an additional 4 h and 48 min of footage was available through manual examination of the Ganz DVR hard drive, compared to the official DiskTools.exe application.

In criminal investigations any additional time recoverable may result in crucial evidence, which had the investigator relied on the official application would not have been recovered. The findings were communicated in a written report with the extracted footage available in multiple formats for ease of playback.

## 5.2  Case Study 2—Swann DVR4-2500

Scenario: Investigators obtained a partially corrupted hard drive from a Swann DVR4-2500. This DVR supports 4 cameras, records in the common H.264 codec, and became available in 2009.

The hard drive within this particular DVR is 500 GB in size. Due to the corruption of the hard drive in the system, the DVR does not recognize the hard drive and therefore video was not able to be reviewed from the device itself. Although this hard drive was smaller in size than Case Study 1, being only 500 GB, the data it contained was corrupted and not recognised by the device. Within the context of Big Data it was impossible to establish the number of cameras that had been recorded, the length of time/recording periods, or the level of data recoverable. The Volume of data was an issue in that it was essentially an unknown entity, the required evidence may have taken up all of the 500 GB or 1 mb of the space. In either event the full 500 GB would need to be examined to establish the evidence.

With regards to the Variety outlined in the Introduction, this case study had identified the H.264 codec, however no information was known as to file system, book keeping, whether audio was present and if so was it located with the video recording or filed separate etc. The examination would have to deal with the Variety of data as well as the Volume.

Using the General Methodology of DVR Forensic Analysis model proposed above the case study starts with Step 1: Background Investigation.

**Step 1: Background Investigation**

An online search quickly resulted in the product manual and several key information was learned from this step. Firstly, that the Swann DVR4-2500 records in the common H.264, otherwise known as MPEG-4 AVC, and that it supports a maximum of 4 channels. Investigators who seized this DVR noted that two cameras were connected at the time of seizure. Within the DVR itself it contained a standard 500 GB hard drive.

No official tool for viewing the DVR hard drive on a computer was found. It is suspected that even if an official tool is available, the tool may not recognize the drive just as the DVR itself did not recognize the drive.

**Step 2: Evidence Preservation**

The hard drive was removed and connected to a write-blocker (read-only interface) to commence acquisition. The operating system, Windows 7, had trouble detecting the drive and a continuous clicking noise can be heard from the drive. Forensic software also had trouble detecting the drive. The drive was connected to a DeepSpar data recovery disk imager where a clone of the drive to another drive of identical storage capacity was attempted. The DeepSpar detected numerous bad sectors but successfully recovered approximately 90% of data. A forensic image was then acquired from the cloned drive. This image was made the working copy while the original drive and the cloned drive were kept as evidence.

**Step 3: Forensic Analysis**

The forensic image was analysed using the forensic software EnCase and HxD.

The drive contained two partitions, both of which are EXT2, a Linux file system. The partitions are named 'hda1' and 'hda2'. The second partition 'hda2' was small, at 6 GB while the first partition 'hda1' was comprised of the bulk of the data, at 459.8 GB.

The first partition 'hda1' contained no files or folders, and completely consisted of unallocated space. The second partition 'hda2' contains folders with naming convention that correspond with the year and month (yyyy-mm) that the folder was created, e.g. '2011-11' and '2012-07'.

It was apparent that the folder naming indicates the year and the month. Within each of these folders are files with naming convention that is consistent with the day (##.log) that the file was created, e.g. 16.log = the 16th day of the month. These .log files are relatively small in size, from less than 500 kb to just over 1 MB.

This consistent naming convention suggests that data within the each .log file corresponds to the date that the naming convention indicates. For example, in Fig. 7, the file 16.log within folder 2012-07 should contain data for 2012-07-16, or July 16th,

2012. In other words, it is likely that the data in these .log files contain bookkeeping data.

Data within the .log files isn't as easily deciphered. Figure 8 shows a portion of data within an example .log file, 16.log from the folder 2012-07. There is 6 bytes of data followed by 0x 00 00 01 00 00 00 01, and then followed by 7 more bytes of 0x00. This is then followed by ASCII "/log/x" which can be easily seen throughout the .log file. This is lastly followed by 126 bytes of 0x00 before the pattern repeats.

The 5 bytes of data initially appear to be some sort of count, perhaps for the timecode of the video and the location of the video. The first 4 series of the 5 bytes of data in the example file are as follows:

1. 0x 0C 07 10 0E 24 1C ... /log/x ...
2. 0x 0C 07 10 0E 25 29 ... /log/x ...
3. 0x 0C 07 10 0E 25 31 ... /log/x ...
4. 0x 0C 07 10 0E 25 34 ... /log/x ...

Attempts to interpret and apply the 5 bytes of data as locations, in terms of potential physical sector and physical offsets, for the corresponding video evidence on the hard drive were unsuccessful.

| | # | Name | Logical Size | Last Written |
|---|---|---|---|---|
| ☐ | 1 | 16.log | 209,316 | 07/16/12 08:55:18AM |
| ☐ | 2 | 17.log | 336,084 | 07/17/12 08:57:10AM |
| ☐ | 3 | 18.log | 223,452 | 07/18/12 08:57:14AM |
| ☐ | 4 | 19.log | 163,412 | 07/19/12 08:55:40AM |
| ☐ | 5 | 20.log | 242,300 | 07/20/12 08:59:55AM |
| ☐ | 6 | 21.log | 243,820 | 07/21/12 08:57:34AM |
| ☐ | 7 | 22.log | 92,276 | 07/22/12 08:49:07AM |
| ☐ | 8 | 23.log | 225,276 | 07/23/12 08:56:47AM |
| ☐ | 9 | 24.log | 204,756 | 07/24/12 08:55:47AM |
| ☐ | 10 | 25.log | 391,108 | 07/25/12 08:53:20AM |
| ☐ | 11 | 26.log | 311,308 | 07/26/12 08:57:08AM |
| ☐ | 12 | 27.log | 170,708 | 07/27/12 08:59:53AM |
| ☐ | 13 | 28.log | 222,996 | 07/28/12 08:58:11AM |
| ☐ | 14 | 29.log | 83,004 | 07/29/12 08:59:06AM |
| ☐ | 15 | 30.log | 175,572 | 07/30/12 08:59:44AM |
| ☐ | 16 | 31.log | 237,740 | 07/31/12 08:57:40AM |

Tree panel: Cases / Home / Entries / Bookm; Home / File Extents; Entries / recovered drive / hda1 / hda2 / 2011-11 / 2012-07 / 2012-08 / 2012-09 / 2012-10 / 2012-11 / 2012-12 / 2013-01 / 2013-02 / 2013-03 / 2013-04 / 2013-05 / 2013-06 / 2013-07 / 2013-08 / Hard Links / lost+found / Lost Files. Top tabs: Table, Report, Gallery, Timeline, Disk, Code.

**Fig. 7** hda2—files and folders

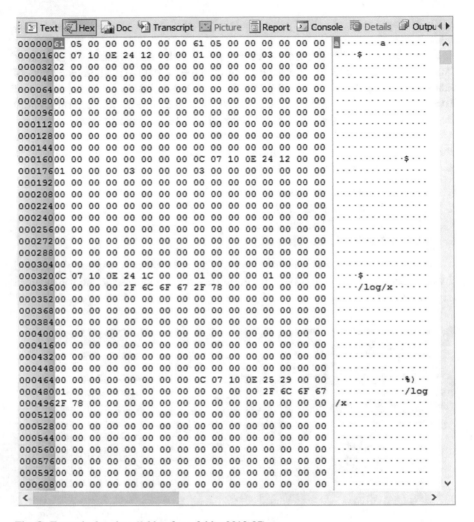

**Fig. 8** Example .log data (16.log from folder 2012-07)

Examining the consistency of the hex data reveal a pattern of hexadecimal to decimal interpretation that coincides with the date of the data set. In this case:

1. 0x 0C 07 10 0E 24 1C: decimal 12 07 16 14 36 28
   = date 2012-07-16 time 14:36:28
2. 0x 0C 07 10 0E 25 29: decimal 12 07 16 14 37 41
   = date 2012-07-16 time 14:37:41

While the coding of timestamp in the log data was successfully interpreted, they are currently not very useful. Further examination of the .log file did not reveal any additional insight. Examination of the drive thus far indicates that the video data is stored in 'hda1' due to its sheer size compared to 'hda2'.

```
15900F10   6E BD C3 53 E8 5A 16 B5 33 FB E4 09 06 64 EC 47   n½ÃSèZ.µ3ûä..dìG
15900F20   24 A6 E1 F4 3A 6E 00 00 00 31 30 64 63 48 32 36   $¦áô:n...10dcH26
15900F30   34 B7 51 00 00 10 00 00 00 85 B7 5E 16 00 00 00   4·Q........‥.^....
15900F40   00 0D 09 0B 10 10 21 00 00 1E 00 00 00 60 01 00   ......!......`..
15900F50   00 00 00 00 01 67 42 E0 14 DB 05 87 C4 00 00 00   .....gBà.Û.‡Ä...
15900F60   01 68 CE 30 A4 80 00 00 00 01 06 E5 01 66 80 00   .hÎ0¤€.....å.f€.
15900F70   00 00 01 65 B8 00 02 CC D0 27 14 2B C4 47 28 33   ...e,..ÌÐ'.+ÄG(3
15900F80   8B 6A 0B 20 00 21 2F 2D B8 C5 33 AA 1A E8 F9 F9   ‹j. .!/-¸Å3ª.èùù
15900F90   41 9E 69 A3 4D 98 50 B1 63 BB 7F 81 37 AD AC 00   Aži£M˜P±c»..7.¬.
```

**Fig. 9** Sample search hit for 0x 00 00 00 01

As a test, several video and file recovery programs such as C4ALL and PhotoRec were ran against the raw data of 'hda1' with no success.

The background investigation revealed that this DVR records in the H.264 video codec, also known as MPEG-4 AVC (Advanced Video Codec). Standard MPEG codec uses the header 0x 00 00 00 01 ??, where ?? is the stream ID (Fig. 9).

A search for 0x 00 00 00 01 ?? resulted in some patterned clustering of hits:

0x 00 00 00 01 67
0x 00 00 00 01 68
0x 00 00 00 01 06
0x 00 00 00 01 65

Data carved with the starting point at 0x 00 00 00 01 67 was recognized by MediaInfo as AVC format video at Baseline@L2 profile with a resolution of 352 × 240 pixels. The carved data was played successfully with Media Player Classic.

When the MPEG header 0x 00 00 00 01 67 was removed, MediaInfo and Media Player Classic no longer recognize the file and it was not playable.

What stand out in the carved video is the hard-coded timestamp and channel number, which states 09/11/2013 16:16:33 with the channel number of 2. Utilizing what was learned in the examination of bookkeeping data, namely, the timestamp coding by this particular DVR, we find this timestamp starting at 16 bytes just prior to 0x 00 00 00 01 67.

As seen in Fig. 10, the coded timestamp in this sample carve is 0x 0D 09 0B 10 10 21. Converting the hexadecimal number to decimal, we get 13 09 11 16 16 33, or 2013-Sept-11 16:16:33 h, which is the hardcoded timestamp seen in Fig. 10.

```
15900F10   6E BD C3 53 E8 5A 16 B5 33 FB E4 09 06 64 EC 47   n½ÃSèZ.µ3ûä..dìG
15900F20   24 A6 E1 F4 3A 6E 00 00 00 31 30 64 63 48 32 36   $¦áô:n...10dcH26
15900F30   34 B7 51 00 00 10 00 00 00 85 B7 5E 16 00 00 00   4·Q........‥.^....
15900F40   00 0D 09 0B 10 10 21 00 00 1E 00 00 00 60 01 00   ......!......`..
15900F50   00 00 00 00 01 67 42 E0 14 DB 05 87 C4 00 00 00   .....gBà.Û.‡Ä...
15900F60   01 68 CE 30 A4 80 00 00 00 01 06 E5 01 66 80 00   .hÎ0¤€.....å.f€.
15900F70   00 00 01 65 B8 00 02 CC D0 27 14 2B C4 47 28 33   ...e,..ÌÐ'.+ÄG(3
15900F80   8B 6A 0B 20 00 21 2F 2D B8 C5 33 AA 1A E8 F9 F9   ‹j. .!/-¸Å3ª.èùù
15900F90   41 9E 69 A3 4D 98 50 B1 63 BB 7F 81 37 AD AC 00   Aži£M˜P±c»..7.¬.
```

**Fig. 10** Video stream header timestamp

Inspecting the rest of the video stream header, we note 0x 31 30 64 63 48 32 36 34, or ASCII 10dcH264, the interpretation of which becomes clear with further examination.

Continuing the video data carve based on the previous criteria along with the preview of the video through Media Player Classic, and inspection of the video stream header, it was observed that video carved showing channel 1 corresponds with the video stream header 00dcH264 while channel 2 corresponds with 10dcH264. Further testing verified that this was consistent, therefore it is established that the second byte before dcH264 indicates the channel, where 0x 00 = channel 1 and 0x 01 = channel 2. No other channels were found on this drive. This is consistent with the background investigation information that only two cameras were connected to this DVR. Here, we extrapolate that 0x 02 = channel 3 and 0x 03 = channel 4.

Examining for further MPEG header data revealed numerous occurrences of 0x 00 00 00 01 61 as seen in Fig. 11. This video stream header differed from the previous one as it was smaller and includes 11dcH264 as opposed to 10dcH264.

Data carved with the starting point of 0x 00 00 00 61 was not recognized by MediaInfo or Media Player Classic. Recognizing the pattern of these two types of occurrences of video stream header, which is one occurrence of 10dcH264 which is playable and has a timestamp, to many occurrences of 11dcH264 which isn't playable, it is concluded that 10dcH264 corresponds to an I-frame while 11dcH264 corresponds to P-frame or B-frames. These I, P and B frames are in reference to video codecs.

The next logical step was to carve for a larger chunk of video. The carving schema starts with an I-frame. When it encounters the next I-frame, a check is done on the channel number and timestamp (year, month, day, and hour only). If it matches the base I-frame, the carve would continue. This new carving schema resulted in smooth usable video. Figure 12 proves an overview of findings made so far.

One advantage to the discovery of channel and timestamp information in the video data header itself is that virtually all of the video data on the drive can be accounted for. This is in contrast to relying on bookkeeping data such as log files as they will not track deleted or otherwise lost video.

Further, the header information is coded into every I-frame. I-frames are generally coded frequently. This means that even if the video header of a video stream data cannot be recovered, the video stream data not recovered with the technique in this paper would only be until the next I-frame. Tests on this DVR show that the I-frame interval is between 10 and 30 s.

The smaller the I-frame interval, the more granular the corresponding video can be skipped to. For example, YouTube uses an interval of 2 s while ESPN uses an

```
15A37FA0   2F 65 5F F2 53 3E B0 F8 31 31 64 63 48 32 36 34   /e_òS>°ø11dcH264
15A37FB0   21 07 00 00 00 00 00 00 DE 52 43 17 00 00 00 00   !........ÞRC.....
15A37FC0   00 00 00 01 61 FA 02 31 42 BC 5C 64 87 49 60 F8   [....aú.1B¼\d‡I`ø
15A37FD0   E2 A8 F2 29 0B FF 84 0A CD 98 48 E3 E4 0C 04 07   â¨ò).ÿ„.Í˜Hãä...
15A37FE0   86 9E 72 EA DC 92 CC A4 1A DC 4F 21 12 E2 F8 EB   †žrêÜ'Ì¤.ÜO!.âøë
```

**Fig. 11** Sample of MPEG header 0x 00 00 00 01 61

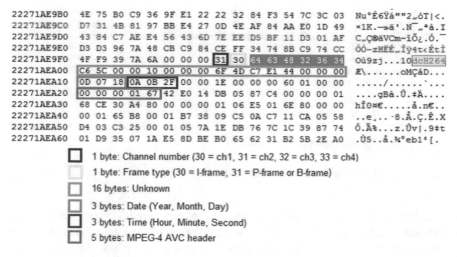

```
22271AE9B0   4E 75 B0 C9 36 9F E1 22 22 32 84 F3 54 7C 3C 03   Nu°É6Ýá""2„óT|<.
22271AE9C0   D7 31 4B 81 97 BB E4 27 0D 4E AF 84 AA E0 1D 49   ×1K.—»ä'.N⁻„ªà.I
22271AE9D0   43 84 C7 AE E4 56 43 6D 7E EE D5 BF 11 D3 01 AF   C„Ç®äVCm~îÔ¿.Ó.⁻
22271AE9E0   D3 D3 96 7A 48 CB C9 84 CE FF 34 74 8B C9 74 CC   ÓÓ-zHËÉ„Îÿ4t<ÉtÌ
22271AE9F0   4F F9 39 7A 6A 00 00 00 31 30 64 63 48 32 36 34   Où9zj...10dcH264
22271AEA00   C6 5C 00 00 10 00 00 00 6F 4D C7 E1 44 00 00 00   Æ\......oMÇáD...
22271AEA10   0D 07 18 0A 0B 2F 00 00 1E 00 00 00 60 01 00 00   ...../......`...
22271AEA20   00 00 00 01 67 42 E0 14 DB 05 87 C4 00 00 00 01   ....gBà.Û.‡Ä....
22271AEA30   68 CE 30 A4 80 00 00 00 01 06 E5 01 6E 80 00 00   hÎ0¤€.....å.n€..
22271AEA40   00 01 65 B8 00 01 B7 38 09 C5 0A C7 11 CA 05 58   ..e,..·8.Å.Ç.Ê.X
22271AEA50   D4 03 C3 25 00 01 05 7A 1E DB 76 7C 1C 39 87 74   Ô.Ã%...z.Ûv|.9‡t
22271AEA60   01 D9 35 07 1A E5 8D BE B0 65 62 31 B2 5B 2E A0   .Ù5..å.¾°eb1ª[.
```

☐ 1 byte: Channel number (30 = ch1, 31 = ch2, 32 = ch3, 33 = ch4)

☐ 1 byte: Frame type (30 = I-frame, 31 = P-frame or B-frame)

☐ 16 bytes: Unknown

☐ 3 bytes: Date (Year, Month, Day)

☐ 3 bytes: Time (Hour, Minute, Second)

☐ 5 bytes: MPEG-4 AVC header

**Fig. 12** Video stream header interpretation

interval of approximately 10 s. The .log files, as well as folder structure and naming convention appear to indicate the date range of video evidence available on the hard drive. The dates indicated are November 2011 and July 2012 to August 2013. Here we test whether there exist video evidence beyond the dates indicated in the bookkeeping data. This can be done one of two ways, by reviewing the carved video from the newly developed carving schema or by examining the data once again.

Figures 13 and 14 both show examples of video data for date range that is beyond the dates indicated by the bookkeeping data. In fact, video from as far as December of 2013 was located on the hard drive.

**Step 4: Presentation**

A forensic report was prepared detailing actions taken including receiving custody of the DVR, evidence continuity, and forensic integrity of evidence, tasks conducted, findings, and their limitations.

```
5F860FF460   86 D9 63 4D 19 09 57 8E 8C BE 61 6F 60 C3 AC A5   †ÙcM..WŽŒ¾ao`Ã¬¥
5F860FF470   8D 09 BB 07 9F 9E 2B 66 E4 B0 A7 DC 07 F1 0D 36   ...».Ÿž+fä°§Ü.ñ.6
5F860FF480   15 4A 9B B2 7F 68 7C 85 ED 2D 84 A7 BA 83 5F B6   .J>².h|.í-„§°f ¶
5F860FF490   87 BD 80 F5 D4 FF 2A D6 49 9B 9B 3E BE F3 C9 FB   ‡½€õÔÿ*ÖI>>>¾óÊû
5F860FF4A0   6B BE E8 1B 63 98 E0 00 31 30 64 63 48 32 36 34   k¾è.c˜à.10dcH264
5F860FF4B0   03 65 00 00 10 00 00 00 33 D3 95 3E 65 00 00 00   .e......3Ó•>e...
5F860FF4C0   0D 09 07 17 1F 1F 00 00 1E 00 00 00 60 01 00 00   ............`...
5F860FF4D0   00 00 00 01 67 42 E0 14 DB 05 87 C4 00 00 00 01   ....gBà.Û.‡Ä....
5F860FF4E0   68 CE 30 A4 80 00 00 00 01 06 E5 01 1C 80 00 00   hÎ0¤€.....å..€..
5F860FF4F0   00 01 65 B8 00 00 83 84 02 71 42 B1 F9 C8 15 4C   ..e,..ƒ„.qB±ùÈ.L
```

☐ 0x 0D 09 07:   2013-09-07

**Fig. 13** Video data indicating a date of September 7, 2013

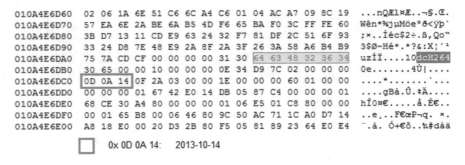

Fig. 14   Video data indicating a date of October 14, 2013

In this case the video data of several hours was recovered. This raw video stream does not contain information such as the total length of the video, locations of key frames and other relevant information about the video as a whole. Due to this lack of information, the raw video stream is only playable on a very small number of players, none of which include players that stakeholders such as other investigators, prosecutors or defence is likely to have, such as Windows Media Player and VLC. In addition, even with players that are able to play this video stream, seeking, skipping or fast forwarding of the video is not possible.

To remedy this, the raw video stream was processed through the multimedia video processor FFMPEG. Settings were carefully chosen to ensure the video stream was not re-encoded, which could lead to a loss of quality. The resulting video file was tested and was compatible with Windows Media Player and VLC, and had the capability to skip to any part of the video.

The original raw video stream evidence was kept on file and preserved as original evidence.

## 5.3   Case Study 3—Operation Baron

Scenario: The St. Pauls Carnival happens in Bristol on the first Saturday of July most years. It is an Afro-Caribbean festival that has been going since 1968, consisting of a masquerade precession of ornate and colourful costumes and floats made by the local schools, cultural associations and performance groups. At night, the festival turns to a street party where the roads are closed and the pubs and bars are open late. In the early hours of the 3rd of July 2011 in the Black Swan and Coach House pubs on Stapleton Road in Bristol, two rival North London gangs found themselves in each other's company. Arguments and obscenities were exchanged between the groups before threats of guns being used. At 4:05 a.m. members of the two gangs exchanged at least 21 shots across Stapleton Road, one shot hit a 21-year-old male, killing him. Further to this, a female walking by was hit and wounded as was another male. The gang members all fled the scene.

CCTV footage from multiple angles and locations was recovered by the investigation team. The challenge posed in this case study was how to present this evidence to the Court, a jury of lay people with varying technical skills and understanding, in such a way that it would understood and evidence the criminal offences alleged. Such challenges are a constant issue with presenting Big Data, namely the volume and variety of the CCTV data seen in these investigations.

### Step 1: Background Investigation
The lead investigator located CCTV footage of the initial interaction and some of the shootings. From this the investigation was able to establish there were at least 3 gunmen—two who had emerged from a side street opposite the Coach House pub and had then run off in different directions, and the third gunman who appeared to leave in a vehicle coming from the direction of the Coach House Pub.

The initial CCTV footage recovery strategy was to track forward and back the movements of the three gunmen, this quickly revealed that a number of cars were involved—three of these cars were Ford Pumas, which because of their unique styling made tracking them around the city before and after the offence relatively easy.

### Step 2: Evidence Preservation
The preservation was conducted in a manner consistent with those proposed by the General Methodology of Forensic Analysis proposed within this paper and as described in Case Studies 1 and 2. As this case study is focused primarily on Step 4: Presentation we have not described Step 2 in detail.

### Step 3: Analysis
The Analysis was conducted in a manner consistent with those proposed by the General Methodology of Forensic Analysis proposed within this paper and as described in Case Studies 1 and 2. As this case study is focused on primarily Step 4: Presentation we have not described Step 3 in detail.

### Step 4: Presentation
The investigation had established multiple CCTV sources, all of which were obtained and examined through the process. The challenge was how to present that volume of data to the court in the simplest manner possible but without losing its importance and integrity.

A storyboard was used in the format of still images with a careful explanation of what is in each image underneath. This case was the first time this technique was used to full effect by Avon and Somerset Constabulary, with an Officer talking the jury through it page by page from the witness box.

As the two defendants came from opposing sides (they had shot at each other) the storyboards were separated and one was effectively created for the movement of group 1 and a separate one for group 2. There were clearly overlaps in the storyboards, but it made sense to be able to tell the story from the either side.

A chronological schedule was also produced that showed the real time for each clip and an explanation of what was happening. Each clip had an entry on the schedule

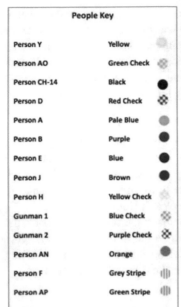

| People Key | | |
|---|---|---|
| Person Y | Yellow | |
| Person AO | Green Check | |
| Person CH-14 | Black | ● |
| Person D | Red Check | |
| Person A | Pale Blue | ● |
| Person B | Purple | ● |
| Person E | Blue | ● |
| Person J | Brown | ● |
| Person H | Yellow Check | |
| Gunman 1 | Blue Check | |
| Gunman 2 | Purple Check | |
| Person AN | Orange | ● |
| Person F | Grey Stripe | ⦀ |
| Person AP | Green Stripe | ⦀ |

**Fig. 15** The colour key chart and the colours used on the footage, from the report presented to the jury at Bristol Crown Court

and each clip was numbered. The jury used this as the various CCTV sources were played to them and they were encouraged to write additional information onto it.

Various edits were made to tweak the wording before the trial and it was agreed with defence so there were no major tweaks required during the trial. This must be avoided as anything put in front of the Jury that then needs removing/editing can cause the need for a re-trial with a different jury. CCTV led police to the offenders' cars which in turn led them to more CCTV and allowed them to get images of them that could identify them alongside intel and other data. CCTV showed why the shooting happened as it showed an earlier argument in the Coach House pub.

The CCTV told the whole story of the incident; it was such a chaotic scene with 50+ people nearby, 20+ bullets fired from three guns and cars and people going in many different directions. Without the clear explanation and the use of the model the court would never have really understood what had gone on (Figs. 15, 16, 17 and 18).

## 6   Conclusion and Further Work

The general methodology of DVR forensic analysis proposed in this paper was used in the analysis of a previously unknown Ganz and Swann DVR with great success.

**Operation Baron**

At **03:38:32** Person AP follows Person E into the rear bar area.

At **03:38:48** Person AP and Person Y from Group 2 come together in the rear bar area. Person AP is lost to sight as he moves towards the left of the image and the members from Group 1 and Group 2 come together.

**Fig. 16** A page from the CCTV report showing the interaction between the 2 gangs at the Coach House pub

The methodology was designed after a comprehensive review of prior research, taking into account the different approaches taken and their successes and challenges. Numerous other considerations and avenues of analysis were added after the successful video data recovery from the Swann DVR. On the third step: forensic analysis, which was the focus of this methodology, 3 core groups were discussed: file system,

**Operation Baron**

At **04:16:19** Person AN is also seen from this camera angle walking along St Marks Rd in the direction of Berwick Rd. A light toned marking is also seen on the sides of the dark toned footwear worn by Person AN.

**Fig. 17** Gunman 1 walking Bristol's back streets after the incident

 **Operation Baron**

**View of Gunman 3 from BCC Cam 168**

At **04:05:45** Gunman 3 (yellow arrow) is seen to run back towards Vehicle E from BCC Cam 168. The front drivers side headlight can be seen in the image (yellow circle). The image quality is not clear enough to establish if Gunman 3 gets into Vehicle E. Gunman 3 is not be seen to pass in front of the illuminated section believed to be the public phone box (yellow box).

**Fig. 18** The moment gunman 2 entered the Alfa Romeo

bookkeeping data and video data. Some considerations, steps and options within the core groups were discussed. In Case Study 1 the Ganz DVR was found to use a proprietary file system, which by conducting the examination as shown in this paper, resulted in the file system being reversed engineered with no initial knowledge, applications or directions available. Further, it was reversed engineered to a sufficient degree to allow for the identification and retrieval of video footage from any specified camera for any specified date recorded without the use of any proprietary applications. In Case Study 2, the Swann DVR used a standard file system (EXT2). However, the bookkeeping data was stored and coded in a proprietary way. The file and folder structure of the small partition 2, along with the successfully interpreted timestamp within the log files, provided a clear indication of video likely available on the drive. In both Case Studies 1 and 2, the analysis of available footage was shown to be beyond the date range indicated by the bookkeeping data or the manufacturers own supplied software. It is believed, and demonstrated with the successful analysis of the Ganz and Swann DVRs, that the developed general methodology can be applied to the analysis of any other unknown DVR. In Case Study 3, we demonstrated how Big Data can be presented in a logical and understandable manner, in a tried and tested case within the UK court system. In terms of big data, we explored both the volume and variety aspects. In dealing with the volume, Step 1: Background Information has allowed for the identification of the exact footage timeframe to be identified. It also allows for specific intelligence and open source information to be established which may reduce the timeframe of the overall examination. Furthermore, such background information promotes and affords specific searches for forensic acquisition once Step 3: Forensic Analysis has been completed and the system reverse engineered. The general methodology has also proven to be valid across two different vendors, of significant variance in their storage methods, file systems, bookkeeping and logging methods. It is proposed that the methodology is not restricted to any one vendor, therefore it addresses the Variety aspect of Big Data, and allows it to be discarded to a certain degree.

Further avenues of research includes expanding upon the methodology to include further principles and techniques of possible analysis within the core groups of file system, bookkeeping data and video data. The challenge when doing so will be to maintain the generality of its application to potentially any DVR or similar device, and not be too specific or vendor specific.

# References

1. The Telegraph: One surveillance camera for every 11 people in Britain. Retrieved from http://www.telegraph.co.uk/technology/10172298/One-surveillance-camera-for-every-11-people-in-Britain-says-CCTV-survey.html (2013)
2. IMS Research: World Market for CCTV and Video Surveillance Equipment Report (2010)
3. Donegen, W.S.: Case study: forensic analysis of a Samsung digital video recorder. Digit. Investig. 5(1–2) (2008)

4. Poole, N.R., Zhou, Q., Abatis, P.: Analysis of CCTV digital video recorder hard disk storage system. Digit. Investig. **5**(3–4) (2009)
5. Ariffin, A., Choo, K.K.R., Slay, J.: Digital camcorder forensics. In: Proceedings of the Eleventh Australasian Information Security Conference (2013)
6. Tobin, L., Shosha, A., Gladyshev, P.: Reverse engineering a CCTV system, a case study. Digit. Investig. **11**(3) (2014)
7. Han, J., Jeong, D., Lee, S.: Analysis of the HIKVISION DVR file system, digital forensics and cyber crime. In: 7th International Conference, ICDF2C, 2015 (2015)
8. Yang, F., Li, R., Wu, C.: Basic principle and application of video recovery software for "Dahua" and "Hikvision" brand. In: ICITCE 2014—International Conference on Information Technology and Career Education (2015)
9. Gomm, R., LeKhac, N.-A., Scanlon, M., Kechadi, T.: An analytical approach to the recovery of data from 3rd party proprietary file systems. In: The 15th European Conference on Cyber Warfare and Security (ECCWS 2016), Munich, Germany, July 2016, pp. 117–126
10. Regional Computer Forensics Laboratory: Best Practices for the Retrieval of Video Evidence for Digital CCTV Systems (2006)
11. Han, J., Jeong, D., Lee, S.: Analysis of the HIKVISION DVR file system. In: Lecture Notes of the Institute for Computer Sciences, Social-Informatics and Telecommunications Engineering, LNICST (2015)
12. Wang, W.: Digital video forensics. Thesis, Computer and Information Systems Department, Nathan Smith Building, Hanover (2009). ISBN 978-1-109-16419-0
13. Walter, C.: Kryder's law. Sci. Am. (2005)
14. Faheem, M., Le-Khac, N.-A., Kechadi, M.-T.: Smartphone forensics analysis: a case study for obtaining root access of an android Samsung S3 device and analyse the image without an expensive commercial tool. J. Inf. Secur. **5**(3), 83–90 (2014). https://doi.org/10.4236/jis.2014.53009
15. Alabdulsalam, S., Schaefer, K., Kechadi, M.-T., Le-Khac, N.-A.: Internet of things forensics: challenges and case study. In: Gilbert, P., Sujeet, S. (eds.) Advances in Digital Forensics XIV. Springer Berlin Heidelberg, New York (2018). https://doi.org/10.1007/978-3-319-99277-8_3
16. Boztas, A., Riethoven, A.R.J., Roeloffs, M.: Smart TV forensics: digital traces on televisions. Digit. Investig. **12**(Supplement 1), S72–S80 (2015)
17. Jacobs, D., Choo, K.K.R., Kechadi, M.-T., Le-Khac, N.-A.: Volkswagen car entertainment system forensics. In: 16th IEEE International Conference on Trust, Security and Privacy in Computing and Communications (TrustCom-17), Sydney, Australia, Aug 2017, pp. 699–705. https://doi.org/10.1109/Trustcom/BigDataSE/ICESS.2017.302
18. Le-Khac, N.-A., Jacobs, D., Nijhoff, J., Bertens, K., Choo, K.-K.R.: Smart vehicle forensics: challenges and case study. Future Gener. Comput. Syst. (2018). https://doi.org/10.1016/j.future.2018.05.081
19. Roder, A., Choo, K.-K.R., Le-Khac, N.-A.: Unmanned aerial vehicle forensic investigation process: DJI phantom 3 drone as a case study. In: 13th Annual ADFSL Conference on Digital Forensics, Security and Law, Texas, USA, May 2018

**Richard Gomm** received an MSc from UCD, Ireland in Forensic Computing and Cyber Crime Investigations in 2014 and is a current PhD student. With over 15 years Law Enforcement he is currently an Investigations Officer with the Garda Síochána Ombudsman Commission, Ireland. His research interests include CCTV Forensics, Crime Scene Forensic Digital Reconstruction and Law Enforcement responses to Big Data.

**Ryan Brooks** received an MSc in Forensic Computing and Cyber Crime Investigation from University College Dublin, Ireland in 2017. He is currently a Digital Evidence Recovery Officer at the

National Crime Agency in the United Kingdom. He is an EnCase Certified Examiner, a GIAC Certified Forensic Examiner and is currently working towards obtaining his GIAC Network Forensic Analyst certification studying SANS 572 - Advanced Network Forensics.

**Kim-Kwang Raymond Choo** received the Ph.D. in Information Security in 2006 from the Queensland University of Technology, Australia. He currently holds the Cloud Technology Endowed Professorship at the University of Texas at San Antonio (UTSA). In 2015, he and his team won the Digital Forensics Research Challenge organized by the Germany's University of Erlangen-Nuremberg. He is the recipient of the 2019 IEEE Technical Committee on Scalable Computing (TCSC) Award for Excellence in Scalable Computing (Middle Career Researcher), 2018 UTSA College of Business Col. Jean Piccione and Lt. Col. Philip Piccione Endowed Research Award for Tenured Faculty, British Computer Society's 2019 Wilkes Award Runner-up, 2019 EURASIP Journal on Wireless Communications and Networking (JWCN) Best Paper Award, Korea Information Processing Society's Journal of Information Processing Systems (JIPS) Survey Paper Award (Gold) 2019, IEEE Blockchain 2019 Outstanding Paper Award, Inscrypt 2019 Best Student Paper Award, IEEE TrustCom 2018 Best Paper Award, ESORICS 2015 Best Research Paper Award, 2014 Highly Commended Award by the Australia New Zealand Policing Advisory Agency, Fulbright Scholarship in 2009, 2008 Australia Day Achievement Medallion, and British Computer Society's Wilkes Award in 2008. He is also a fellow of the Australian Computer Society, an IEEE senior member, and co-chair of IEEE Multimedia Communications Technical Committee's Digital Rights Management for Multimedia Interest Group.

**Nhien-An Le-Khac** is a lecturer at the School of Computer Science (CS), University College Dublin (UCD), Ireland. He is currently the program director of MSc program in Forensic Computing and Cybercrime Investigation (FCCI)—an international program for the law enforcement officers specializing in cybercrime investigations. He is also the co-founder of UCD-GNECB Postgraduate Certificate in fraud and e-crime investigation. Since 2008, he is a research fellow in Citibank, Ireland (Citi). He obtained his Ph.D. in Computer Science in 2006 at the Institut National Polytechnique de Grenoble (INPG), France. His research interest spans the area of Cybersecurity and Digital Forensics, Data Mining/Distributed Data Mining for Security, and Fraud and Criminal Detection. Since 2013, he has collaborated on many research projects as a principal/co-PI/funded investigator. He has published more than 150 scientific papers in peer-reviewed journal and conferences in related research fields, and his recent edited book has been listed the Best New Digital Forensics Book according to the Book Authority.

**Kien Wooi Hew** received his MSc in Forensic Computing and Cybercrime Investigation from University College Dublin in 2016. He is a police officer with 14 years of service with the Royal Canadian Mounted Police (RCMP). He is currently a senior digital evidence specialist at Digital Forensic Services (formerly Technological Crime Unit), RCMP. He has deep hands-on practical experience having conducted hundreds of digital forensics analysis on criminal investigations and is a recognized expert witness in court who have provided expert testimony on over 12 cases. His experience and knowledge is enriched by wide cooperation with law enforcement agencies from all over the world while working on international cybercrime investigations. His research interests include cybersecurity and digital forensics as it relates to the proliferation of connected devices including IoT devices, the utilization of open source data and data mining for crime detection, investigation and prevention; as well as, the capacity and effectiveness of the varying law enforcement models on countering cybercrime and the exponential increased demand for digital forensics.

# Forensic Investigation of PayPal Accounts

**Lars Standare, Darren Hayes, Nhien-An Le-Khac, and Kim-Kwang Raymond Choo**

**Abstract** PayPal, Inc. is one of the leading international online payment method providers, with more than 218 million active customer accounts across the globe. PayPal not only appeals to consumers who wish to purchase goods online, it is also of interest to criminals in a variety of ways. When it comes to criminal investigations, it is critical to determine who committed the crime and how the case can be proven in court. When a criminal investigation relates to PayPal, the questions to be answered include: Which PayPal account was used by the suspect, which computer should be seized? How can we prove criminality? This chapter is geared towards digital investigators, who are interested in digital evidence related to PayPal accounts, used with a Web browser. Herein, we provide an overview of the techniques that PayPal actually uses to identify their customers, which goes beyond online user credentials. More specifically, this chapter highlights evidence related to PayPal accounts, which can be found on an acquired hard disk image file. This in turn should help to determine if a PayPal account was in fact used and identify which account was used. This research focuses on a behavioural analysis of PayPal, using the Mozilla Firefox Web browser, in an effort to monitor and identify ways to determine how a PayPal account was utilized. Furthermore, we have detailed the examination and analysis of acquired image files, involving different use cases of PayPal, to illustrate these indicators and subsequently analyse the findings.

**Keywords** PayPal · Digital forensics · Network forensics · Financial fraud

L. Standare (✉)
Rhineland-Palatinate Police University, Buechenbeuren, Germany

D. Hayes
Pace University, New York, USA

N.-A. Le-Khac
University College Dublin, Dublin, Ireland

K.-K. R. Choo
University of Texas, San Antonio, USA

© The Editor(s) (if applicable) and The Author(s), under exclusive license to Springer 141
Nature Switzerland AG 2020
N.-A. Le-Khac and K.-K. R. Choo (eds.), *Cyber and Digital Forensic Investigations*,
Studies in Big Data 74, https://doi.org/10.1007/978-3-030-47131-6_7

# 1  Introduction

Electronic commerce is a growing market especially when it comes to retail sales globally. Electronic payment methods are, according to a study [1], the preferred payment method worldwide. A statistical forecast for the U.S. online retail market indicates a rise in sales from $360 billion, starting in 2016 and rising to more than $603 billion in 2021 [2]. From an alternative perspective, money laundering [3], fraud and online crime has also increased, while traditional crime has transitioned from the real world into the virtual sphere as the Financial Times reported in 2017, in reference to U.K. crime statistics [4].

Currently, the leading online payment provider, based on market share [5], with more than 81% and more than 763,000 web pages offering their service, is PayPal, Inc. (hereafter referred to as "PayPal"), which is an American company founded in 1998. The company offers online payment services that allows individuals and businesses to transfer funds electronically. PayPal has 218 million active customer accounts and processed 1.9 billion payment transactions in the third quarter of 2017 and a posted revenue of $3.24 billion [6]. Thus, PayPal's strong market position and growth is attractive to criminals who attempt to benefit from their success.

On the other hand, law enforcement continually encounters numerous criminal cases that involve one or more PayPal accounts. There are primarily two types of cases: (1) where a PayPal account was registered with the name, address, bank account or credit card details for someone else or (2) the accounts were originally created by a legitimate person but were subsequently stolen and used. In both scenarios, the goal is to identify the suspect who fraudulently used a specific PayPal account and subsequently uncover evidence that links the suspect to the crime. One way to determine this is by forensically analysing the suspect's electronic device to find evidence related to the use of PayPal's website and services.

PayPal uses a variety of technologies to identify their customers. Nevertheless, there are practical problems associated with these investigations, especially when it comes to finding evidence and analysing Web browser activity. More specifically there are challenges associated with finding PayPal transactions, while determining if there are artefacts that identify the use of a specific PayPal account.

This research will focus on the use of the Mozilla Firefox[1] (hereafter referred to as "Firefox") running on a Microsoft Windows 10 (here in after referred to as "Windows") operating system (here in after referred to as "OS").

# 2  Related Work

There are two different foci for this research. One is that of the forensic artefacts created by Web browsers and the other is the privacy/security related issues associated with Web browsers and tracking techniques.

---

[1] https://www.mozilla.org.

From a forensics perspective, Oh et al. researched forensic artefacts created by Web browsers, including browsing history, browser cache, and cookie files and suggested that investigators must analyse the information generated from each browser, using the same timeline, to extract more significant information rather than parsing out information. To achieve this, he introduced a new tool called Web Browser Forensic Analyzer (WEFA) to collect artefacts from different Web browsers [7].

Pereira researched Firefox 3 Internet history and identified where the browser stored its Internet history in SQLite databases and also explained how deleted SQLite records could be retrieved [8]. Nalawade et al. compared different tools to retrieve Web browser artefacts and discovered their benefits and limitations [9]. Rathod focussed his research on the Google Chrome Web browser and demonstrated how to obtain information about the last accessed date and time, search items, visited URLs and deleted data [10].

Some researchers have focussed on the Web browsers that provide private mode features and have shown how these features still leave Web browsing [11–14]. Said et al. for example, was able to extract evidence from private browsing sessions, using RAM analysis [15], and Hedberg's research led to similar findings by recovering artefacts for Google Chrome, Mozilla Firefox and Microsoft Internet Explorer from within the hard disk drive and memory of the system [16]. Other research, including Donny et al. looked at portable Web browsers and discovered that the best way to retrieve evidence is to obtain the data from RAM or working memory [17].

Le-Khac et al. researched particular Web browsers that support privacy preserving Web browsing and found out that they still leave evidence in RAM or on the local drive about the user's browsing activities, like keyword searches, websites visited, and viewed pictures [18, 19].

With the HTML5 standard there were new features added to access client storage and resources, like Web Storage and IndexedDB, and these have also been examined by researchers.

Matsumoto et al. demonstrated the new forensic possibilities to retrieve Web Storage artefacts from RAM [20] or from the local disk [21]. Mendoza et al. also researched Web Storage and developed a new tool, called Browser Storage Extractor (BrowStEx), to aggregate data from Web Storage files [22].

Researchers around Kimak focussed on the client-side database, called IndexedDB, which is also one of the newer functionalities in HTML5, to provide an offline functionality with reduced load on the Web server and determined that IndexedDB stored data in an unencrypted state transparently, which was external to the actual database [23]. He proposed a security model to secure the data on a client's machine [24].

Boucher et al. [25] a framework to guide computer forensic examiners in their quest to determine if data is local or synced via Web browser apps because of the ever-increasing proliferation of Internet connected devices. The end user benefits of this seamless synchronization of data between devices. However, the synchronization of data between devices translates to a challenge for computer forensic examiners as a device may contain evidence that synced from another device that cannot be found.

Some researchers, like Soltani et al. have taken a data privacy approach to the use of Flash cookies, which track users [26]. Flash cookies, and their ability to respawn HTTP cookies, were researched by McDonald et al. [27]. Subsequently, their interaction with HTML5 storage concepts for respawning were later researched by [28, 29]. Samy Kamkar provided an overview of where data can be placed on the local machine, by webpages, to track the client and how these techniques can be used to create a persistent cookie called 'evercookie' [30].

Eckersly [31] studied whether Web browsers could be uniquely identified via "device fingerprinting" and what techniques are used to achieve this. The outcome of this study led to the creation of a website to check your own Web browser fingerprint.[2]

It appears that there no research has focussed on Web browser forensics that is related to the use of PayPal. This chapter will focus on digital forensics techniques used to examine Web browser artefacts associated with PayPal accounts. A variety of tracking techniques are also important to understand when determining what Web browser sessions exist to ascertain whether a specific Internet service was used.

## 3 Problem Statement

Modern websites use different techniques to track and identify their visitors because of security and commercial reasons. Those techniques may leave traces on a local machine which could be later analysed and used to reconstruct the user's activities. PayPal has stated that they make use of various techniques to track or identify a customer. By browsing the webpages of PayPal's website, login into an existing PayPal account and performing other PayPal related activities via a web browser there will be very likely traces left behind of these tracking activities on the used computer for further analysis. Therefore, in this chapter, we aim to answer the following research questions: Is the use of a PayPal account, accessed via the Firefox web browser, detectable after forensically imaging a hard disk drive and what are the indicators of use?

## 4 Adopted Approach

This section deals with a description of our research methodology to identify indicators which prove that a PayPal account has been used. To achieve this, we have selected two different approaches; (a) the web browser monitoring approach and (b) the computer forensic approach, as displayed in the figure. Experiments were carried out in a special experiment environment.

The usage of PayPal is divided into different steps and each step will first be analysed in the monitoring approach to identify indicators of the usage of a PayPal

---

[2]https://panopticlick.eff.org.

account by performing a behavioural analysis on the web browser. The computer forensic approach will then examine acquired disk image files from the experiments to show if the identified indicators can be found and analysed to prove that a PayPal account has been used.

## 4.1 Web Browser Monitoring Approach

The first used approach is to monitor the web browser's activities in each step, as shown in Fig. 1, with the built-in developer tools[3] from Firefox. This will primarily be achieved using the Storage Inspector[4] tool to analyse the different data types, stored or changed by Firefox, on the computer system in addition to the Network Monitor[5] tool that will capture the network interactions between the web browser and the selected web server. These tools enabled us to monitor and manipulate the stored and transferred data for behavioural analysis. They are built-in and can be displayed by typing '*Shift + F9*' within Firefox [32].

Figure 2 displays the Network Monitor Tool from Firefox in action which gives an overview of the HTTP requests and transferred files from the web browser interaction with the web server during a browser session.

Another tool from Firefox is the Storage Inspector, which provides an overview of stored cookies or other storage functions, including Web Storage's Local Storage and Session Storage (Fig. 3).

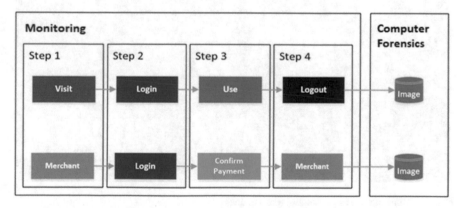

**Fig. 1** Adopted approach process

---

[3]https://developer.mozilla.org/en-US/docs/Tools.

[4]https://developer.mozilla.org/en-US/docs/Tools/Storage_Inspector.

[5]https://developer.mozilla.org/en-US/docs/Tools/Network_Monitor.

**Fig. 2** Firefox network monitor tool (dark theme)

**Fig. 3** Firefox storage inspector tool (dark theme)

## 4.2 Computer Forensics Approach

The second approach is a computer forensic analysis of acquired disk image files to identify usage of PayPal. With the collected information, from the first approach, we can search for the identified indicators to check if they can be found in a forensic analysis of a disk image file. This would enable us to see if a PayPal account was used and analyse if there can be more indicators found within.

In both scenarios, a virtualisation software is used on a host machine so as to have a clean environment in a virtualised OS with an installed web browser who has not visited a web page after its installation. The snapshot function of the virtualisation software was used to revert the state of the virtual machine after each experiment to the origin state.

The following subsections describe the experimental environment and process used in the experiments.

## 4.3  Experimental Procedure

This section describes the followed procedures in the various experiments during the two different approaches.

The experiments always started with the virtual machine in the origin state, which is a saved virtual machine state after the OS and the web browser was first installed and it has not been used to visit a web pages after its installation.

Beginning with the virtual machine in the origin state in the monitoring approach the separate steps as shown in Fig. 1 was monitored with the developer tools and the different changes (for example creating, altering or deleting cookies) made by the web browser in each step were recorded. These records were later analysed to identify indicators for the usage of PayPal.

After identifying possible indicators for the use of PayPal, three different kinds of experiments are conducted:

(1)  Only visiting PayPal's website.
(2)  Visiting PayPal's website with login, browsing through and logout of a PayPal account.
(3)  Using PayPal as a payment method in online shopping.

After each web browser session, we acquired a forensic disk image of the virtual hard disk within the virtual machine as demonstrated in the next figure.

A computer forensic examination would normally follow after a forensic imaging process of the powered off hard disk drive with a write blocker to protect the evidence from modification by the connected OS. This research project used a virtual machine to ensure that the carried-out experiments began with a clean environment. To ensure that the snapshot function wouldn't affect a forensic imaging process of a suspended or powered off virtual machine we have chosen to conduct the forensic image process within the running machine as a live imaging process with an application with minimal footprint.

The forensic disk image was made with the Access Data FTK Imager command line version which was stored on a portable hard disk drive. The portable hard disk drive was connected to the virtual machine via a USB 3.0 connection from the host machine (Fig. 4).

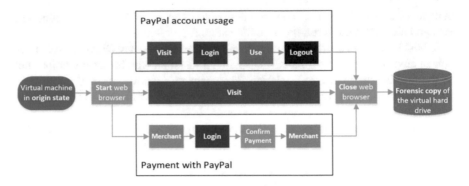

**Fig. 4** Experiment procedure: forensic disk image creation

To automate the whole experimental process, we created a batch script which was used to collect system information and timestamps in log files and to automatically start the web browser to load the web page www.paypal.de and to acquire the disk image file at the end of the experiment which was then stored on the connected portable hard disk drive.

## 5　Experimental Results and Analysis

This section describes the outcomes for the applied approaches to solve the research problem. It is divided into two main subsections, beginning with the outcomes for the monitoring approach which is followed by the outcome in the computer forensic approach.

### 5.1　Web Browser Monitoring

The following subsections describe the major results from the carried-out experiments where indicators were found during the monitoring approach separated by the different steps of visiting PayPal web site, log in, use and log out of PayPal accounts.

#### 5.1.1　Monitoring Website Visits

**Web History Findings**
A customer will initially visit PayPal's website to log into his PayPal account. PayPal has a wide selection of country-specific web pages and which all follow the same structure as shown below

https://www.paypal.com/[ab]/webapps/mpp/home

where *[ab]* is replaced by a two-digit long code for each specific country like *mx* for Mexico, *us* for the USA or *de* for Germany [33].

A common way to browse to PayPal's country-specific web page is to enter a country-specific URL directly into the address bar of the web browser. To get to the German web page for example you would possibly enter paypal.de into the address bar and hit 'ENTER'. The monitoring approach indicates the following connection establishment after hitting the 'ENTER' button:

(1)  paypal.de
(2)  https://www.paypal.de/
(3)  https://www.paypal.com/de/home/
(4)  https://www.paypal.com/de/webapps/mpp/home.

Beginning from the user entered URL under number 1. The web browser is first directed to the URL under number 2., then redirected to the URL number 3. And finally, to the URL under number 4., which is the starting web page for German customers.

**Cookie Findings**
While visiting the webpage of PayPal the web browser interacts with PayPal's web server and stores cookies on the customer's computer system. The Firefox Storage Inspector tool displays the stored cookies belonging to the current domain, as shown in Fig. 5.

**Web Storage Findings**
The Storage Inspector tool also allows us to see if a website uses newer storage techniques, like Web Storage. By visiting PayPal's web page, the web browser stores key/value pairs, as shown in Fig. 6, with the key names *44907* and *47364*, which are

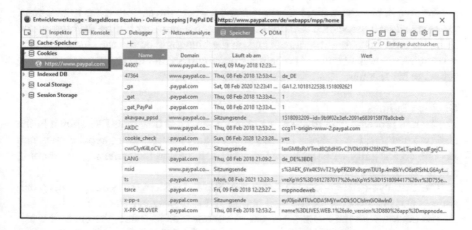

**Fig. 5**  PayPal monitoring: stored cookies when visiting PayPal's web page

**Fig. 6** PayPal monitoring: stored web storage entries when visiting PayPal's web page

identically named like two cookies as shown in Fig. 5. The key named *47364* seems to contain the cookie Expires attribute, as a JavaScript date object in milliseconds, since NSPR epoch, midnight (00:00:00), 1 January, 1970 Coordinated Universal Time (UTC) [34] and the same value of the identically named cookie.

### 5.1.2 Monitoring the Sign in Process

**Web History Findings**
To get access to a PayPal account, the customer must first sign in. This can be achieved by visiting PayPal's website and click on the log in button which will direct the web browser to the sign in webpage under

https://www.paypal.com/signin?country.x=DE&locale.x=de_DE

The URL of the *signin* page is followed by variables after the "?" symbol to specify the language settings for the login process; for example like *country.x=DE&locale.x=de_DE* for German. Without this additional information the sign in page would be displayed in English.

The sign in process is divided into two steps where the customer has to first enter the email address associated with the PayPal account and after hitting the blue button he has to enter his password and clicked on the blue login button the sign in process is completed when all entered credentials were correct and the customer has finally access to his account.

**Cookie Findings**
The sign in process alters the existing cookies and adds new ones like shown in the next two pictures: Fig. 7 shows the new subdomain entry https://c.paypal.com which has one cookie entry with the same Domain attribute and the name *sc_f* containing a long string as its value.

The most eye-catching cookie is created after the customer has entered his email address and hit 'ENTER' or the next button and is highlighted in the Fig. 8.

The cookie with the name *login_email* contains, as its value, the entered email address of the customer, in unencrypted format, where the @ symbol is replaced by %40 for encoding reasons with character interpretation by the web server.

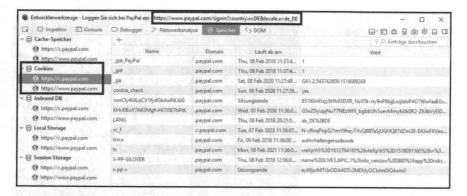

**Fig. 7** PayPal monitoring: stored cookies of PayPal sign in process part one

**Fig. 8** PayPal monitoring: stored cookies of PayPal sign in process part two

Figure 9 shows the details of this cookie with its name: value pair used in this example. Further research about this cookie indicated that it always contains the most recently entered email address which was used in the sign in process and it will be kept by Firefox at least until the Expire attribute deadline is reached or another email address is entered.

Several conducted sign-in procedures, also with email addresses where no PayPal account is linked to, have shown that this cookie value is filled with the string which the customer has most recently entered during the sign in process no matter if this string contains a legitimate email address as shown in Fig. 10.

**Fig. 9** PayPal monitoring: stored cookie login_email during PayPal sign in process

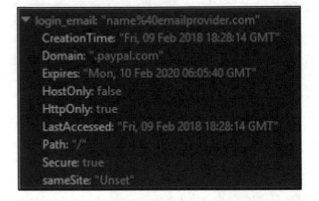

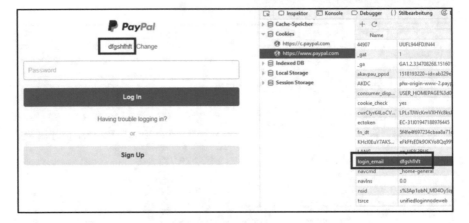

**Fig. 10** PayPal monitoring: experimenting with the login_email cookie

While the *login_email* cookie contains the most recently entered email address from the sign in process, further experiments show that this cookie is not an identifying session management cookie used to remember that this client is logged in.

**Web Storage Findings**

In addition to the existing key/value Local Storage entries from the web page visit, prior to the sign in process a new key/value pair is added under the new subdomain entry https://c.paypal.com, as shown in the next figure. The key is named *sc-lst* and its value is the same as the one from the cookie named *sc_f* under the same domain (Fig. 11).

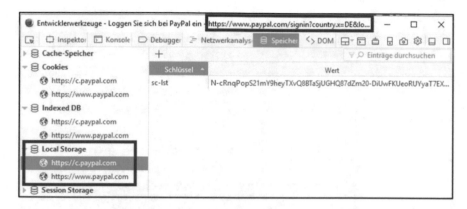

**Fig. 11**   PayPal monitoring: sign in web storage changes

## 5.2   *Monitoring the Account Usage*

**Web History Findings**

When a customer has successfully completed the sign in process, the web browser will be directed to the following URL: https://www.paypal.com/myaccount/home which is the overview web page of the customer's PayPal.

After the customer has reached the overview page he can browse within his account to different sections which all start with the same URL https://www.paypal.com/myaccount/ and end as follows:

a.   https://www.paypal.com/myaccount/home
b.   https://www.paypal.com/myaccount/settings
c.   https://www.paypal.com/myaccount/activity
d.   https://www.paypal.com/myaccount/wallet
e.   https://www.paypal.com/myaccount/transfer.

Now the customer can change his account settings under */settings*, see his account activities under */activity* or transfer funds to another PayPal account under */transfer*.

**Cookie Findings**

While the customer is logged in, the web browser has again altered and stored different cookies, which still contain the *login_email* cookie but now contain a legitimate email address of the currently used PayPal account as shown in Fig. 12.

**Web Storage findings**

In the Local Storage fields, the former created entry shown in Fig. 13 disappeared after the successful login and left over the values from the web page visit, as shown in the next figure.

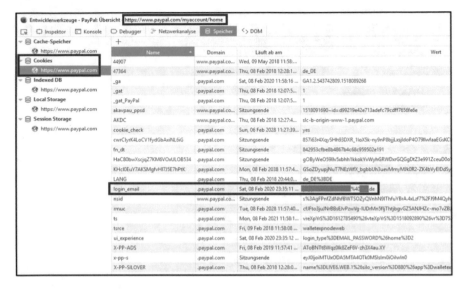

**Fig. 12** PayPal monitoring: stored cookies after login

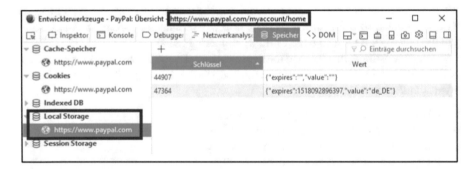

**Fig. 13** PayPal monitoring: stored web storage after login

### 5.2.1 Monitoring the Logout Process

**Web History Findings**

The log out process starts after hitting the log out button which directs the web browser, initially accessing https://www.paypal.com/signout and then is redirected to https://www.paypal.com/de/webapps/mpp/offers?,[6] which is a web page with different commercials by PayPal.

---

[6]The ? symbols is followed by a long string which seems to contain to be transferred information for the requested web server.

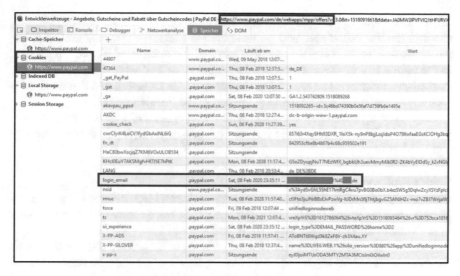

**Fig. 14** PayPal monitoring: stored cookies after logout

### Cookie Findings

After the logout process is finished, many of the created cookies are still stored depending on their Expires or Max-Age attributes. The experiments showed that the session cookies didn't survive a restart of the web browser but the other cookies were still stored like the *login_email* cookie with the email address formerly used for login as shown in the Fig. 14.

### Web Storage Findings

Once the user logs out, the Storage Inspector tool still display the stored key/value pairs, which were created on the web page visit. The value for the key *44907* didn't change and the key *47364* only changed the assigned number of the expired variable.

### 5.2.2    Monitoring Payments with PayPal

One of the main functions of PayPal is the ability to use it as a payment method when buying goods from supported merchants online [6].

Using PayPal as a payment method usually begins when the customer has chosen the desired goods within an online store and at the end of the ordering process he will be directed to PayPal's landing web page to log in and confirm the payment.

### Web History Findings

After the successful completion of the sign in process an overview page with the URL

https://www.paypal.com/webapps/hermes?flow=1-P&ulReturn=true&
country.x=DE&hermesLoginRedirect=xoon&locale.x=de_DE&token=EC-

0FJ15628VS360925V&useraction=commit&country.x=DE&locale.x=de_DE#/ checkout/review was loaded and we were able to check the payment and complete the payment process. After the payment successfully completed we were redirected to the online store, where we originally started from and where the payment was marked as confirmed. The analysis of the recorded URLs has shown that a indicator for a successful login can be found at the end of the following URL

https://www.paypal.com/webapps/hermes?[…]/checkout/review where */check-out/review* indicates that the final overview web page was displayed after a successful login.

**Cookie Findings**
During the payment process, there were many cookies added or altered. The majority of these cookies were similar to the cookies which were created while a customer goes through the normal process of login and logout.

The cookies named *HaC80bwXscjqZ7KM6VOxULOB534*, *X-PP-ADS*, *rmuc* indicated a successful login and could also be found while the customer is logged in, during the payment process. One unknown cookie was created after a successful login with the name *ectoken* who could also be identified as an indicator for the usage of a PayPal account which contained as its value the token value who could also be found in the URL during the payment process.

An analysis of all recorded cookies during the payment process resulted in an overview with the cookie names and the time when they were seen in the different steps from visiting PayPal's landing web page (column named *landing*), signing in with step one where the customer has to enter his email address (column named *signin1*), signing in with step two where the customer has to enter his password (column named *signin2*), the next step is when the customer is successfully logged in (column named *logged_in*) and the last column shows if the cookie will survive when the web browser gets closed (column named *persistent*). The X symbol represents that this cookie was seen at that time in the monitoring approach with the Storage Inspector. If the field is left blank then this cookie was not seen at this stage.

Four cookies of interest are marked, three in green for those who will survive a restart of the web browser and the one marked in yellow is a session cookie who should be deleted after the browsing session has ended (Table 1).

**Web Storage Findings**
Monitoring the Web Storage, during the payment process, showed that changes to the Local Storage were quite similar to the regular website login process. After a successful login, when the overview page is loaded, two unknown values were added under the domain https://www.paypal.com to the Local Storage as shown below:

| key | value |
|---|---|
| *family_device_test* | *["Z7HTHDGF2439N"]* |
| *public_device_test* | *false* |

**Table 1** Monitoring results: table of recorded cookie names with payment (full list)

| Cookie_name/Step | Session Cookie? | PP Payment Method Usage | | | | |
|---|---|---|---|---|---|---|
| | | landing | signin1 | signin2 | logged_in | persistent |
| 47364 | no | X | X | X | X | X |
| _ga | no | | | | X | X |
| _gat_PayPal | no | | | | X | X |
| akavpau_ppsd | yes | X | X | X | X | |
| AKDC | no | X | X | X | X | X |
| cookie_check | no | | X | X | X | X |
| LANG | no | X | X | X | X | X |
| nsid | yes | X | X | X | X | |
| ts | no | X | X | X | X | X |
| tsrce | no | X | X | X | X | X |
| x-pp-s | yes | X | X | X | X | |
| X-PP-SILOVER | no | X | X | X | X | X |
| KHclOEuY7AKSMgfvHl7J5E7hPtK | no | | X | X | X | X |
| sc_f | no | | X | X | X | X |
| ui_experience | no | | X | X | X | |
| fn_dt | yes | | X | X | | |
| login_email | no | | X | X | X | X |
| X-PP-ADS | no | | | | X | X |
| HaC80bwXscjqZ7KM6VOxULOB534 | yes | | | | X | |
| rmuc | no | | | | X | X |
| X-PP-K | no | | X | X | X | X |
| AV894Kt2TSumQQrJwe-8mzmyREO | yes | | | | X | |
| x-csrf-jwt | no | X | X | X | X | X |
| ectoken | no | | | | X | X |
| CONSENT | no | | | X | | |
| NID | no | | | X | | |
| PYPF | no | | | X | | |
| DPz73K5mY4nlBaZpzRkjl3ZzAY3QMmrP | yes | | | | X | |
| id_token | yes | | | | X | |

These two records could also act as an indicator for the successful login and are maybe used by PayPal as a tracking method to identify the customer by the stored string in the *family_device_test* value.

### 5.2.3 Analysis

This section explains the results obtained from the monitoring experiments within the former subsections.

**Web History Findings**

Monitoring URLs, while visiting PayPal's website, loging in and loging out of an existing account, indicated usage of a PayPal account, which can be identified by the following URLs in the displayed order:

(1) https://www.paypal.com/signin[…]
(2) https://www.paypal.com/myaccount/[…]

(3)   https://www.paypal.com/signout.

After the customer has successfully signed in under the URL number 1, where […] stands for a specific string for different language settings for the sign in process, he will be logged into his account, which is recognizable by the URL number 2. where […] is replaced by the visited subsections of the PayPal account like */home* for the main account overview page, */transfer* for the transferring funds page or */settings* for the account settings page.

Finally, the URL under number 3. indicates the logout of the PayPal account.

While these records do not indicate that a specific account was used, because they could also be typed in manually or be selected from the web browser, bookmarks or web history records, they do set a timeframe in which someone has possibly used a PayPal account.

**Cookie Findings**

PayPal makes use of cookies, as previously discussed. While monitoring the different steps, from visit to logout, we recorded each stored cookie with its value for later analysis.

An analysis of all recorded cookies, during the normal account usage process, via PayPal's website, resulted in an overview table containing the cookie names and the time when they were seen in the different steps from visiting PayPal's website (column named *visit*), signing in with step one where the customer has to enter his email address (column named *signin1*), signing in with step two where the customer has to enter his password (column named *signin2*), the next step is when the customer is successfully logged in (column named *logged_in*) and the last column shows if the cookie will survive when the web browser gets closed after the logout of the account (column named *persistent*). The X symbol represents that this cookie was seen at that time in the monitoring approach with the Storage Inspector. If the field is left blank, then this cookie was not seen at this stage.

There are three cookies that are only created when a user is finally logged in and which could be used as indicators for the use of a PayPal account in later computer forensic analysis. These three cookies are marked, two in green for those who will survive a gracefully closing of the web browser and the one marked in yellow is a session cookie who should be deleted after the browsing session has ended. The three cookie names are:

*HaC80bwXscjqZ7KM6VOxULOB534*

This is a session cookie that contains a value, which is changed each time a new interaction with PayPal's web servers takes place, like the reloading of the account web page or browsing to another section page of the account. When the customer logs out of his account the value of this cookie gets deleted.

Further experiments showed that, if this cookie is manually deleted after the customer has logged in, then this will result in an automated logout by the next interaction with PayPal's web servers, similar to reloading a web page. Furthermore, altering the cookie value manually will result in a logout. This cookie seems to be used for session management to keep track of the current login status of the used web browser (Table 2).

**Table 2** Monitoring results: table of recorded cookie names (without payment)

| Cookie_name/Step | Session Cookie? | PP Web Site Usage | | | | |
|---|---|---|---|---|---|---|
| | | visit | signin1 | signin2 | logged_in | persistent |
| 44907 | no | X | X | X | X | X |
| 47364 | no | X | X | X | X | X |
| _ga | no | X | X | X | X | X |
| _gat | no | X | X | X | X | X |
| _gat_PayPal | no | X | X | X | X | X |
| akavpau_ppsd | yes | X | X | X | X | |
| AKDC | no | X | X | X | X | X |
| cookie_check | no | X | X | X | X | X |
| cwrClyrK4LoCV1fydGbAxiNL6iG | yes | X | X | X | X | |
| LANG | no | X | X | X | X | X |
| nsid | yes | X | X | X | X | |
| ts | no | X | X | X | X | X |
| tsrce | no | X | X | X | X | X |
| x-pp-s | yes | X | X | X | X | |
| X-PP-SILOVER | no | X | X | X | X | X |
| KHclOEuY7AKSMgfvHl7J5E7hPtK | no | | X | X | X | X |
| sc_f | no | X | X | | | X |
| ui_experience | no | | X | | X | X |
| fn_dt | yes | | X | X | | |
| login_email | no | | X | | X | X |
| X-PP-ADS | no | | | | X | X |
| HaC80bwXscjqZ7KM6VOxULOB534 | yes | | | | X | |
| rmuc | no | | | | X | X |

### X-PP-ADS

This cookie is set to expire after one year beginning from the most recent interaction with PayPal's web servers. The value of this cookie contains a string which stays the same as long as the user does not log out and in again, because a new login will result in a new value. This cookie may be used for session management by PayPal.

### rmuc

This cookie is set to expire after ten years, beginning from the most recent interaction with PayPal's web servers. The value of this cookie contains a string, which remained the same throughout the experiments. This cookie could be used by PayPal as an identifier for the used PayPal account within the experiments.

With the information about these three cookies, it is now possible to tell that the email address within the value of the cookie named *login_email* was used for a successful login into this PayPal account by comparing the cookie related timestamps when they were created or last accessed.

### Web Storage Findings

Monitoring the Web Storage used by PayPal's website, during the monitoring approach, showed that PayPal is using the Local Storage function. This storage function is able to be accessed from web pages from the same web site but from different browser windows and the stored data will last after the session was terminated (50). A Local Storage entry with the key name *sc-lst* was created during the sign in process whose value contained the same string as the cookie value from the cookie named *sc_f* but both got deleted after successful login. The other two key entries *44907* and *47364* seemed to be a copy of the cookies with the identical names and their values got updated when the related cookies were altered.

These Local Storage entries were still stored after the logout like the related cookies. The existence of the following Local Storage database entries indicate a successful login during a payment process:

| key | value |
|-----|-------|
| *family_device_test* | *[ "Z7HTHDGF2439N" ]* |
| *public_device_test* | *false* |

These records seem to have no related time values and other artefacts like cookies or web history entries may needed to tell when the login took place.

### IndexedDB Findings

During the monitoring approach, there was no usage of the IDB storage function detectable, by the Storage Inspector tool. To verify how the Storage Inspector tool worked, we created a HTML file named index.html which contained a JavaScript code to create an IndexedDB database (DB) manually with the name *MyDatabase* and adds the following two entries to the DB:

*{id: 12345, name: {first: "John", last: "Doe"}, age: 42}*
*{id: 67890, name: {first: "Bob", last: "Smith"}, age: 35}*

Opening the *index.html* file with Firefox showed that the Storage Inspector tool couldn't find any IDB entries, as shown in Fig. 15.

**Fig. 15** Monitoring IndexedDB experiments

To verify this, we performed additional research related to the possible storage location of the IDB data and found out that the opened *index.html* file created a new folder under

*%USERPROFILE%\AppData\Roaming\Mozilla\Firefox*
*\Profiles\[profilename]*^Error! Bookmark not defined.*.default\storage\default*

with the name *file++++C++Users+PP+Desktop+index.html*, as shown in the Fig. 16.

Within this folder two more subfolders and two files were created. The folder named *idb* contained a file named *1556056096MeysDaabta.sqlite* and a folder which was empty.

A brief check of the Linux built-in tools xxd and file showed that this file is a SQLite 3.x database file (Fig. 17).

Further analysis, with the DB Browser for SQLite showed that this file stores the database named *MyDatabase* but the *name*, *id* or *age* values couldn't be found in the different tables. By viewing the file content in the Linux xxd tool we was able to find some fragments of the searched-for value.

As a final check to make sure that these results weren't caused by the used JavaScript code we crosschecked the same *index.html* file with the Google Chrome

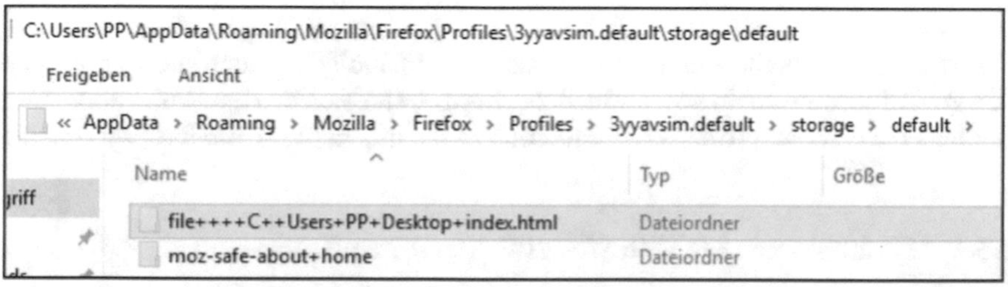

**Fig. 16** IDB: storage folder in Firefox

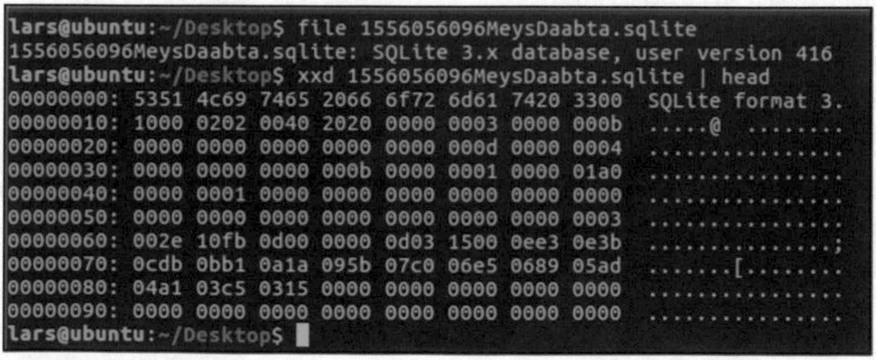

**Fig. 17** IDB: file analysis

web browser v62 and its developer tools which displayed the correct values for the
created database.

The experiments conducted with the IDB functionality show that the Firefox
Storage Inspector tool seem to not display stored IDB data and this must be kept in
mind when monitoring web pages with this tool.

**Cache Findings**
The Storage Inspector possesses a field for cache content but it has never displayed
any files or data during the experiments and therefore were no monitoring records
created. However, the preferences of Firefox showed that the Cache has used hard
disk drive storage as shown in the next figure which will later be analysed in the
computer forensic approach.

**Flash Cookie Findings**
Our version of Firefox blocked Flash content by default and asks the user for permission. During the monitoring approach there were no messages displayed which
asked the user for permission or any other Flash related activity noticed.

## 5.3   Computer Forensic Approach

With the findings in the monitoring approach where indicators for a successful usage
of a PayPal account could be identified the forensic analysis will now build upon these
findings the computer forensic examination of the image files from the experiments.

### 5.3.1   Findings from the Web Site Visit Experiments

This subsection describes the results found with the forensic software Magnet Axiom[7] within the image files acquired in the experiments where only a web
site visit and no login occurred.

**Web History Findings**
With the help of Axiom, the web history from within the SQLite database file
*places.sqlite* could be found and extracted as shown in the Fig. 18.

Figure 18 shows the Axiom findings for three analysed image files where the
web site of PayPal has only been visited. The found URL entries within the three
*places.sqlite* files are consistent with the findings in the monitoring approach.

---

[7](https://www.magnetforensics.com/).

| URL | Title | Last Visited Date/Time | Location | Source |
|---|---|---|---|---|
| http://www.paypal.de/ | | 09.11.2017 12:49:46 | Table: moz_places(id: 9) | VM01FF.img.1.1.E01 |
| https://www.paypal.com/de/home/ | | 09.11.2017 12:49:48 | Table: moz_places(id: 10) | VM01FF.img.1.1.E01 |
| https://www.paypal.com/de/webapps/mpp/home | Bargeldloses Bezahlen - Online Shopping \|... | 09.11.2017 12:49:48 | Table: moz_places(id: 11) | VM01FF.img.1.1.E01 |
| http://www.paypal.de/ | | 09.11.2017 13:16:56 | Table: moz_places(id: 9) | VM01FF.img.2.1.E01 |
| https://www.paypal.com/de/webapps/mpp/home | Bargeldloses Bezahlen - Online Shopping \|... | 09.11.2017 13:16:58 | Table: moz_places(id: 11) | VM01FF.img.2.1.E01 |
| https://www.paypal.com/de/home/ | | 09.11.2017 13:16:58 | Table: moz_places(id: 10) | VM01FF.img.2.1.E01 |
| http://www.paypal.de/ | | 09.11.2017 13:33:47 | Table: moz_places(id: 9) | VM01FF.img.3.1.E01 |
| https://www.paypal.com/de/home/ | | 09.11.2017 13:33:49 | Table: moz_places(id: 10) | VM01FF.img.3.1.E01 |
| https://www.paypal.com/de/webapps/mpp/home | Bargeldloses Bezahlen - Online Shopping \|... | 09.11.2017 13:33:50 | Table: moz_places(id: 11) | VM01FF.img.3.1.E01 |
| http://www.paypal.de/ | | 09.11.2017 13:54:53 | Table: moz_places(id: 9) | VM01FF.img.4.1.E01 |
| https://www.paypal.com/de/home/ | | 09.11.2017 13:54:55 | Table: moz_places(id: 10) | VM01FF.img.4.1.E01 |
| https://www.paypal.com/de/webapps/mpp/home | Bargeldloses Bezahlen - Online Shopping \|... | 09.11.2017 13:54:55 | Table: moz_places(id: 11) | VM01FF.img.4.1.E01 |

**Fig. 18** Forensic findings: web history when visiting PayPal's website

The experiments also showed that Firefox is storing the visited URLs with an *id* number in order of their first visit within the *moz_places* table of the *places.sqlite* database which can be seen Fig. 18, in the column '*Location*'.

**Cookie Findings**
The findings in the forensic analysis were consistent with the records in the monitoring approach and cookies from other websites, like mediaplex.com and doubleclick.net were found, which seem to belong to third-party companies.

**Web Storage Findings**
The *webappsstore.sqlite* file could be examined with the Axiom built-in hex viewer tool. The two keys, named *44907* and *47364*, were found within the database and contained the known values from the monitoring approach.

### 5.3.2 Forensic Findings After PayPal Account Usage

This subsection describes the found forensic artefacts with Axiom within the image files acquired in the experiments where a successful login into a PayPal account occurred.

**Web History Findings**
After using a PayPal account there were more web history artefacts found as expected. Figure 19 displays the findings by Axiom in two different image files where a successful login and logout with a PayPal account occurred. The login can be identified by the URL of the sign in page, beginning with https://www.paypal.com/signin followed by the account overview page with the URL https://www.paypal.com/myaccount/home.

Figure 20 is an example from an image file, where, after the login, different sections (like the settings or transfer section) of the account were visited. During the monitoring approach we discovered that these actions, within the account, can be

| URL | Title | Last Visited Date/Time ^ | Location | Source |
|---|---|---|---|---|
| http://www.paypal.de/ | | 10.11.2017 16:05:06 | Table: moz_places(id: 9) | VM01FF.img.6.1.001 |
| https://www.paypal.com/de/home/ | | 10.11.2017 16:05:20 | Table: moz_places(id: 10) | VM01FF.img.6.1.001 |
| https://www.paypal.com/de/webapps/mpp/home | Bargeldloses Bezahlen - Online Shopping \|... | 10.11.2017 16:05:23 | Table: moz_places(id: 11) | VM01FF.img.6.1.001 |
| https://www.paypal.com/signin?country.x=DE&locale.x=de_DE | Loggen Sie sich bei PayPal ein | 10.11.2017 16:07:52 | Table: moz_places(id: 12) | VM01FF.img.6.1.001 |
| https://www.paypal.com/myaccount/home | PayPal: Übersicht | 10.11.2017 16:11:32 | Table: moz_places(id: 13) | VM01FF.img.6.1.001 |
| https://www.paypal.com/ | | 10.11.2017 16:12:43 | Table: moz_places(id: 15) | VM01FF.img.6.1.001 |
| https://www.paypal.com/signout | | 10.11.2017 16:12:43 | Table: moz_places(id: 14) | VM01FF.img.6.1.001 |
| https://www.paypal.com/de/home | Bargeldloses Bezahlen - Online Shopping \|... | 10.11.2017 16:12:44 | Table: moz_places(id: 16) | VM01FF.img.6.1.001 |
| http://www.paypal.de/ | | 10.11.2017 17:01:28 | Table: moz_places(id: 9) | VM01FF.img.7.1.001 |
| https://www.paypal.com/de/home/ | | 10.11.2017 17:01:39 | Table: moz_places(id: 10) | VM01FF.img.7.1.001 |
| https://www.paypal.com/de/webapps/mpp/home | Bargeldloses Bezahlen - Online Shopping \|... | 10.11.2017 17:01:42 | Table: moz_places(id: 11) | VM01FF.img.7.1.001 |
| https://www.paypal.com/signin?country.x=DE&locale.x=de_DE | Loggen Sie sich bei PayPal ein | 10.11.2017 17:05:16 | Table: moz_places(id: 12) | VM01FF.img.7.1.001 |
| https://www.paypal.com/myaccount/home | PayPal: Übersicht | 10.11.2017 17:10:21 | Table: moz_places(id: 13) | VM01FF.img.7.1.001 |
| https://www.paypal.com/signout | | 10.11.2017 17:13:33 | Table: moz_places(id: 14) | VM01FF.img.7.1.001 |
| https://www.paypal.com/ | | 10.11.2017 17:13:40 | Table: moz_places(id: 15) | VM01FF.img.7.1.001 |
| https://www.paypal.com/de/home | Bargeldloses Bezahlen - Online Shopping \|... | 10.11.2017 17:13:47 | Table: moz_places(id: 16) | VM01FF.img.7.1.001 |

**Fig. 19** Forensic findings: web history when using a PayPal account

identified by the URL starting with https://www.paypal.com/myaccount/ following the name of the subfolder of each section.

The */signin* and */signout* parts of the URL are highlighted in yellow and everything between these two web history records is web browsing activity within the used PayPal account.

Figure 20 also demonstrates that a reconstruction of the web history should not only focus on the last visited timestamps, it should also consider the *id* value of the *moz_places* table as displayed in the column *'Location'* and the values for the number of visits as displayed in column *'Visit Count'* because in this example the URL https://www.paypal.com/myaccount/home (highlighted in blue) was first visited after the URL of the sign in page beginning with https://www.paypal.com/signin. This can

| URL | Last Visited Date/Time ^ | Title | Visit Count | Location | Source |
|---|---|---|---|---|---|
| http://paypal.de/ | 06.02.2018 20:33:07 | | 1 | Table: moz_places(id: 9) | VM01FF.img.11.1.001 |
| https://www.paypal.de/ | 06.02.2018 20:33:08 | | 1 | Table: moz_places(id: 10) | VM01FF.img.11.1.001 |
| https://www.paypal.com/de/home/ | 06.02.2018 20:33:08 | | 1 | Table: moz_places(id: 11) | VM01FF.img.11.1.001 |
| https://www.paypal.com/de/webapps/mpp/home | 06.02.2018 20:33:09 | Bargeldloses Bezahlen -... | 1 | Table: moz_places(id: 12) | VM01FF.img.11.1.001 |
| https://www.paypal.com/signin?country.x=DE&local... | 06.02.2018 20:33:16 | Loggen Sie sich bei Pay... | 1 | Table: moz_places(id: 13) | VM01FF.img.11.1.001 |
| https://www.paypal.com/myaccount/activity | 06.02.2018 20:33:48 | PayPal: Aktivitäten | 1 | Table: moz_places(id: 15) | VM01FF.img.11.1.001 |
| https://www.paypal.com/myaccount/transfer | 06.02.2018 20:33:51 | | 1 | Table: moz_places(id: 16) | VM01FF.img.11.1.001 |
| https://www.paypal.com/myaccount/transfer/ | 06.02.2018 20:33:52 | PayPal: Senden | 1 | Table: moz_places(id: 17) | VM01FF.img.11.1.001 |
| https://www.paypal.com/myaccount/settings | 06.02.2018 20:34:03 | | 1 | Table: moz_places(id: 19) | VM01FF.img.11.1.001 |
| https://www.paypal.com/myaccount/settings/ | 06.02.2018 20:34:04 | PayPal: Einstellungen | 1 | Table: moz_places(id: 20) | VM01FF.img.11.1.001 |
| https://www.paypal.com/myaccount/settings/security | 06.02.2018 20:34:15 | PayPal: Einstellungen | 1 | Table: moz_places(id: 21) | VM01FF.img.11.1.001 |
| https://www.paypal.com/myaccount/settings/payme... | 06.02.2018 20:34:17 | PayPal: Einstellungen | 1 | Table: moz_places(id: 22) | VM01FF.img.11.1.001 |
| https://www.paypal.com/myaccount/settings/notific... | 06.02.2018 20:34:18 | PayPal: Einstellungen | 1 | Table: moz_places(id: 23) | VM01FF.img.11.1.001 |
| https://www.paypal.com/myaccount/home | 06.02.2018 20:34:28 | PayPal: Übersicht | 2 | Table: moz_places(id: 14) | VM01FF.img.11.1.001 |
| https://www.paypal.com/myaccount/wallet/balance | 06.02.2018 20:34:38 | PayPal: E-Börse | 1 | Table: moz_places(id: 24) | VM01FF.img.11.1.001 |
| https://www.paypal.com/myaccount/wallet | 06.02.2018 20:34:45 | PayPal: E-Börse | 2 | Table: moz_places(id: 18) | VM01FF.img.11.1.001 |
| https://www.paypal.com/signout | 06.02.2018 20:34:49 | | 1 | Table: moz_places(id: 25) | VM01FF.img.11.1.001 |
| https://www.paypal.com/de/webapps/mpp/offers?v... | 06.02.2018 20:34:50 | Angebote, Gutscheine u... | 1 | Table: moz_places(id: 26) | VM01FF.img.11.1.001 |

**Fig. 20** Forensic findings: web history when using a PayPal account

| login_email | paypal_dissertation01%40n3ll.de | 06.02.2018 20:34:45 | 06.02.2018 20:33:31 | 07.02.2020 08:11:04 | .paypal.com |
| rmuc | cfJFto3juJNrBBsEJvPzwVg-IUDrMn5fljThtjbgvGZ5ANJH... | 06.02.2018 20:34:45 | 06.02.2018 20:33:41 | 06.02.2028 20:33:33 | .paypal.com |
| X-PP-ADS | AToBHhF6Wh5qnDzNKddM4h4LNqRORcU | 06.02.2018 20:34:45 | 06.02.2018 20:33:41 | 06.02.2019 20:33:34 | .paypal.com |

**Fig. 21** Forensic findings: found login indicating cookies

be identified by comparing the *id* numbers which are stored in ascending order after each first visit of a URL with the *last_visit_date* timestamp and the *visit_count* value in the *moz_places* table to see if a URL has been visited more than once and its ID value is lower than the other URL entries when sorted by the *last_visit_date* timestamps.

**Cookie Findings**

An analysis of *cookies.sqlite* files with Axiom confirmed the results gained in the monitoring approach. Cookies that showed a successful login (Fig. 21) could be found in all of images examined where a PayPal account login occurred.

A comparison of the created timestamps of the cookies, see Fig. 21, with the last visit timestamps from the web history also confirms the assumption gained in the monitoring approach that these cookies are created after a successful login because the cookies named *rmuc* and *X-PP-ADS* were created at 20:33:41 and, according to Fig. 21, the login must have taken place between 20:33:16 and 20:33:48.

**Web Storage Findings**

Axiom was also able to display the *webappsstore.sqlite* database in some kind of database viewer as shown in the Fig. 22.

The meaning of these entries could not be identified but the value belonging to the key named *sc-lst* had the same value as the cookie named *sc_f* and could work as a backup for the cookie value.

| webappsstore2 | | | |
| --- | --- | --- | --- |
| originKey | scope | key | value |
| noc.lapyap.www.:https:443 | moc.lapyap.www.:https:443 | 44907 | {"expires":"","value":""} |
| noc.lapyap.c.:https:443 | moc.lapyap.c.:https:443 | sc-lst | JwMg1bchcDa1LR7KcmWwbj5-bKb1B8jDTgFk1. |
| noc.lapyap.www.:https:443 | moc.lapyap.www.:https:443 | loglevel | SILENT |
| noc.lapyap.www.:https:443 | moc.lapyap.www.:https:443 | 47364 | {"expires":1517951092488,"value":"de_DE"} |

**Fig. 22** Forensic findings: web storage entries after account usage

### 5.3.3    Forensic Findings After Payment with PayPal

This subsection describes the results found with Axiom, within the image files, acquired during the experiments, where a payment with PayPal occurred.

**Web History Findings**
In the experiments that we conducted, PayPal has been used as a payment method for buying goods in online stores from different merchants. In all cases, after the desired goods were bought the merchant used a payment service provider, who facilitated the payment and directed us to PayPal's web site to login and confirm the payment and redirected us to the merchant's web site to confirm the successful payment. The next figure shows an excerpt of this type of payment process, which was uncovered using Axiom in the *places.sqlite* file of Firefox afterwards with the columns: *visited_URL; date/time_of last_visit; title*.

Figure 23 shows in detail the web history from the checkout of the merchant's web page beginning with the URL https://www.rheinwerk-verlag.de/checkout/preview/.

The next entry is the involved payment service provider with the URL beginning with

https://secure.ogone.com/ncol/prod/orderstandard_utf8.asp[…].

This URL contained details like the merchant customer's name and email address along with other information about the transaction which was paid with PayPal as highlighted in the top red box.

After this the web page of PayPal was visited as seen in the monitoring approach where the login into the PayPal account can be identified by the URL:

https://www.paypal.com/webapps/hermes?[…]/checkout/login.

The next webpage, after the successful login, where the payment overview and confirmation were made can be identified by the URL:

https://www.paypal.com/webapps/hermes?[…]/checkout/review.

Both are highlighted by the blue box, displayed in Fig. 23.

After the payment was made, the confirmation was first sent back by the payment service provider to the merchant's webpage, where the payment was confirmed.

The visited URLs contained comprehensive information about the entire payment related to the PayPal account. The red highlighted box at the top contained information about the merchant's customer name and email address, while the red box at the bottom contained the name of the PayPal account holder that was used for the payment, the IP address and PayPal-related email address. With this information it is possible to tell when which PayPal account was used for payment with which IP address and under which name.

The highlighted green box contains the following substring:

*PayerID = Z7HTHDGF2439N*

The assigned string *Z7HTHDGF2439N* could also be found in the monitoring approach within the active Local Storage, by PayPal, in the payment process.

| | | |
|---|---|---|
| https://www.rheinwerk-verlag.de/checkout/preview/ | 11.02.2018 15:41:58 | Rheinwerk Verlag GmbH - Sicher online einkaufen |
| https://secure.ogone.com/ncol/prod/orderstandard_utf8.asp?cn=Lars+Nell&brand=PAYPAL&currency=EUR&declineurl=https%3A%2F%2Fwww.rheinwerk-verlag.de%2Fcheckout%2Fpaypal%2F%3Fstatus%3Ddeclined%26token%3De82f37e96b61ea485c951dbf7430ea4e1c51c84b&operation=SAL&orderID=1147717&paramplus=sessionid%3D3f70564780bb4e749458bd7770955fca%26customer_email%3D█████████████%40█.de%26total%3D4490%26order_no%3D1147717%26basket_id%3D447717%26date_frozen%3D2018-02-11+15%3A42%3A18.755297%2B00%3A00&language=de_DE&PSPID=GalileoPressPro&amount=4490&SHASign=bba2b7ff0d90efb0e3a52e2b209912c4f153cd00&accepturl=https%3A%2F%2Fwww.rheinwerk-verlag.de%2Fcheckout%2Fpaypal%2F%3Fstatus%3Daccepted%26token%3Dd2e75396b1897b0640ed39c7eb63c61e6daaebcf&pm=PAYPAL | 11.02.2018 15:42:27 | Zahlungsbestätigung |
| https://www.paypal.com/DE/cgi-bin/webscr?cmd=_express-checkout&token=EC-0FJ15628VS360925V&useraction=commit | 11.02.2018 15:42:30 | PayPal |
| https://www.paypal.com/DE/cgi-bin/webscr?cmd=_express-checkout&token=EC-0FJ15628VS360925V&useraction=commit#/checkout/signup | 11.02.2018 15:42:39 | PayPal-Kaufabwicklung - Eröffnen Sie ein PayPal-Konto! |
| https://www.paypal.com/webapps/hermes?country.x=DE&hermesLoginRedirect=xoon&locale.x=de_DE&token=EC-0FJ15628VS360925V&useraction=commit | 11.02.2018 15:47:47 | Loggen Sie sich bei PayPal ein |
| https://www.paypal.com/webapps/hermes?flow=1-P&ulReturn=true&country.x=DE&hermesLoginRedirect=xoon&locale.x=de_DE&token=EC-0FJ15628VS360925V&useraction=commit&country.x=DE&locale.x=de_DE | 11.02.2018 16:02:05 | PayPal |
| https://www.paypal.com/webapps/hermes?flow=1-P&ulReturn=true&country.x=DE&hermesLoginRedirect=xoon&locale.x=de_DE&token=EC-0FJ15628VS360925V&useraction=commit&country.x=DE&locale.x=de_DE#/checkout/landing | 11.02.2018 16:02:08 | PayPal |
| https://www.paypal.com/webapps/hermes?flow=1-P&ulReturn=true&country.x=DE&hermesLoginRedirect=xoon&locale.x=de_DE&token=EC-0FJ15628VS360925V&useraction=commit&country.x=DE&locale.x=de_DE#/checkout/login | 11.02.2018 16:02:08 | PayPal-Kaufabwicklung - Einloggen |
| https://www.paypal.com/webapps/hermes?flow=1-P&ulReturn=true&country.x=DE&hermesLoginRedirect=xoon&locale.x=de_DE&token=EC-0FJ15628VS360925V&useraction=commit&country.x=DE&locale.x=de_DE#/checkout/review | 11.02.2018 16:02:13 | PayPal-Kaufabwicklung - Zahlungsdetails überprüfen |
| https://secure.ogone.com/ncol/prod/order_APaypalExpress_FlowHandler_UTF8.asp?RET=0&PAYID=3925468549&CasePaypal=3&Hash=B68329BB1CB32A95993CED0EF7A3477CF9BFC12D&token=EC-0FJ15628VS360925V&PayerID=Z7HTHDGF2439N | 11.02.2018 16:10:49 | Zahlungsbestätigung |
| https://www.rheinwerk-verlag.de/checkout/thank-you/ | 11.02.2018 16:11:03 | Rheinwerk Verlag GmbH - Sicher online einkaufen |
| https://www.rheinwerk-verlag.de/checkout/paypal/?status=accepted&token=d2e75396b1897b0640ed39c7eb63c61e6daaebcf&orderID=1147717&currency=EUR&amount=44%2E9&PM=PAYPAL&ACCEPTANCE=&STATUS=9&CARDNO=paypal%2DXXXXXXXX%2Dde&ED=&CN=█████++█&TRXDATE=02%2F11%2F18&PAYID=3925468549&NCERROR=0&BRAND=PAYPAL&IP=█%2E█%2E█%2E█&sessionid=3f70564780bb4e749458bd7770955fca&customer%5Femail=█████████████%40█%2Ede&total=4490&order%5Fno=1147717&basket%5Fid=447717&date%5Ffrozen=2018%2D02%2D11+15%3A42%2E755297%2B00%3A00&SHASIGN=685DE081654B8874BD16C7301127F4FC4EC5AE68 | 11.02.2018 16:11:03 | |
| https://www.rheinwerk-verlag.de/ | 11.02.2018 16:12:15 | Rheinwerk – Der Verlag für IT, Design und Fotografie |

**Fig. 23** Forensic findings: web history findings after payment with PayPal

Further research about this string determined that this string could be found in payments from different merchants. It seems that this string is used by PayPal as an identifier for the active PayPal account used in our experiments.

**Cookie Findings**

An examination of the acquired image files, using Axiom, for PayPal payments also found three login-related cookies, along with other cookies. These were located within the *cookies.sqlite* Firefox database. The recorded cookie from the monitoring approach, named *ectoken*, could also be found within the database, as displayed in the Table 3.

This cookie seems to be used in the payment process, as an identifier for the current processed payment, between the involved payment service provider and PayPal. It can also be found within the visited URLs during the payment process.
*HaC80bwXscjqZ7KM6VOxULOB534*

We conducted a keyword search for the other login, indicating a session cookie named *HaC80bwXscjqZ7KM6VOxULOB534*, and we were able to find this

**Table 3** Forensic findings: indicating cookies after payment

| Experiment number | Host | Name | Value |
|---|---|---|---|
| 20.1.E01 | .paypal.com | login_email | ██████%40██de |
| 20.1.E01 | .paypal.com | X-PP-ADS | AToB1usJWlnW-GYwMwTZT7aD57Gexyo |
| 20.1.E01 | .paypal.com | rmuc | cfJFto3juJNrBBsEJvPzwVg-IUDrMn5fljThtjbgvGZ5ANJHZc-mo7vZBJ7WrijaWMBHvLJlbcA_X3EXEHRgKcOL6EJRBOR2lePvwIZdrn9c1kHb3Lx9gkrz6HF6dY4j1Y68dm |
| 20.1.E01 | www.paypal.com | ectoken | EC-0ME0548223652902V |
| 12.1.E01 | .paypal.com | login_email | ██████%40██.de |
| 12.1.E01 | .paypal.com | rmuc | cfJFto3juJNrBBsEJvPzwVg-IUDrMn5fljThtjbgvGZ5ANJHZc-mo7vZBJ7WrijaWMBHvLJlbcA_X3EXEHRgKcOL6EJRBOR2lePvwIZdrn9c1kHb3Lx9gkrz6HF6dY4j1Y68dm |
| 12.1.E01 | .paypal.com | X-PP-ADS | AToBdix7WoZBawjrxrMuWKVakMyxq40 |
| 12.1.E01 | www.paypal.com | ectoken | EC-7T228075C5489973U |
| 14.1.E01 | .paypal.com | login_email | ██████%40██.de |
| 14.1.E01 | .paypal.com | rmuc | cfJFto3juJNrBBsEJvPzwVg-IUDrMn5fljThtjbgvGZ5ANJHZc-mo7vZBJ7WrijaWMBHvLJlbcA_X3EXEHRgKcOL6EJRBOR2lePvwIZdrn9c1kHb3Lx9gkrz6HF6dY4j1Y68dm |
| 14.1.E01 | .paypal.com | X-PP-ADS | AToB9miAWk43aQ6p7CPRZEuBGtyLzPw |
| 14.1.E01 | www.paypal.com | ectoken | EC-0FJ15628VS360925V |

cookie stored within two different files in different image files. The file named *sessionstore.jsonlz4* could be found in the folder

*\AppData\Roaming\Mozilla\Firefox\Profiles\[profilename]..default\* and the other file, named *previous.jsonlz4*, was stored within

*\AppData\Roaming\Mozilla\Firefox\Profiles\[profilename].default\sessionstore-backups\* (Figs. 24 and 25)

The session cookies stored, within these two files, could be restored via the Firefox session restore function which restores a previous browsing session [35]. By using a newly installed Firefox web browser, and replacing the original files with the recently discovered files, from an image file, the session cookies can be recovered with the restore function, as shown in the Fig. 26.

**Web Storage Findings**

As previously mentioned, the 13-digit string, containing only upper case letters and numbers, *Z7HTHDGF2439N*, seems to be used by PayPal as an account identifier and was also found in the monitoring approach within the Local Storage of Firefox. Therefore, several files from the conducted payment experiments named

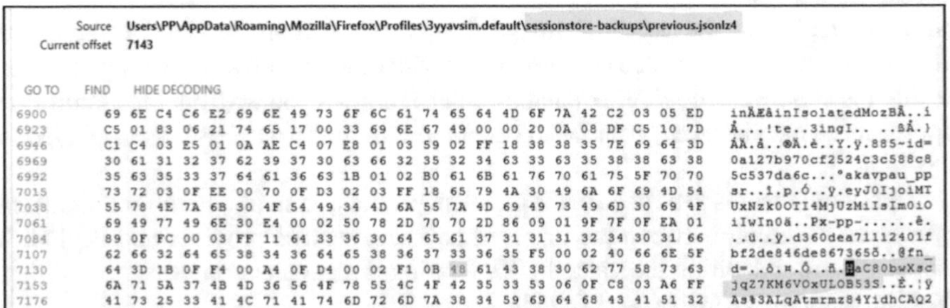

**Fig. 24** Forensic findings: found session cookies in previous.jsonlz4

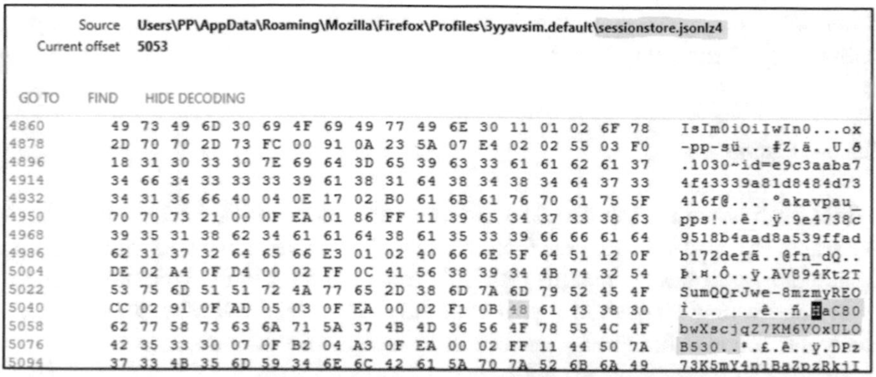

| Source | Users\PP\AppData\Roaming\Mozilla\Firefox\Profiles\3yyavsim.default\sessionstore.jsonlz4 |
| Current offset | 5053 |

GO TO    FIND    HIDE DECODING

```
4860      49 73 49 6D 30 69 6F 69 49 77 49 6E 30 11 01 02 6F 78    IsIm0iOiIwIn0...ox
4878      2D 70 70 2D 73 FC 00 91 0A 23 5A 07 E4 02 02 55 03 F0    -pp-sü...‡Z.ä..U.ð
4896      18 31 30 33 30 7E 69 64 3D 65 39 63 33 61 61 62 61 37    .1030~id=e9c3aaba7
4914      34 66 34 33 33 33 39 61 38 31 64 38 34 38 34 64 37 33    4f43339a81d8484d73
4932      34 31 36 66 40 04 0E 17 02 B0 61 6B 61 76 70 61 75 5F    416f@....°akavpau_
4950      70 70 73 21 00 0F EA 01 86 FF 11 39 65 34 37 33 38 63    pps!..ê..ÿ.9e4738c
4968      39 35 31 38 62 34 61 61 64 38 61 35 33 39 66 66 61 64    9518b4aad8a539ffad
4986      62 31 37 32 64 65 66 E3 01 02 40 66 6E 5F 64 51 12 0F    b172defã..@fn_dQ..
5004      DE 02 A4 0F D4 00 02 FF 0C 41 56 38 39 34 4B 74 32 54    Þ.¤.Ô..ÿ.AV894Kt2T
5022      53 75 6D 51 51 72 4A 77 65 2D 38 6D 7A 6D 79 52 45 4F    SumQQrJwe-8mzmyREO
5040      CC 02 91 0F AD 05 03 0F EA 00 02 F1 0B 48 61 43 38 30    Ì....ê..ñ.█aC80
5058      62 77 58 73 63 6A 71 5A 37 4B 4D 36 56 4F 78 55 4C 4F    bwXscjqZ7KM6VOxULO
5076      42 35 33 30 07 0F B2 04 A3 0F EA 00 02 FF 11 44 50 7A    B530..'.£.ê..ÿ.DPz
5094      37 33 4B 35 6D 59 34 6E 6C 42 61 5A 70 7A 52 6B 6A 49    73K5mY4nlBaZpzRkjI
```

**Fig. 25**  Forensic findings: found session cookies in sessionstore.jsonlz4

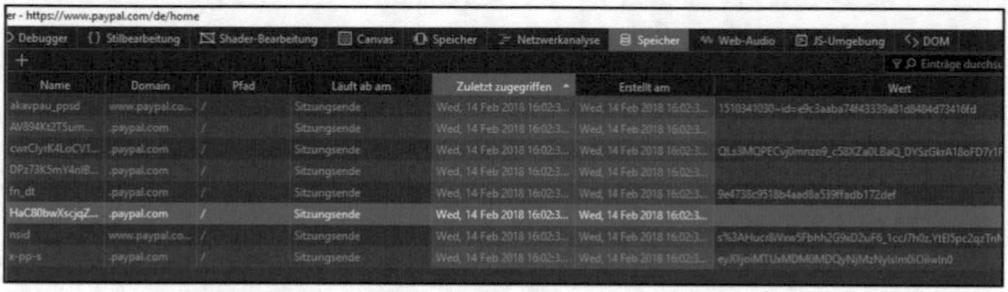

**Fig. 26**  Forensic findings: restored session cookies

*webappsstore.sqlite* were examined with Axiom and its built-in database viewer as demonstrated in the Fig. 27 and the same string could be found in each file where a payment with PayPal had occurred.

Axiom is currently not supporting Web Storage within their artefact search and therefore we had to export the *webappsstore.sqlite* files for further analysis. The SQLite files that were opened with SQLite3 and the database entries from the only

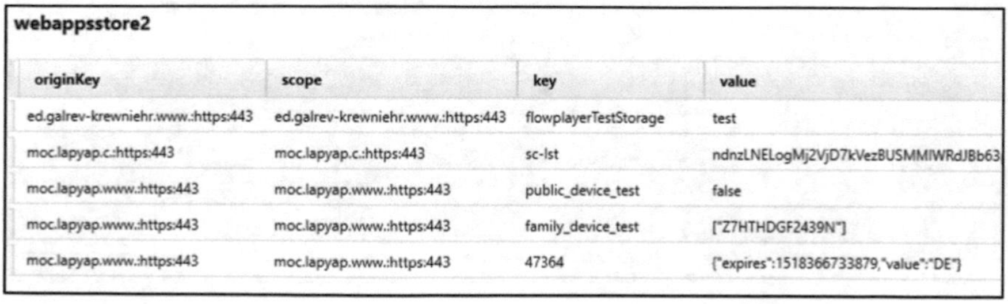

**webappsstore2**

| originKey | scope | key | value |
|---|---|---|---|
| ed.galrev-krewniehr.www.:https:443 | ed.galrev-krewniehr.www.:https:443 | flowplayerTestStorage | test |
| moc.lapyap.c.:https:443 | moc.lapyap.c.:https:443 | sc-lst | ndnzLNELogMj2VjD7kVezBUSMMMIWRdJBb63 |
| moc.lapyap.www.:https:443 | moc.lapyap.www.:https:443 | public_device_test | false |
| moc.lapyap.www.:https:443 | moc.lapyap.www.:https:443 | family_device_test | ["Z7HTHDGF2439N"] |
| moc.lapyap.www.:https:443 | moc.lapyap.www.:https:443 | 47364 | {"expires":1518366733879,"value":"DE"} |

**Fig. 27**  Forensic findings: web storage entries after payment with PayPal

found table named *webappsstore2*, were exported to a .csv file with the SQLite built-in export function.

### 5.3.4 Cache

While the monitoring approach did not find any cache entries, because the records were not displayed with the used tool, the forensic examination with Axiom was able to find indicators based upon the results from the findings in the Web Storage.

By performing a keyword search with the string *Z7HTHDGF2439N* there were many hits found in the Firefox cache entries which were stored under

*%USERPROFILE%\AppData\Roaming\Mozilla\Firefox*
*\Profiles\[profilename]*[8]*.default\\cache2\entries\*

Upon further examination of these findings, we discovered that they were part of a URL from *doubleclick.net* and all were beginning with the URL

https://ad.doubleclick.net/ddm/activity/[...], as shown in the next figure.

Figure 28 also shows that the stored URLs within the cache files not only contained the possible identification string for a specific PayPal account (highlighted in blue), it also contained information about the visited web page of PayPal as highlighted in these two URLs where the first example was exported out of an image file where a login into a PayPal account occurred (Fig. 29).

The second example was found within an image file, where a PayPal payment was made. This also contains a different identification string for a possible second PayPal account, which could identify the recipient PayPal account (Fig. 30).

Related timestamps, for each cache entry, could be retrieved by the OS file metadata timestamps, collected by Axiom, from the filesystem of the image file. All

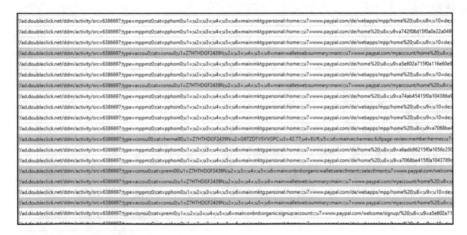

**Fig. 28** Forensic findings: web cache findings for doubleclick

---

[8][profilename] *is replaced by a random string created by Firefox for each installation.*

https://ad.doubleclick.net/ddm/activity/src=6386697;type=accou0;cat=consu0;u1=Z7HTHDGF2439N;u2=;u3=;u4=
;u5=;u6=main:walletweb:summary::main:::;u7=www.paypal.com/myaccount/home%20;u8=;u9=a74ab45415f0a10
4386a99e6fffc24b;u10=de;u11=;u12=;u13=;u14=;u15=;u16=;u17=;u18=;u19=;u20=;dc_lat=;dc_rdid=;tag_for_chi
ld_directed_treatment=;ord=1?

**Fig. 29**   Forensic findings: web cache entries of doubleclick (sample 1)

https://ad.doubleclick.net/ddm/activity/src=6386697;type=consu00;cat=herme00;u1=Z7HTHDGF2439N;u2=G6TZ
DTVSVVGPC;u3=42.77;u4=EUR;u5=;u6=main:ec:hermes::fullpage-
review:member:hermes:;u7=www.paypal.com/webapps/hermes%20;u8=ec:hermes:;u9=b6c0056e15f0ac02b3ae
0e12ffff2106;u10=de;u11=;u12=;u13=;u14=;u15=;u16=;u17=;u18=;u19=;u20=;dc_lat=;dc_rdid=;tag_for_child_dir
ected_treatment=;ord=1?

**Fig. 30**   Forensic findings: web cache entries of doubleclick (sample 2)

image files acquired from the experiments contained similar cache entries depending on how PayPal was used. These findings show that cache files can be used to prove that a PayPal account was used. Furthermore, they can indicate which account subpages were visited after a login or when someone made a payment with PayPal, including which PayPal account, and identified with a 13-digit string.

**Flash Cookies**
The forensic examination of all acquired disk images with Axiom could not find any evidence that Flash was used by PayPal or any other web site.

**IndexedDB Findings**
A search within the folders as found out in the experiments could not find any indicators for the usage of IDB by PayPal.

# 6   Conclusion

This chapter was led by the question if the use of a PayPal account via the Firefox web browser is detectable in digital forensics of the hard disk drive and what are the indicators for the usage? As showed in previous sections, a wide range of different indicators for the usage were found to indicate a usage of a PayPal account in forensic examination. The outcome of this research is practical useful for forensic examiners who want to prove that a PayPal account was used and they are now able to identify these indicators and how to interpret them. The results for the new storage functions Web Storage and IndexedDB also enhanced the knowledge about these quite new functions and where their forensic artefacts can be found. The findings about the Web Storage usage by PayPal also enhances the awareness for these storage functions because more and more companies will possibly adopt these new possibilities to their portfolio of tracking and monitoring techniques in the near future and forensic experts should continue to monitor these technologies.

Further research will focus on crypto-currencies forensics [36, 37] to investigate PayPal accounts as currently cryptocurrency exchanges have offered their users options of payment via PayPal transfers [12].

# References

1. Statista.com.: Preferred payment methods of online shoppers worldwide as of March 2017. [Online] Ipsos, April 2017. Accessed Nov 2019. https://www.statista.com/statistics/508988/preferred-payment-methods-of-online-shoppers-worldwide/
2. Statista.com.: US retail e-commerce sales forecast. [Online] October 2017. Accessed Nov 2019. https://www.statista.com/statistics/272391/us-retail-e-commerce-sales-forecast/
3. Le-Khac, N.-A., Markos, S., Kechadi, M.-T.: Towards a new data mining-based approach for anti money laundering in an international investment bank. In: International Conference on Digital Forensics and Cyber Crime (ICDF2C 2009), Springer LNICST 31, 30 Sept–2 Oct, Albany, New York, USA (2009)
4. Warrell, H.: Financial Times. [Online] 19 Jan 2017. Accessed Nov 2019. https://www.ft.com/content/03e8674e-de47-11e6-9d7c-be108f1c1dce (2017)
5. Datanyze: Online payments market share. [Online] Nov 2017. Accessed Nov 2019. https://www.datanyze.com/market-share/payments/ (2017)
6. PayPal Inc.: About Us (USA). [Online] 2017. Accessed Nov 2019. https://www.paypal.com/us/webapps/mpp/about (2017)
7. Oh, J., Lee, S., Lee, S.: Advanced evidence collection and analysis of web browser activity. Dig. Invest. 8, S62–S70 (2011)
8. Pereira, M.T.: Forensic analysis of the Firefox 3 internet history and recovery of deleted SQLite records. Dig. Invest. 5(3), 93–103 (2009)
9. Nalawade, A., Bharne, S., Mane, V.: Forensic Analysis and Evidence Collection for Web Browser Activity (2016)
10. Rathod, D.M.: Web browser forensics: Google Chrome. Int. J. Adv. Res. Comput. Sci. 8(7) (2017)
11. Gaurav, A., Bursztein, E., Jackson, C., Boneh, D.: An analysis of private browsing modes in modern browsers. In: Proceedings of the 19th USENIX conference on Security 2010. USENIX Security'10, Washington DC: USENIX Association (2010)
12. Houben, R., Snyers, A.: Cryptocurrencies and Blockchain: Legal Context and Implications for Financial Crime, Money Laundering and Tax Evasion. Policy Department for Economic, Scientific and Quality of Life Policies, European Parliament (2018)
13. Chivers, H.: Private browsing: a window of forensic opportunity. Dig. Invest. 11, 20–29 (2014)
14. Tsalis, N., Mylonas, A., Nisioti, A.: Exploring the protection of private browsing in desktop browsers. Comput. Secur. 67, 181–197 (2017)
15. Huwida, S., Noora, A., Al Awadhi, I.: Forensic analysis of private browsing artifacts. In: 2011 International Conference on Innovations in Information Technology (2011)
16. Hedberg, A.: The Privacy of Private Browsing (2013)
17. Ohana. D.J., Narasimha, Shashidhar: Do private and portable web browsers leave incriminating evidence? A forensic analysis of residual artifacts from private and portable web browsing sessions. EURASIP J. Inf. Secur. (2013)
18. Warren, C., El-Sheikh, E., Le-Khac, N.-A.: Privacy preserving internet browsers—forensic analysis of browzar In: Daimi, K., et al. (eds.) Computer and Network Security Essentials, 18 pp. Springer, Berlin (2017). https://doi.org/10.1007/978-3-319-58424-9_21
19. Reed, A., Scanlon, M., Le-Khac, N.-A.: Forensic analysis of epic privacy browser on windows operating systems. In: 16th European Conference on Cyber Warfare and Security, Dublin, Ireland (2017)

20. Matsumoto, S., Sakurai, K.: Acquisition of evidence of WebStorage in HTML5 web browsers from memory image. In: Ninth Asia Joint Conference on Information Security (2014)
21. Matsumoto, S., Onitsuka, Y., Kawamoto, J., Sakurai, K.: Reconstructing and visualizing evidence of artifact from firefox session storage. In: Yi, J., Rhee, K.H. (eds.) Information Security Applications (2015)
22. Mendoza, A., Kumar, A., Midcap, D., Cho, H., Varol, C.: BrowStEx: A tool to aggregate browser storage artifacts for forensic analysis. In: Elsevier, B.V. (ed.) Digital Investigation, vol. 14, pp. 63–75 (2015)
23. Kimak, S., Ellman, J., Laing, C.: Some Potential Issues with the Security of HTML5 IndexedDB (2014)
24. Kimak, S., Ellman, J.: The role of HTML5 IndexedDB, the past, present and future. In: The 10th International Conference for Internet Technology and Secured Transactions (ICITST-2015) (2015)
25. Boucher, J., Le-Khac, N.-A.: Forensic framework to identify local vs synced artefacts. J. Dig. Invest. **24**(1), S68–S75 (2018). https://doi.org/10.1016/j.diin.2018.01.009
26. Soltani, A., Canty, S., Mayo, Q., Thomas, L., Hoofnagle, C.J.: Flash Cookies and Privacy. s.n., Berkeley, USA (2009)
27. McDonald, A., Cranor, M., Faith, L.: A Survey of the Use of Adobe Flash Local Shared Objects to Respawn HTTP Cookies. s.n., Carnegie (2011)
28. Mika, A., et al.: Flash Cookies and Privacy II: Now with HTML5 and ETag Respawning. s.n., Berkeley (2011)
29. Acar, G., et al.: The Web Never Forgets: Persistent Tracking Mechanisms in the Wild. s.n., Leuven Belgium (2014)
30. Samy, K.: samy.pl. [Online] 11 Oct 2010. Accessed: 30 Oct 2019. https://samy.pl/evercookie/
31. Eckersley, P.: How unique is your web browser? In: Atallah, M.J., Hopper, N.J. (eds.) Privacy Enhancing Technologies. PETS 2010. Lecture Notes in Computer Science, vol. 6205. Springer, Berlin (2010)
32. mozilla.org.: Firefox Developer Tools. [Online] 2018. Accessed: 05 Feb 2018. https://developer.mozilla.org/en-US/docs/Tools
33. PayPal Inc.: PayPal Worldwide. [Online] Accessed: 08 Feb 2019. https://www.paypal.com/de/webapps/mpp/country-worldwide
34. Mozilla.: PRTime. [Online]. Accessed: 14 Feb 2019. https://developer.mozilla.org/en-US/docs/Mozilla/Projects/NSPR/Reference/PRTime
35. support.mozilla.org.: How do I restore my tabs from last time? [Online]. Accessed: 14 Feb 2019. https://support.mozilla.org/en-US/kb/how-do-i-restore-my-tabs-last-time
36. Koerhuis, W., Kechadi, T., Le-Khac, N.-A.: Forensic Analysis of Privacy-Oriented Cryptocurrencies. Elsevier (2020). DOI:https://doi.org/10.1016/j.fsidi.2019.200891
37. Zollner, S., Choo, K.K.R., Le-Khac, N.-A.: An automated live forensic and postmortem analysis tool for bitcoin on windows systems. IEEE Access **7** (2019). https://doi.org/10.1109/ACCESS.2019.2948774

**Lars Standare** received a Master of Science from University College Dublin Ireland in Forensic Computing and Cybercrime Investigation (FCCI) in 2018. He is currently working as a lecturer at the Rhineland-Palatinate Police University, Germany. His research interests include Digital Forensics in Cybercrime Investigations.

**Darren Hayes** received a PhD, in Industrial Engineering, from Sapienza University, Rome, Italy, in 2019. Additionally, he received a Doctorate in Professional Studies at Pace University, New York, USA, in 2008. Hayes has also completed a number of certifications in digital forensics. He is currently Associate Professor at Pace University. At Pace, he has developed a computer forensics program, which includes an undergraduate degree concentration a number of courses at the graduate level. Hayes has been principal and co-principal investigator on more than $3 million in US government grants related to digital forensics and cyber security. Hayes is a prolific author with numerous peer-reviewed publications in computer security, intelligence and digital forensics. He is the author of a number of books, including A Practical Guide to Digital Forensics. He is listed as one of the Top 10 Computer Forensics Professors, by Forensics Colleges. Hayes founded the Digital Forensics Research Laboratory at Pace where he and his team continually conducts research in support of law enforcement in the USA and in Europe. Hayes was selected as the recipient of the 2020 Homeland Security Investigations New York Private Sector Partnership Award. Hayes has also been a Visiting Professor at Sapienza University, Rome, Italy, University College Dublin, New Jersey City University and New York University. Hayes has also consulted on numerous civil and criminal investigations and has been declared an expert witness in US federal court.

**Nhien-An Le-Khac** is a lecturer at the School of Computer Science (CS), University College Dublin (UCD), Ireland. He is currently the program director of MSc program in Forensic Computing and Cybercrime Investigation (FCCI)—an international program for the law enforcement officers specializing in cybercrime investigations. He is also the co-founder of UCD-GNECB Postgraduate Certificate in fraud and e-crime investigation. Since 2008, he is a research fellow in Citibank, Ireland (Citi). He obtained his Ph.D. in Computer Science in 2006 at the Institut National Polytechnique de Grenoble (INPG), France. His research interest spans the area of Cybersecurity and Digital Forensics, Data Mining/Distributed Data Mining for Security, and Fraud and Criminal Detection. Since 2013, he has collaborated on many research projects as a principal/co-PI/funded investigator. He has published more than 150 scientific papers in peer-reviewed journal and conferences in related research fields, and his recent edited book has been listed the Best New Digital Forensics Book according to the Book Authority.

**Kim-Kwang Raymond Choo** received the Ph.D. in Information Security in 2006 from the Queensland University of Technology, Australia. He currently holds the Cloud Technology Endowed Professorship at the University of Texas at San Antonio (UTSA). In 2015, he and his team won the Digital Forensics Research Challenge organized by the Germany's University of Erlangen-Nuremberg. He is the recipient of the 2019 IEEE Technical Committee on Scalable Computing (TCSC) Award for Excellence in Scalable Computing (Middle Career Researcher), 2018 UTSA College of Business Col. Jean Piccione and Lt. Col. Philip Piccione Endowed Research Award for Tenured Faculty, British Computer Society's 2019 Wilkes Award Runner-up, 2019 EURASIP Journal on Wireless Communications and Networking (JWCN) Best Paper Award, Korea Information Processing Society's Journal of Information Processing Systems (JIPS) Survey Paper Award (Gold) 2019, IEEE Blockchain 2019 Outstanding Paper Award, Inscrypt 2019 Best Student Paper Award, IEEE TrustCom 2018 Best Paper Award, ESORICS 2015 Best Research Paper Award, 2014 Highly Commended Award by the Australia New Zealand Policing Advisory Agency, Fulbright Scholarship in 2009, 2008 Australia Day Achievement Medallion, and British Computer Society's Wilkes Award in 2008. He is also a fellow of the Australian Computer Society, an IEEE senior member, and co-chair of IEEE Multimedia Communications Technical Committee's Digital Rights Management for Multimedia Interest Group.

# Digital Forensic Approaches for Cloud Service Models: A Survey

Sebastian Schlepphorst, Kim-Kwang Raymond Choo, and Nhien-An Le-Khac

**Abstract** Cloud computing has become one of the fastest-growing IT infrastructures in the world. Criminals are aggressively expanding the use of digital technologies for illegal activities. As a consequence, the rise of cybercrimes in cloud systems exacerbates the problem of scale for digital forensic practitioners. Traditional digital forensic approaches such as the data acquisition of electronic devices, personal computer forensics, live data forensics, network investigations and forensics, mobile phone forensics, and are not sufficient or only partly applicable for the investigation, acquisition and analysis of evidence from cloud computing platforms. The evaluation of digital forensic techniques for cloud service models is still a challenge due to the lack of efficient criteria. Therefore, in this chapter, we first define the criteria for evaluating existing digital forensic approaches for the three main cloud service models: Infrastructure as a Service (IaaS), Platform as a Service (PaaS) and Software as a Service (SaaS). After that, we will review, analyse and compare each digital forensic approach in order to display existing gaps that need further solutions.

**Keywords** Cloud forensics · Cloud computing · Cloud service model · Infrastructure as a service · Platform as a service · Software as a service · Investigation · Digital forensics · Survey

## 1 Introduction

Cloud computing can be considered as a new paradigm for delivering Information Communications Technology (ICT) to organisations that do not need to invest in hardware, software and network infrastructures. Cloud computing encompasses

S. Schlepphorst (✉)
Federal Office for Information Security (BSI), Bonn, Germany
e-mail: sebastian.schlepphorst@ucd.connect.ie

K.-K. R. Choo
University of Texas, San Antonio, USA

S. Schlepphorst · N.-A. Le-Khac
University College Dublin, Dublin, Ireland

N.-A. Le-Khac and K.-K. R. Choo (eds.), *Cyber and Digital Forensic Investigations*,
Studies in Big Data 74, https://doi.org/10.1007/978-3-030-47131-6_8

any subscription-based or pay-per-use service that can extend an organisations ICT capabilities. A recent study by International Data Corporation (IDC), "Quantitative Estimates of the Demand for Cloud Computing in Europe and the Likely Barriers to Up-take," illustrates that the adoption of cloud computing is on the rise. Among the 1056 organisations surveyed, 32.7% have fully adopted the cloud computing in more than one business area, and 13.4% have fully adopted it in a single business area [1].

The National Institute of Standards (NIST) defined cloud computing as a model for enabling ubiquitous, convenient, on-demand network access to a shared pool of configurable computing resources (e.g., networks, servers, storage, applications, and services). These resources can be rapidly provisioned and released with minimal management effort or service provider interaction [2]. In general, cloud computing divides into three cloud service models, which are known as Infrastructure as a Service (IaaS), Platform as a Service (PaaS) and Software as a Services (SaaS). In total, cloud computing appears in four different deployment models, Public Cloud, Private Cloud, Hybrid Cloud and Community Cloud. All cloud service providers give their customers different layers of access control according to their cloud service model and relating cloud-computing resource (Fig. 1).

Besides the implementation of cloud technology in companies and millions of independent households, criminals are aggressively expanding the use of digital technologies for illegal activities. Cloud computing solutions provide an additional opportunity for criminals to use and commit crimes with the help of cloud computing platforms. Cybercrime such as data theft, internet fraud, computer sabotage, data tampering, business espionage, phishing, dissemination and possession of child exploitation material and other conceivable crimes could be committed with the help

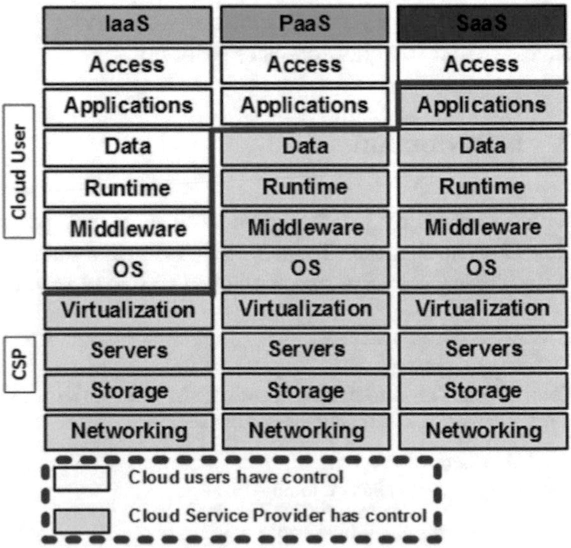

**Fig. 1** Access control within three cloud computing service models

of cloud infrastructure and poses a significant challenge for law enforcement and their investigators. In a recent cybercrime statistics from 2018, the German Federal Criminal Police Office stated that cloud computing expects to gain relevance for law enforcement and criminals due to further development and distribution of cloud computing technology [3].

Unfortunately, cloud computing is impacting significantly in the way digital forensic investigations are conducted [4, 5]. Cloud forensics represents a new category and is until today, one of the most challenging disciplines in the recognised science of digital forensics [6]. Investigators are faced with the challenges of evidence identification, accessing, acquiring, analysing and examining data from and within cloud-based environments. However, most of the cloud forensics solutions and frameworks proposed in the literature are analysing cloud as one platform [7]. Consequently, surveys in the literature on cloud forensics mostly focused on cloud storage artefacts. In recent publications, only Dykstra and Sherman [8] focused on one of the cloud computing service models: Infrastructure as a Service (IaaS), by forensic acquisition and analysis. Until now, there has been very little in the literature on the evaluation of approaches for the two other cloud service models; the Platform as a Service (PaaS), and Software as a Service (SaaS) in the context of digital forensics.

As a result of a fast-growing cloud computing infrastructure, the above three cloud service models are merging. A single examination of the cloud service model (IaaS) in the context of digital forensics will not be sufficient. Many cloud service providers are already using and combining the three cloud service models in their products. IaaS, PaaS and SaaS must be linked to and remain in consideration of forensic analysis. Therefore, in this chapter, we firstly define criteria used to evaluate digital forensic approaches for different cloud service models, and then we review and compare each forensic approach for the three cloud service models IaaS, PaaS and SaaS.

The rest of the chapter is organised as follows: Sect. 2 describes previous survey articles. Section 3 details the survey methodology and the evaluation criteria catalogue. We review and compare existing digital forensic approaches on IaaS, PaaS and SaaS in Sects. 4, 5, 6 and 7. Section 8 provides a discussion analysis of these approaches and a conclusion. Finally, we outline some future work in Sect. 9.

## 2 Related Work

Ruan et al. [4] presented results and analysis of a survey about cloud forensics and critical criteria for cloud forensic capability. In their work, they polled digital forensic experts and practitioners and received a total of 156 responses. Their survey identified three categories: (1) the category concerning the background of cloud computing and digital forensics, (2) the category concerning cloud forensics research and techniques and (3) the category concerning critical criteria for forensics capability. A majority of responders agreed that: (i) Cloud forensics is an application of digital forensics to cloud computing and that it is a mixture of traditional computer forensics, digital device forensics and network forensics; (ii) Tools, procedures, staffing, agreements, policies and guidelines must be developed to support forensic investigations in the cloud environment. The authors concluded that a forensic architecture for cloud computing environments is required and needs to be developed.

In 2013, Ruan et al. [9] repeated their survey of 257 respondents and compared the results with that obtained in 2011. In comparison to 2011, most of the results were consistent. The conducted survey led to a working definition for cloud forensics and confirmed that cloud forensics shows significant challenges for digital forensics whose challenges were summarised in a list. Furthermore, the survey also showed that it is crucial to establish definitions for cloud forensic capabilities that should include a set of toolkits and cloud investigations procedures.

Ruan et al.'s surveys from 2011 and 2013 do not deliver solutions for the general problems in the application of digital forensics in the cloud computing environment. However, they provide an initial evaluation of digital forensic experts and practitioners opinions in the context of cloud computing and digital forensics. This, in turn, enables framing the general challenges, opportunities and difficulties in the application of digital forensics in the cloud computing environment.

Authors in [10] focused their work on existing literature based on the term cloud forensics and grouped the literature into three classes: (1) survey-based, (2) technology-based and (3) forensics-procedural-based. They also reviewed cloud forensics and their standardisations and created a list of digital forensic tools that were applied in digital forensic solutions. The list also indicated tools that were used in test environments for cloud forensics. The authors also provided two mind maps that visualise these three types, the standardisations, research gaps, digital forensic tools testing and simulation environments.

The survey-based category covered the challenges of cloud computing and digital forensics viewed from technical and legal aspects of cloud computing, digital forensics, results of conducted questionnaires and surveys. The latter explained the results obtained by Ruan et al. [9] and Al Fahdi et al. [11]. They concluded that most of the literature focused on the IaaS cloud service model and deduced that no research on the technical findings in the cloud service models PaaS and SaaS had been carried out. From their statistical analysis, they outlined that many kinds of literature focused on IaaS while the remaining did not specify a particular cloud service model. More than half of survey-based literature concentrated on technical issues while the rest focused on legal aspects.

In the technology-based category, the authors reviewed the topic of distributed computing and virtualisation in the context of digital forensics and cloud computing. In their discussion, the authors found that there are two solutions for the application of digital forensics in cloud computing environments. 58% of the literature suggested add-on-solutions for the application of digital forensics, while 42% of the literature suggested by-design solutions. Cloud service providers could be obliged by governments to consider digital forensics in their cloud computing environment. Another of their statistical result showed that 50% of the proposed solutions targeted distributed virtualisation, 25% targeted non-distributed virtualisation and another 25% targeted distributed computing.

The procedural-based category reviewed literature in the context of the digital forensic process model of [12] including the phases of identification, preservation, collection, examination, analysis and presentation with reporting stages. Employing their own conducted statistical analysis, the authors pointed out that the majority of the literature focused on the analysis and collection phases. The other phases; identification, reporting, examination and preservation, were barely mentioned. Their statistical analysis also suggested that the majority of solutions proposed in the literature concentrated on evidence that resided in the cloud environment. Only a small amount of solutions made use of the resources available in the cloud environment to conduct digital forensics.

The survey article [10] is one of the first that searched the literature with the term cloud forensics and provided a summary and consideration of the topic. The authors found and acknowledged that most of the literature focused on IaaS while there had been no research carried out on forensic strategies for PaaS and SaaS up to 2014. Furthermore, the authors found that over half of the suggested solutions were add-on-solutions rather than by-design solutions. The paper reviewed existing literature for solutions in a different context in which the reader can find a good overview of the digital forensics in cloud computing. They provided state-of-the-art research and concluded that more research efforts are required in cloud forensics.

In [13] authors conducted a study on cloud forensics challenges and solutions based on 22 published articles. In their work, the authors described and showed five tables comparing current existing methodologies, stages of digital forensic models, the complexity of the methodology stages, cloud forensic challenges and the existing solutions for cloud forensics. This work catered a detailed description of the state-of-the-art on the existing literature in cloud forensics and presented a review of all frameworks, methodologies and a discussion about their functions, advantages, and drawbacks. It also offered a detailed analysis in which all stages and phases of the methodologies were classified according to their complexity. All existing cloud forensic challenges were considered separately, and the cloud forensic solutions were addressed concerning the three cloud service models; IaaS, PaaS, and SaaS. From these tables, it becomes clear that almost any challenge for the cloud service model IaaS has a suggested solution. This demonstrates that the IaaS cloud service model has been the priority while there remains work to do to come up with comprehensive forensic solutions for PaaS and SaaS cloud service models. The authors concluded that the key forensic challenges for the PaaS and SaaS models are the scarcity of forensic tools for acquisition, the physical inaccessibility and the volatile data with live investigation. Therefore, it is imperative to define digital forensic processes for data acquisition.

## 3 Methodology and Evaluation Criteria Catalogue

In this section, we describe the methodology of the survey and the general evaluation criteria catalogue. For the survey, we created 14 individual evaluation criteria points.

All evaluation criteria points can be categorised into three main categories which are namely:

- The digital forensic approach in cloud computing (1, 2, 3)—In this category, we range the literature into three possible solutions for performing digital forensics in cloud computing. With the help of these criteria points, we will be able to display the boundaries and differences of the latest digital forensic approaches in the reviewed literature.
- General Digital Forensic Model (4, 5, 6, 12, 13, 14)—In this category, we use the results from an unpublished survey which compared forty-four digital forensic models and frameworks from the past literature. With the help of the derived general digital forensic model and its six phases, we want to display to what extent the reviewed literature fulfils a specific phase of the general digital forensic model.
- Indicative Criteria (7, 8, 9, 10, 11)—These points display additional criteria for the application of digital forensics. With the help of the additional criteria, we want to evaluate whether the digital forensic approach in the reviewed literature takes additional essential points into account.

After that, we searched literature that concentrated on digital forensic approaches for individual cloud computing platforms and the three cloud service models IaaS, PaaS and SaaS. We reviewed and summarised the individual approaches. In the end, we compared the reviewed literature with the help of a comparison table. Here we examined whether the reviewed literature meets the individual criteria of the catalogue. With the comparison, we want to give an overview of the three cloud service models and digital forensic approaches. This comparison also aims to uncover existing gaps between the three different cloud service models (Table 1).

## 4  Digital Forensics in IaaS

Cloud computing services that match the characteristics of the cloud service model Infrastructure as a Service offer the most access control to the cloud customer. In [16] authors stated that cloud service providers provide IaaS cloud services in the form of:

- Backup and recovery of file systems and raw data stores on servers and desktop systems
- Computing server resources for running cloud-based systems which can be dynamically provisioned and configured
- Content Delivery Networks (CDNs) which store content and files to improve the performance and cost of delivering content for web-based systems
- Managing cloud infrastructure platforms
- Massively scalable storage capacity that can be used for applications, back-ups, archival, and file storage.

**Table 1** Evaluation criteria catalogue for digital forensic approaches in cloud computing

| No. | Evaluation criteria | Explanation |
| --- | --- | --- |
| 1 | Cloud service provider forensics (by-design solution) | Cloud service provider forensics is the application and implementation of digital forensic approaches by the responsible cloud service provider who runs and owns the cloud computing platform. These digital forensic approaches or solutions generally concentrate on implementation in the cloud environment design. These solutions or approaches are also called "By-Design Solution" or also referenced as "cloud forensic readiness" [14, 15] |
| 2 | Cloud user forensics (add-on solution) | Cloud user forensics is the application of digital forensic approaches by the individual cloud user or investigator. These digital forensic approaches concentrate on the existing interfaces of a cloud computing platform and provide the user or investigator with the ability to perform an investigation without a change of the cloud environment. These solutions or approaches are also called "Add-on Solution" |
| 3 | Cloud client-side forensics | Cloud client-side forensics is the application of digital forensic approaches that concentrate on locating cloud user artefacts on end-user devices like mobile phones, computers and other devices |
| 4 | Planning and preparation | In the planning and preparation phase, the investigators plan and prepare the entire ongoing digital forensic investigation. Besides, the preparation of technical equipment for the future acquisition and collection phase of an electronic device, the investigators organise the necessary search warrant for the scene. Once the search target is identified, the planning and preparation go into more in-depth actions such as formulating an approach strategy, considering the aim of the search warrant, the expecting of possible electronic devices and the general awareness-raising of the search team |

(continued)

**Table 1** (continued)

| No. | Evaluation criteria | Explanation |
|-----|---------------------|-------------|
| 5 | Identification and recognition | In the identification and recognition phase, the investigators generate a general overview and identify all electronic devices at the scene. After the identification and recognition, the investigators prepare the future seizing of the identified evidence and determine the order of seizure. Before this can happen, the investigators must document the situation and immediately determine the status of an electronic device whether it is still running, suspended or turned off. A live running system can request the performing of immediate live forensics and request prioritisation. Turned off electronic devices are documented, seized and transported to the forensic laboratory for further forensic acquisition and analysis |
| 6 | Collection and acquisition | The collection and acquisition phase in the digital forensic process model is paramount for the success of an investigation. This phase follows once the investigators have identified potential electronic devices with relevant data and evidence for the investigation. Therefore, investigators use and apply forensic tools and techniques to collect and acquire volatile and non-volatile data from the electronic device for the following examination phase |
| 7 | Imaging | The imaging phase involves forensic copying of the desired electronic devices or data. The forensic copying creates a technical one to one copy of the desired electronic device or data in the form of an image. This image can afterwards be used for the examination phase of the digital forensic process. In order to validate the acquired image, forensic imaging programs use cryptographic hash functions to check, compare and validate the acquired with the original electronic device |

(continued)

**Table 1** (continued)

| No. | Evaluation criteria | Explanation |
|-----|---------------------|-------------|
| 8 | Volatile data | In digital forensics, there are two types of data sources that can be seized and acquired during an investigation; namely: Persistent data sources such as hard discs, flash memory devices and solid-state discs. This type of data content remains permanently and can be acquired at a later time. Volatile data sources like random-access memory. This type of data content is volatile and can only be acquired while the computer system is running. Random-access memory data can provide fruitful additional information alongside the persistent data sources and support the investigation |
| 9 | Forensic tools | Forensic tools are often used to acquire data from the evidence during the collection phase of the digital forensic process. A forensic tool fulfils the term "forensic" if the tool can acquire the desired data completely, reliably, replicable without inducing changes and alteration of the evidence data source. In live data forensic, an investigation within a running computer environment using forensic tools might change the evidence data in the form of data contamination and alteration. In this particular case, all forensic tools must keep data contamination and alteration to a minimum. Forensic tools must record and provide logs of performed actions and cryptographic hash functions in order to maintain the chain of custody and its auditability |

(continued)

The following section gives an overview of the latest literature and digital forensic approaches for the cloud service model IaaS.

From [17] they were the first authors that provided and defined a general digital forensic approach that would work with the cloud service models IaaS, PaaS and SaaS. The authors worked with the method of isolation and identified six conditions that must be met for a successful digital investigation. In the end, the authors applied and evaluated the conditions with the IaaS cloud computing platform Nimbula Director.

**Table 1**    (continued)

| No. | Evaluation criteria | Explanation |
|-----|---------------------|-------------|
| 10 | Logs and integrity | Logs play a crucial role during an investigation. The main task of logs is to record the performed actions during an investigation from the beginning to the end in a separate log file. Logs can give information about the performed and carried out activities of the investigators. Furthermore, logs can provide information about the beginning of an investigation, the success, errors that have occurred, further performed actions and the end of the investigation. The acquired logs are essential for the later reconstruction and presentation of the acquired data and evidence in front of a jury in court. Furthermore, logs can prove that the chain of custody was kept in during the investigation |
| 11 | Artefacts | The word artefact is widely used within digital forensics and describes a specified evidence source or data structure in which a specific performed action of a user, computer system or a third party can be concluded. Many computer forensic researchers study operating systems and their applications to make conclusions on performed actions that had been previously carried out on the electronic device. Information and findings gathered can be summed by a clear rule and behavioural explanation that forms the general artefact and their description. Artefacts include information about their location on the electronic device, way of appearance, meaning, behaviour and possible conclusions. Digital forensic examiners can use the artefact description for future digital forensic investigations and identify or explore performed actions on the electronic device |

(continued)

Authors in [18] tested with forensic tools, the remote acquisition of data in the cloud computing platform Amazon Elastic Cloud 2 (Amazon EC2) from the cloud service provider Amazon Web Services (AWS). This cloud computing platform fulfilled the characteristics of the cloud service model IaaS and gave the examiners adequate access control to the underlying cloud computing components. At the beginning of their work, the authors explained the six layers of the IaaS cloud computing

**Table 1** (continued)

| No. | Evaluation criteria | Explanation |
|-----|---------------------|-------------|
| 12 | Examination and data analysis | In the examination and analysis phase, the investigators examine and analyse the acquired electronic devices for traces and artefacts with forensic tools. Here the investigators prepare and process the found data structure, separate insignificant data structures from the further analysis and fulfil previously defined investigation targets. If necessary, the investigators prove or disprove previously declared hypotheses |
| 13 | Presentation and reporting | After the examination phase, the investigators summarise the evidence, results, confirmed or unconfirmed hypotheses in a final report. The final report provides the information to the court and informs the judge, defence and prosecutor about the performed digital investigation and forensic analysis |
| 14 | Preservation | The preservation phase is overarching and considered during the whole digital forensic process. This phase contains the element of the chain of custody, which is the process of validating and proving that the acquired evidence or data was not altered, changed or manipulated during the digital forensic investigation process and its different phases. This chain of custody provides the auditability of performed actions during the investigation and determines the validity of brought in evidence during court proceedings |

platform, their potential forensic acquisition type and the necessary trust in the components such as guest operating system, hypervisor, host operating system, hardware and network. After that, the authors performed three experiments in the forensic acquisition of data in the Amazon cloud computing platform. Those experiments were conducted with forensic tools such as EnCase, FTK, FTK Imager, Fastdump, Memoryze, Volume Block Copy, the direct agent injection of EnCase Servlet and FTK Agent into the virtual machine and the cloud service provider tool Amazon Web Services Export. All forensic tools were able to acquire forensic data successfully in a certain amount of time. Dykstra and Sherman outlined that most widely used forensic tools are capable of performing forensics acquisition from the cloud computing platform Amazon EC2. However, they also claimed that the integrity of the acquired forensic data could only partly be verified during the experiments. Furthermore, the authors also emphasised that a high level of necessary trust must be granted to the cloud computing platform and its different layers, hence advised refraining from using commonly used forensic tools such as EnCase and FTK. They, therefore, recommended tools that can be run by the cloud management plane provided by the cloud service provider. In their experiment, the authors performed the data acquisition with Amazon EC2, Amazon Web Services that showed the best balance between acquisition speed and trust in the cloud computing platform components. Both authors noted that their experiments should be performed with other cloud computing platforms such as Microsoft Azure and Google App Engine to find and discover parallels with these platforms.

In [8] authors developed in another paper a framework called Forensic Open Stack Tools (FROST) that contains three new forensic tools for the cloud computing platform OpenStack. OpenStack is an open-source software platform for cloud computing which is mostly deployed and run as IaaS. In their work, the authors described the design and implementation of a forensic cloud management plane within the cloud computing platform of OpenStack. The first tool in FROST can acquire images from the target virtual machine's disk and validate the integrity of the acquired images by a cryptographic checksum. The second tool, the investigator can retrieve logs of all API requests made to the cloud service provider and validate their integrity. The third tool can retrieve the OpenStack firewall logs for the target virtual machine. The authors outlined that FROST can be integrated into an OpenStack cloud computing platform and additionally be expanded for other forensic capabilities.

Dykstra's dissertation [19] was based on his previous works [8, 18, 20, 21] and dealt with issues in the application of digital forensics in the cloud service model IaaS. Dykstra addressed three problems in his dissertation, namely criminal and civil actions in cloud computing, capabilities of recent forensics and existing policies and laws concerning cloud computing in the United States of America. As for the first problem, Dykstra explained why criminal actions are difficult to prosecute by analysing two hypothetical case studies and illustrating the digital forensic challenges in acquisition, the chain of custody, trust and forensic integrity. The second problem he addressed forensic tools like EnCase and FTK and discussed why these tools are not recommendable for cloud acquisition. Besides, Dykstra provided and described the design and implementation of the forensic framework FROST in the cloud computing platform OpenStack. He evaluated it with two experiments and proved the integrity, completeness, and accuracy of the forensic data acquired by FROST. In the third problem, Dykstra analysed existing policies and laws of the United States of America by cloud computing and offered guidance on how to author a search warrant for cloud computing data.

Authors in [22] provided a practical forensic solution model that can be implemented into the OpenNebula cloud computing platform which falls under the cloud service model IaaS. In their work, the authors displayed a model and approach that concentrates on a forensic acquisition of cloud-hosted virtual machines that are the subject of an attacker intrusion. In order to achieve a successful acquisition of the virtual machine, the authors suggested the previous preparation and modification of the cloud environment relating to the hypervisor and the running virtual machines by a set of open-source tools. Once an intrusion occurs the previously installed IDS on the virtual machines or the hypervisor alerts a running daemon service of the cloud environment which is responsible for the acquisition process. With the help of the tool Google Rapid Response (GRR), the affected virtual machine is saved in the form of a snapshot. In the next step, the snapshot and related data are saved on a separate server in the cloud computing platform, which further can be accessed by an investigator. Furthermore, the authors evaluated the model with different attacking scenarios. The authors outlined that their model overcame a couple of main challenges for the cloud service model IaaS. In their conclusion section, the authors described that their proposed solution could also be applied for other cloud service models like PaaS and SaaS. The authors see a possibility for the usage with Google Rapid Response server if the Google Rapid Response client software is installed on a PaaS or SaaS cloud service.

In [23] authors developed a forensic framework for the cloud computing platform OpenStack with the help of existing forensic tools. The framework provides the acquisition capability of live snapshots of virtual discs, random access memory, image evidences, packet captures and log evidence from a virtual machine. The authors also evaluated the solution in OpenStack successfully and emphasised that the thin client code can easily be extended by other cloud computing platforms like OpenNebula or Eucalyptus.

Banas [24] was inspired by the work of Dykstra et al. [8] and invented a new digital forensic framework with the functionality of memory and disk image extraction from the hypervisor. Furthermore, the author added a different approach to check and verify acquired images afterwards. Additionally, he implemented and evaluated the digital forensic framework within the OpenStack cloud computing platform successfully.

Digambar [25] designed in his dissertation a general digital forensic framework for the cloud investigation and identified the challenges and requirements for digital forensics in the cloud computing environment. He focused on the digital forensic method and proposed an architecture for cloud data acquisition and analysis. Digambar also tested his digital forensic framework in the cloud computing platform OpenStack. The tools that were designed for the acquisition of a virtual machine will run on the investigator's workstation while the data segregation tool will run on the cloud hosting servers. To acquire a forensic image of the running virtual machine within the cloud computing platform, Digambar suggested following the approach and concepts from [8] or using traditional file transfer applications like WinSCP or PuTTY. The applied acquisition of the virtual machine saves a copy in a file directory in the cloud computing platform OpenStack that the investigator must collect for further analysis in the next step. Furthermore, the cryptographic hash value will be calculated over the acquired data to preserve forensic integrity. According to the acquisition of the virtual machine's memory, Digambar suggested injecting traditional memory acquisition tools like FTK Imager or LiME (Linux Memory) into the target virtual machine. A segregated log data is collected and performed by the investigator's workstation while the log data is saved in the controller node of OpenStack. After that, Digambar continued with the examination of the previously acquired evidence from the target virtual machine in the form of a disk image and a memory image. In this examination, he analysed the data with partly self-written analysis programs and examined in general the file system metadata, windows registry files, physical memory examination with the volatility framework and performed keyword searching within the memory image.

In [26] authors provided a novel model for the data acquisition of an IaaS cloud service that prevents the involvement of the cloud service provider as far as possible. This approach follows the idea of an agent-based approach that would be installed in every hosted virtual machine. If a digital investigation of a virtual machine is needed, the pre-installed agents send the requested information to a central Cloud Forensic Acquisition and Analysis System (Cloud FaaS) for a further examination and analysis. The pre-installed agent consists:

- Non-volatile memory agent—This agent acquires the hard drive which is associated with a virtual machine
- Volatile memory agent—This agent acquires the random access memory of the virtual machine
- Network traffic agent—This agent logs and stores ingressing and egressing network traffic
- Activity log agent—This agent saves system and application logs.

Authors in [27] proposed a digital forensic framework for the cloud computing platform Eucalyptus, which mainly provides IaaS cloud services such as virtual machines. In their framework, the authors offer a couple of command-line commands that could transfer a snapshot of a target virtual machine to a persistent storage service, which is called Walrus. In the next step, the investigator can use the Euca2ool to collect a snapshot from the storage service Walrus for further investigation.

In [28], authors provided a solution for the live forensic acquisition of virtual machines that are hosted with the server virtualisation software called Proxmox. Proxmox is a virtualisation software which allows the user a quick deployment of a cloud computing platform with several computer resources that can be subsumed under the cloud service model IaaS. In the first step, the authors installed the Proxmox virtualisation software on hardware infrastructure. After that, the authors added one virtual machine with Windows 10 and one virtual machine with Linux Ubuntu to the cloud computing platform Proxmox. Once the infrastructure was prepared, the authors performed a live forensic acquisition of one virtual machine with a previously prepared bash script that is saved on a separate USB stick. By inserting the USB Stick into the computer that hosts the whole cloud computing platform Proxmox, the investigator can complete the acquisition through a guided menu. In the end, the authors were able to examine the acquired structures of virtual machines with forensic tools like Autopsy. Besides, the authors tested and evaluated their acquisition technique with some case scenarios successfully.

## 5 Digital Forensics in PaaS

Cloud computing services that match the characteristics of the cloud service model Platform as a Service give the customer control over the application layer within the cloud environment. Authors in [16] stated that most cloud service providers enable PaaS services as follows:

- Business intelligence in the form of platforms for the creation of applications like dashboards, reporting systems and data analysis
- Databases which offer scalable relational database solutions or scalable non-SQL data stores
- Development and testing platforms for the development and testing cycles of application development
- Integration applications in the cloud and within the enterprise
- Application deployment in which platforms are suited for general-purpose application development. These services provide databases and web application runtime environments.

For the cloud service model PaaS only the following authors partly addressed PaaS in their research work.

Authors in [29] focused in their work on the general technical aspects of digital forensics in cloud environments and assessed the possibility of performing traditional digital investigation techniques. They discussed possible solutions and methodologies for the cloud service models IaaS, PaaS and SaaS. The authors found out that one advantage of the PaaS cloud service model is that the application layer is under the control of the cloud customer. The cloud customer has the power to determine how the application layer interacts with other dependencies such as storage entities and databases. Cloud customers can interact with the cloud computing platform that runs PaaS cloud services in the form of a previously prepared Application Programming Interface (API). The API can extract and collect system states and specific application logs from the PaaS cloud services.

In [30] authors discussed the challenges of cloud computing investigation and named possible solutions. For the cloud service model PaaS, the authors state that cloud service providers usually provide an API for their clients, which allows them to develop customised applications. The authors assume that obtaining the checking of system status and log files will not be possible due to the reason that the client access is only limited to the API or other pre-defined interfaces. Furthermore, the authors assume that vital evidence of PaaS and SaaS cloud services could reside on the client-side user interface in the form of web browser artefacts or temporary data structures. Besides, the authors also assume that the acquisition in the form of a forensic image of a PaaS or SaaS cloud service will not be possible. This is due to the point that the client does not have any direct access to the storage media.

Ruan [31] generally addressed in her dissertation the background of cloud computing and digital forensics. She analysed the related challenges and opportunities of cloud forensics and provided three different models that are based on survey results and further carried out analysis. According to PaaS, the author outlined that a user of a PaaS cloud platform has control over the application layer and their associated artefacts. Furthermore, she identified that a PaaS interface layer could provide software building blocks like libraries, Java virtual machines and databases for developing application software within the cloud environment. Existing artefacts and traces on the application layer can be similar to traditional development environments and appear in the form of log files, source code and account information. The author concluded that investigations in the development environment are unlikely to be carried out and needed further analysis in future work.

Authors in [32] compared the past research and methodologies in the cloud computing forensic field and addressed the gaps and loopholes in previously suggested frameworks. They also identified that a PaaS cloud platform model that provides forensic-friendly functions does not exist. In their future work, the authors propose a cloud design model that offers easy access to cloud-specific evidence data by the minimisation of non-essential logs, the maximisation of reliability and by preserving the chain of custody at each stage.

## 6   Digital Forensics in SaaS

Cloud computing services that match the characteristics of the cloud service model Software as a Service only give the customer access to the offered resources within the cloud environment. Authors in [16] stated that cloud service providers provide SaaS cloud services in the form of:

- Email and office productivity in the form of applications for email, word processing, spreadsheets and presentations
- Billing applications that manage customer billing based on usage and subscriptions to products and services
- Customer Relationship Management (CRM) applications that range from call centre applications to sales force automation
- Collaboration tools which allow users to collaborate in workgroups, within enterprises and across enterprises
- Content management for the production of and access to content for web-based applications
- Document management in which applications can manage documents, enforce document production workflows and provide workspaces for groups or enterprises to find and access documents
- Financial applications that manage financial processes ranging from expense processing and invoicing to tax management
- Human resources software which can manage human resources functions within companies
- Sales applications that are specifically designed for sales functions such as pricing and commission tracking

- Social networks software which establishes and maintains a connection among users that are tied in one or more specific types of interdependency
- Enterprise Resource Planning (ERP) Integrated computer-based system used to manage internal and external resources, including tangible assets, financial resources, materials, and human resources.

The following section gives an overview of the latest literature and digital forensic approaches for the cloud service model SaaS.

Chung et al. [33] conducted a study on digital forensic investigation of cloud storage services that meet the characteristics of SaaS. In their work, they investigated and analysed the artefacts of cloud storage services, namely Amazon S3, Google Docs and Evernote on general devices like Windows, Mac, iPhone and Android smartphones. In their research, the authors provided a digital forensic framework which includes a procedure for investigations in cloud storage services. Furthermore, the study presented the traces left in computers and smartphones that can access the above-mentioned cloud storage services. The authors also evaluated their digital forensic framework with a crime scenario in which a suspect had leaked a document to the competitor by using the cloud storage service Dropbox. In the end, the authors outlined that the examination of artefacts in cloud storage services cannot be only limited to the computer and must be expanded to all possible devices such as smartphones since they also have access to the cloud storage service.

Following their work [34], the research question if cloud environments can be analysed with traditional digital forensics procedures. In their research, the authors conducted five scenarios in which they accessed Dropbox, Google Docs, PicasaWeb and Flickr via the web browser. In the fifth scenario, the authors analysed a Dropbox client installation with a locally synchronised folder. In the result, the authors were able to find local logs and temporary files that were saved via the used web browser on the computer. Furthermore, the authors found a local copy of folder and data structures in a specified synchronised Dropbox folder on the computer without the need of accessing the SaaS cloud service at all.

Authors in [35, 36] examined in their research, the three cloud storage services Dropbox, Microsoft SkyDrive, Google Drive and their related artefacts that are left on the client-side computer or mobile phone. In their research, the authors conducted several experiments with different virtual machines that used the above-mentioned cloud storage services with the accessing Windows client software, Internet Explorer, Mozilla Firefox, Google Chrome and Apple Safari. Furthermore, the authors concentrated on an in-depth analysis of volatile random access memory, the related data structures and folders on the hard drive and the related network traffic in the form of packet capture. Besides, Quick and Choo also conducted the same experiments with the mobile phone iPhone 3G and used for analysis purposes the mobile phone forensic software XRY. Besides, the author proposed a cyclic digital forensic framework which enlarges the traditional digital forensic process model for additional phases like Commence (Scope) and Prepare.

Hale [37] examined in his research paper that digital artefacts are left during the use of an Amazon Cloud Drive on a computer. He presented methods to identify file transfers from and to an Amazon Cloud Drive on a computer. Hale analysed Amazon Cloud Drive artefacts from unallocated spaces and provided two Perl scripts that automate the investigation process of Amazon Cloud Drive artefacts.

Authors in [38] performed a forensic examination of two popular public cloud services Microsoft OneDrive and Amazon Cloud Drive. They proposed a data preservation process framework and found out that there are various artefacts such as log files, database files, system configuration and setup files which were stored on the examined computer system. Besides, they also found out that username and password for a Microsoft OneDrive user account were stored in a collected RAM dump. Furthermore, username and password for an Amazon Cloud Drive user account were recoverable through the analysis of network traffic packets.

In [39] authors examined user activity artefacts with a Microsoft OneDrive client application on two selected Android smartphones. The authors reviewed the digital forensic model of [36] and proposed a new digital forensic model with small changes. Furthermore, the authors carried out the analysis of thirteen individual user scenarios like data operations in the form of installing, uploading, downloading, uninstalling, signing-on and signing off. Within the thirteen user scenarios, the authors were able to attain eleven individual user artefacts. Only two user artefacts could not be obtained. Therefore the authors identified and recommended the use of live log forensics, which is associated with memory forensics in order to obtain the necessary artefact information.

Authors in [40] also described a data acquisition approach in the cloud computing environment. They presented different aspects related to the challenges of data acquisition in the cloud computing environment as well as legal perspectives related to cloud service providers. In their first case study, the authors showed how to locate the data centre where the computer host of the investigator's interest is running and proposed an efficient approach to tackle this main challenge. In the second case study, the authors demonstrated a forensic acquisition, analysis and examination with the popular cloud storage platform Amazon Simple Storage Service (Amazon S3) from Amazon Web Services. The authors analysed occurring artefacts such as logs, images, and internet history that are left on the end-user device when using the Amazon S3 user account. In their case study the authors used virtual machines for simulating user activities with a Amazon S3 user account and focused on persistent data and volatile data sources for finding the relevant artefacts. The authors also presented an acquisition possibility with a tool called CloudBerry Drive. With the help of the tool and the account credentials, an investigator will be able to mount, access an Amazon S3 cloud service account locally as a network drive and perform acquisition of the cloud service data.

## 7   Comparison of Digital Forensic Approaches

The previously reviewed digital forensic approaches for the different cloud service models IaaS, PaaS and SaaS, are displayed in the following comparison table. The

**Table 2** Comparison of digital forensic approaches in cloud service models

|  | IaaS | PaaS | SaaS |
|---|---|---|---|
| Cloud service provider forensics (by-design solution) | (3) (4) (5) (6) (7) (9) (10) | | |
| Cloud user forensics (add-on solution) | (1) (2) (8) (11) | | (16) (17) (20) (22) |
| Cloud client-side Forensics | | | (16) (17) (18) (19) (20) (21) (22) |
| Planning and preparation | (1) (2) (3) (4) (5) (6) (7) (8) (9) (10) (11) | | (16) (17) (18) (19) (20) (21) (22) |
| Identification and recognition | (1) (2) (3) (4) (5) (6) (7) (8) (9) (10) (11) | | (16) (17) (18) (19) (20) (21) (22) |
| Collection and acquisition | (1) (2) (3) (4) (5) (6) (7) (8) (9) (10) (11) | | (16) (17) (18) (19) (20) (21) (22) |
| Imaging | (1) (2) (3) (4) (5) (6) (7) (8) (9) (10) (11) | | (16) (17) (18) (19) (20) (21) (22) |
| Volatile data | (1) (5) (6) (7) (8) (9) | | (16) (19) (20) (21) (22) |
| Forensic tools | (2) (3) (4) (5) (6) (7) (8) (9) (11) | | (16) (17) (18) (19) (20) (21) (22) |
| Logs and integrity | (1) (2) (3) (4) (5) (6) (7) (8) (9) (11) | | (16) (17) (18) (19) (20) (21) (22) |
| Artefacts | (3) (4) (5) (6) (7) (8) (9) | (12) (13) (14) (15) | (16) (17) (18) (19) (20) (21) (22) |
| Examination and data analysis | (6) (8) (11) | | (16) (17) (18) (19) (20) (21) (22) |
| Presentation and reporting | (8) | | |
| Preservation | (1) (2) (3) (4) (5) (6) (7) (8) (9) (10) (11) | | (16) (17) (18) (19) (20) (21) (22) |

following comparison table shows if the literature meets the individual evaluation criteria that are defined in Sect. 3 (Tables 2 and 3).

# 8   Discussion and Conclusion

The comparison table in the previous pages showed that a couple of authors had contributed a digital forensic approach to the cloud service models IaaS and SaaS. While the majority of researchers had focused on solving the cloud service model IaaS, followed by SaaS, the PaaS is still underrepresented.

**Table 3** Quick reference table

| (1) = Delport and Olivier [17] | (12) = Birk and Wegener [29] |
|---|---|
| (2) = Dykstra and Sherman [18] | (13) = Damshenas et al. [30] |
| (3) = Dykstra and Sherman [8] | (14) = Ruan [31] |
| (4) = Dykstra [19] | (15) = Shirude et al. [32] |
| (5) = Reichert et al. [22] | (16) = Chung et al. [33] |
| (6) = Saibharath and Geethakumari [23] | (17) = Marturana et al. [34] |
| (7) = Banas [24] | (18) = Quick [35], Quick and Choo [36] |
| (8) = Digambar [25] | (19) = Hale [37] |
| (9) = Alqahtany et al. [26] | (20) = Easwaramoorthy et al. [38] |
| (10) = Rani et al. [27] | (21) = Satrya et al. [39] |
| (11) = Sudyana and Lizarti [28] | (22) = Le-Khac et al. [40] |

For IaaS, the majority of authors have provided a "By-Design Solution" that would request an active implementation of digital forensic functions in the cloud design and cloud computing platform by the responsible cloud service provider. Only a few authors have suggested a digital forensic approach that would concentrate on an "Add-on Solution". So far, none of those authors, as mentioned above, have focussed on the topic "Cloud Client-Side Forensics" for the cloud service model IaaS. In their work, the researchers found practical forensic approaches and partly tested them with virtual machines hosted on the cloud computing platforms Nimbula Director, Amazon Elastic Cloud 2, OpenStack, OpenNebula, Eucalyptus and Proxmox. It must be noted that the authors have provided a digital forensic approach that fulfils almost all phases of the digital forensic process model. While almost every author contributed a solution for the following linear phases like planning and preparation, identification and recognition, collection and acquisition, preservation and chain of custody only a few authors contributed a solution for the topics examination and data analysis, presentation and reporting. The topics artefacts, imaging, forensic tools, logs and integrity were also in the scope of the researchers.

For SaaS, most authors have focussed on "Cloud Client-Side Forensics" in which the authors mainly described artefacts and traces that are left on end-user devices like computers and mobile phones when using cloud computing platforms and services. Only the authors [33, 34, 38 and 40] focussed on an "Add-on Solution" that would also concentrate on the acquisition of data structures from cloud computing platforms that run resources in the cloud service model SaaS. In total, the researchers reviewed cloud computing platforms or services, namely Amazon S3, Amazon Cloud Drive, Google Docs, Google Drive, Microsoft OneDrive, Microsoft SkyDrive, Dropbox, Evernote, PicasaWeb and Flickr. While almost every author contributed a solution for the following linear phases like planning and preparation, identification and recognition, collection and acquisition, preservation and chain of custody, examination and data analysis none of the authors dealt with the phase of presentation and reporting. The topics artefacts imaging, forensic tools, logs and integrity were also in the scope of the researchers.

For PaaS, only two authors contributed an idea for a digital forensic approach in the form of a "By-Design-Solution". Authors in [32] planed to propose and design a cloud model that provides digital forensic features for PaaS. How the design and the proposed solution is going to be implemented remains to be seen. Furthermore, [17], who provided a digital forensic approach with the method of isolation, proposed that their solution might be transferred to the two other cloud service models PaaS and SaaS. Even if the two authors tested their digital forensic approach with the IaaS cloud computing platform Nimbula Director, it could not be concluded, that the approach is directly transferable to the other two cloud service models.

Besides the two "By-Design-Solutions" the other three following authors concentrated on the topic artefacts and provided some first ideas about what might be acquired from PaaS cloud services. Author in [31] outlined that PaaS cloud services offer different artefact types such as log files, source code, account information, system states and specific application logs, which hugely depend on the used PaaS cloud services and their appearance. In [30] authors argued that the acquisition of a PaaS or SaaS cloud service in the form of a forensic image would not be possible, which is due to the missing direct access to the storage media. The authors in [29] showed in their work that the argument, an acquisition of a PaaS cloud service resource might not be possible due to missing access possibility, might be incorrect. Birk and Wegener outlined that the application layer of a PaaS cloud service model is under the control of the cloud user, which enables interaction and access with the help of an Application Programming Interface (API). This API idea might be a vital point of a still-developing solution, that would enable performing digital forensics within PaaS.

All present research for digital forensics in PaaS shows that a digital forensic approach has not been developed and tested with any corresponding cloud computing platform yet. In summary, this comparison also showed, that the existing digital forensic approaches for the cloud service models IaaS and SaaS are individual solutions for specific cloud computing platforms and their corresponding cloud resources. While IaaS approaches always focussed on the acquisition of complete virtual machines including disc images, random access memory and log files, the SaaS approaches showed the acquisition of data structures that represent cloud services like Dropbox, Amazon Cloud Drive, Google Docs, Google Drive and Amazon S3 or had a focus on artefacts and traces that are left on the end-using devices. On this occasion, it shows again that all three cloud service models provide different layers of access to their customers and cloud application resource examples, as explained in the introduction chapter. While the cloud customer can receive in the cloud service model IaaS the most significant access to the underlying infrastructure, the access in the other two cloud service models PaaS and SaaS are further restricted. This aspect, in turn, also has an impact on a planned investigation of a cloud computing platform. PaaS cloud service resources like a database, a developer environment or an application deployment example might provide access until the application layer level. This level of access might also provide an acquisition possibility assumed by [29] and provide resulting artefact types according to [31]. In the result of this comparison, it must be noted that the cloud service model PaaS will need an individual digital forensic approach that is developed regardless of the other two cloud service models IaaS and SaaS.

This comparison of literature also showed that existing digital forensic approaches can be categorised into the three main categories By-Design-Solution, Add-On Solution and Cloud Client-Side Forensics. The literature showed that many approaches concentrated on a "By-Design-Solution" which would probably give the best possible solution for the main problem of performing digital forensics within cloud environments. However, this solution stands and fails with the necessary implementation of the responsible cloud service provider. While the literature showed a couple of variations on how to design and implement forensic techniques in individual cloud computing platforms, it stands out that this must be implemented before a possible investigation. A subsequent implementation in a cloud computing platform is not a success-promising operation due to the reason that all hosted cloud resources are volatile and irreversible once they are turned off. While cloud service providers could also acquire and hand over the cloud data to the investigators raises the problem that the used methodology might not meet digital forensic principles and endanger the evidential value.

In contrast to the "By-Design-Solution", the "Add-On Solution" may provide a transitional solution until cloud service providers implement digital forensic techniques in the cloud design from the beginning. The comparison of the literature also showed that investigators could also use the already existing interfaces for administrating the cloud computing platform in order to access, acquire and analyse the cloud data of individual cloud computing resources. The key benefit of an "Add-On Solution" is that an alteration of the cloud design, as suggested in By-Design-Solutions, is not necessary. Furthermore, the investigators will depend less on the help of the cloud service provider. The literature that concentrated on Cloud Client-Side Forensics is also a considerable part for a general ongoing investigation. Findings in this subject fall under already recognised traditional computer forensics methodologies and techniques.

## 9 Future Work

According to our observation of the gaps, we plan to develop a new digital forensic model that would suit the cloud service model PaaS. Following this idea, we will first need to identify the occurring challenges for the application of digital forensics in PaaS. Only with the identification of challenges, will we be able to provide a universally applicable digital forensic model that can address the cloud service model PaaS. Due to the point that a "By-Design-Solution" is under the above-mentioned circumstances not promising, we will follow in this research; therefore, the "Add-on Solution". We also plan to evaluate the digital forensic model with PaaS cloud computing resources on current popular cloud computing platforms like OpenStack, Microsoft Azure and Amazon Web Services.

# References

1. Bradshaw, D., Cattaneo, G., Folco, G., Kolding, M.: Quantitative Estimates of the Demand for Cloud Computing in Europe and the Likely Barriers to Up-take. European Commission—DG Information Society, Brussels, Belgium (2012). Available at: https://ec.europa.eu/digital-single-market/en/news/quantitative-estimates-demand-cloud-computing-europe-and-likely-barriers-take-final-report. Accessed: 22.01.2020
2. Mell, P., Grance, T.: The NIST Definition of Cloud Computing, vol. 53, pp. 1–7. National Institute of Standards and Technology Special Publication (2011)
3. Bundeskriminalamt: Cybercrime Bundeslagebild (2018). Available at: https://www.bka.de/SharedDocs/Downloads/DE/Publikationen/JahresberichteUndLagebilder/Cybercrime/cybercrimeBundeslagebild2018.html. Accessed: 22.01.2020
4. Ruan, K., Baggili, I. P., Carthy, J., Kechadi, T.: Survey on cloud forensics and critical criteria for cloud forensic capability: a preliminary analysis. In: Proceedings of the Conference on Digital Forensics, Security and Law, pp. 16 (2011)
5. Chen, L., Le-Khac, N.-A., Schlepphorst, S., Xu, L.: Cloud forensics: model, challenges, and approaches. In: Chen, L., Takabi, H., Le-Khac, N.-A. (eds.) Security, Privacy, and Digital Forensics in the Cloud, pp. 201–216. Wiley, Singapore (2019). https://doi.org/10.1002/9781119053385.ch1
6. Reilly, D., Wren, C., Berry, T.: Cloud computing: forensic challenges for law enforcement. In: Internet Technology and Secured Transactions (ICITST), pp. 1–7. IEEE, London, UK (2010)
7. Le-Khac, N.-A., Plunkett, J., Kechadi, M.-T., Chen, L.: Digital forensic process and model in the cloud. In: Chen, L., Takabi, H., Le-Khac, N.-A. (eds.) Security, Privacy, and Digital Forensics in the Cloud, pp. 239–255. Wiley, Singapore (2019). https://doi.org/10.1002/9781119053385.ch12
8. Dykstra, J., Sherman, A.: Design and implementation of FROST: digital forensic tools for the OpenStack cloud computing platform. Dig. Invest. 10, S87–S95 (2013)
9. Ruan, K., Carthy, J., Kechadi, T., Baggili, I.: Cloud forensics definitions and critical criteria for cloud forensic capability: an overview of survey results. Dig. Invest. 10(1), 34–43 (2013)
10. Almulla, S., Iraqi, Y., Jones, A.: A state-of-the-art review of cloud forensics. J. Dig. Forensics, Secur. Law 9(4), 7–28 (2014)
11. Al Fahdi, M., Clarke, N.L., Furnell, S.M.: Challenges to digital forensics: a survey of researchers & practitioners attitudes and opinions. In: Information Security for South Africa (ISSA 2013), Johannesburg, South Africa, pp. 1–8. IEEE (2013)
12. Kent, K., Chevalier, S., Grance, T., Dang, H.: *Guide to Integrating Forensic Techniques into Incident Response* 121. Gaithersburg: NIST (2006). Available at: http://nvlpubs.nist.gov/nistpubs/Legacy/SP/nistspecialpublication800-86.pdf
13. Simou, S., Kalloniatis, C., Gritzalis, S., Mouratidis, H.: A survey on cloud forensics challenges and solutions: a survey on cloud forensics challenges and solutions. Secur. Commun. Netw. 9(18), 6285–6314 (2016)
14. De Marco, L., Kechadi, T., Ferrucci, F.: Cloud Forensic Readiness: Foundations. Digital Forensics and Cyber Crime Cham. Springer, Berlin (2013)
15. De Marco, L., Le-Khac, N.-A., Kechadi, M.-T.: Digital evidence management, presentation, and court preparation in the cloud: a forensic readiness approach. In: Chen, L., Takabi, H., Le-Khac, N.-A. (eds.) Security, Privacy, and Digital Forensics in the Cloud, pp. 283–299. Wiley, Singapore (2019). https://doi.org/10.1002/9781119053385.ch14
16. Fang, L., Jin, T., Jian, M., Robert, B., John, M., Lee, B., Dawn, L.: NIST Cloud Computing Reference Architecture: Recommendations of the National Institute of Standards and Technology (Special Publication 500-292). CreateSpace Independent Publishing Platform (2011)
17. Delport, W., Olivier, M.: Isolating instances in cloud forensics. In: Peterson, G., Shenoi, S. (eds.) Advances in Digital Forensics VIII, pp. 187–200. Springer, Berlin (2012)
18. Dykstra, J., Sherman, A.: Acquiring forensic evidence from infrastructure-as-a-service cloud computing: exploring and evaluating tools, trust, and techniques. Dig. Invest. 9, S90–S98 (2012)

19. Dykstra, J.: Digital Forensics for Infrastructure-as-a-Service Cloud Computing. Doctor of Philosophy Ph.D., University of Maryland [Online] (2013). Available at: http://www.cisa.umbc.edu/papers/dissertations/dykstradissertation-2013.pdf. Accessed: 22.01.2020

20. Dykstra, J., Sherman, A.T.: Understanding issues in cloud forensics: two hypothetical case studies. In: Annual ADFSL Conference on Digital Forensics, Security and Law, p. 10. Richmond, Virginia (2011)

21. Dykstra, J., Riehl, D.: Forensic collection of electronic evidence from infrastructure-as-a-service cloud computing. Richmond J. Law Technol. **19**(1), 1–48 (2012)

22. Reichert, Z., Richards, K., Yoshigoe, K.: Automated forensic data acquisition in the Cloud. In: 11th International Conference on Mobile Ad Hoc and Sensor Systems (MAAS), pp. 725–730. IEEE, Philadelphia, PA, USA (2014)

23. Saibharath, S., Geethakumari, G.: Design and Implementation of a forensic framework for cloud in OpenStack cloud platform. In: International Conference on Advances in Computing, Communications and Informatics (ICACCI) 2014, New Delhi, India, pp. 645–650. IEEE (2014)

24. Banas, M.: Cloud Forensic Framework For IaaS With Support for Volatile Memory. Master of Science M.Sc, National College of Ireland [Online] (2015). Available at: http://trap.ncirl.ie/2068/1/matusbanas.pdf. Accessed: 22.01.2020

25. Digambar, P.: A Novel Digital Forensic Framework for Cloud Computing Environment. Doctor of Philosophy Ph.D., Birla Institute of Technology and Science, Pilani [Online] (2015). Available at: http://shodhganga.inflibnet.ac.in/bitstream/10603/84911/1/thesis-digambarpovar.pdf. Accessed: 22.01.2020

26. Alqahtany, S., Clarke, N., Furnell, S., Reich, C.: A forensic acquisition and analysis system for IaaS. Cluster Comput. **19**(1), 439–453 (2016)

27. Rani, D.R., Sultana, N., Sravani, P.L.: Challenges of digital forensics in cloud computing environment. Indian J. Sci. Technol. **9**(17) (2016)

28. Sudyana, D., Lizarti, N.: Digital evidence acquisition system on IAAS cloud computing model using live forensic method. Sci. J. Inf. **6**(1), 125–137 (2019)

29. Birk, D., Wegener, C.: Technical issues of forensic investigations in cloud computing environments. Workshop on Cryptography and Security in Clouds (2011). https://doi.org/10.1109/SADFE.2011.17

30. Damshenas, M. et al.: Forensics investigation challenges in cloud computing environments. In: 2012, International Conference on Cyber Security, Cyber Warfare and Digital Forensic (CyberSec), pp. 190–194. IEEE (2012)

31. Keyun, R.: Cybercrime and Cloud Forensics: Applications for Investigation Processes. IGI Global Press (2012)

32. Shirudeet, D., et al.: Cloud Forensics: Drawbacks in Current Methodologies and Proposed Solution. Int. J. Eng. Res. Appl. **7**(2), 79–81 (2017). https://doi.org/10.9790/9622-070203798179

33. Chung, H., Kang, C., Lee, S., Park, J.: Digital forensic investigation of cloud storage services. Dig. Invest. **9**(2), 81–95 (2012)

34. Marturana, F., Me, G., Tacconi, S.: A case study on digital forensics in the cloud. In: International Conference on Cyber-Enabled Distributed Computing and Knowledge Discovery, Sanya, China, pp. 111–116: IEEE (2012)

35. Quick, D.: Cloud Storage Forensic Analysis. Master of Science (Cyber Security and Forensic Computing), University of South Australia, Adelaide, South Australia [Online] (2012). Available at: http://citeseerx.ist.psu.edu/viewdoc/download?doi=10.1.1.465.5292&rep=rep1&type=pdf. Accessed: 22.01.2020

36. Quick, D., Choo, K.-K.R.: Forensic collection of cloud storage data: does the act of collection result in changes to the data or its metadata? Dig. Invest. **10**(3), 266–277 (2013)

37. Hale, J.S.: Amazon cloud drive forensic analysis. Dig. Invest. **10**(3), 259–265 (2013)

38. Easwaramoorthy, S., Thamburasa, S., Samy, G., Bhushan, S.B., Aravind, K.: Digital forensic evidence collection of cloud storage data for investigation. In: Fifth International Conference on Recent Trends in Information Technology (ICRTIT), Chennai, India, pp. 1–6. IEEE (2016)

39. Satrya, G.B., Nasrullah, A.A., Shin, S.Y.: Identifying artefact on Microsoft OneDrive client to support android forensics. Int. J. Electron. Secur. Dig. Forensics **9**(3), 269–291 (2017)
40. Le-Khac, N.-A., Mollema, M., Craig, R., Ryder, S., Chen, L.: Data acquisition in the cloud". In: Chen, L., Takabi, H., Le-Khac, N.-A. (eds.) Security, Privacy, and Digital Forensics in the Cloud, pp. 257–282. Wiley, Singapore. https://doi.org/10.1002/9781119053385.ch13

**Sebastian Schlepphorst** received the MSc in Forensic Computing & Cybercrime Investigation from University College Dublin, Ireland in 2014. He was a cybercrime and digital forensics lecturer in the police academy of North Rhine-Westphalia from 2013 until 2018. Since 2018, he has worked as a senior civil servant in the German National IT Situation Center of the Federal Office for Information Security (BSI). His research interests include digital forensics, cybercrime investigation, cloud computing and cloud forensics.

**Kim-Kwang Raymond Choo** received the Ph.D. in Information Security in 2006 from the Queensland University of Technology, Australia. He currently holds the Cloud Technology Endowed Professorship at the University of Texas at San Antonio (UTSA). In 2015, he and his team won the Digital Forensics Research Challenge organized by the Germany's University of Erlangen-Nuremberg. He is the recipient of the 2019 IEEE Technical Committee on Scalable Computing (TCSC) Award for Excellence in Scalable Computing (Middle Career Researcher), 2018 UTSA College of Business Col. Jean Piccione and Lt. Col. Philip Piccione Endowed Research Award for Tenured Faculty, British Computer Society's 2019 Wilkes Award Runner-up, 2019 EURASIP Journal on Wireless Communications and Networking (JWCN) Best Paper Award, Korea Information Processing Society's Journal of Information Processing Systems (JIPS) Survey Paper Award (Gold) 2019, IEEE Blockchain 2019 Outstanding Paper Award, Inscrypt 2019 Best Student Paper Award, IEEE TrustCom 2018 Best Paper Award, ESORICS 2015 Best Research Paper Award, 2014 Highly Commended Award by the Australia New Zealand Policing Advisory Agency, Fulbright Scholarship in 2009, 2008 Australia Day Achievement Medallion, and British Computer Society's Wilkes Award in 2008. He is also a fellow of the Australian Computer Society, an IEEE senior member, and co-chair of IEEE Multimedia Communications Technical Committee's Digital Rights Management for Multimedia Interest Group.

**Nhien-An Le-Khac** is a lecturer at the School of Computer Science (CS), University College Dublin (UCD), Ireland. He is currently the program director of MSc program in Forensic Computing and Cybercrime Investigation (FCCI)—an international program for the law enforcement officers specializing in cybercrime investigations. He is also the co-founder of UCD-GNECB Postgraduate Certificate in fraud and e-crime investigation. Since 2008, he is a research fellow in Citibank, Ireland (Citi). He obtained his Ph.D. in Computer Science in 2006 at the Institut National Polytechnique de Grenoble (INPG), France. His research interest spans the area of Cybersecurity and Digital Forensics, Data Mining/Distributed Data Mining for Security, and Fraud and Criminal Detection. Since 2013, he has collaborated on many research projects as a principal/co-PI/funded investigator. He has published more than 150 scientific papers in peer-reviewed journal and conferences in related research fields, and his recent edited book has been listed the Best New Digital Forensics Book according to the Book Authority.

# Long Term Evolution Network Security and Real-Time Data Extraction

Neil Redmond, Le-Nam Tran, Kim-Kwang Raymond Choo, and Nhien-An Le-Khac

**Abstract** Long Term Evolution ("LTE") or 4G data services are relied upon by many people to process information such as payments, VOIP and applications on their mobile devices. Today this technology is widely used and many people depend upon its security every day for all manner of tasks. GSM traffic, SMS traffic and data traffic are encrypted by the network to prevent third party access. In this paper, we are going to investigate a methodology to overcome the security restrictions of LTE network to extract data from mobile device in real-time. The outcome of this research not only assists the digital investigators to tackle the challenges of extracting relevant artefacts from suspect devices using LTE network but also allows us to evaluate the possibility of extract important data from LTE security network in real-time. We also test our approach with real-world scenarios and services.

**Keywords** LTE security · Data extraction · Blade RF X115 · Hacking · Interception

## 1 Introduction

In today's "smart world" users are careful about how their personal data is used. This is even more so with the devices people use every day—mobile phones in particular. In the developed world most people have at least one mobile handset (60% of the worldwide population will use a mobile phone at least once a month) [1] and often more than one mobile device.

Furthermore, technology such as 4G is marketed as being data safe, a key selling point given revelations of government access to other forms of communications

N. Redmond (✉)
National Cyber Security Centre, Dublin, Ireland
e-mail: neil_lte@ohr.ie

N. Redmond · L.-N. Tran · N.-A. Le-Khac
University College Dublin, Dublin, Ireland

K.-K. R. Choo
University of Texas at San Antonio, San Antonio, USA

© The Editor(s) (if applicable) and The Author(s), under exclusive license to Springer Nature Switzerland AG 2020
N.-A. Le-Khac and K.-K. R. Choo (eds.), *Cyber and Digital Forensic Investigations*, Studies in Big Data 74, https://doi.org/10.1007/978-3-030-47131-6_9

over recent years. This level of marketing can lull users into a false sense of security, believing that once they are "on-net" that their calls, texts and applications are secure. The use of banking applications and SMS authentication, VoIP [2] platforms and other "secure" products heavily rely on Long Term Evolution (LTE) or 4G network security to function as intended.

LTE is presented by telecom operators as being reliable, safe and designed for data traffic. However, in this paper we study whether in fact the technology is vulnerable to a "man in the middle" attack—which seeks to identify ways of retrieving data for intelligence gathering. Additionally, if the technology is ultimately vulnerable the extent of the vulnerability of end-user data will be examined. In this paper, we propose an approach taken to address the matter of real-time extraction involved a test network simulating full GSM (Global System for Mobile Communication) functionality, one which will provide for the migration of a user off a 4G enabled network and provide the user with a level of service that is comparable to their home network. By providing a quality network the interceptor can facilitate the throughput of data from a target device and remove the normal LTE and GSM cipher streams (effectively deny usage to A5/2 and A5/3 stream algorithms) and use the GSM standard A5/0 with no response expected. In this way short message service texts can be read as clear text when sent from a user device, the recipient of the text can also be determined in clear text and website URLs can be viewed. Additionally, voice calls can be recorded and calling parties identified. The contribution of this paper is two-folds: on one hand, the consideration is where criminals use such approach proposed in this paper for their illegal activities. How is it then possible to access users' information when they are associated with an LTE network? On the other hand, in the context of digital forensics, traditional investigative techniques involve the physical possession of a targeted mobile device and the examination of data that is contained within the device [3]. Such methodologies include screen capture, direct extraction using non-forensic tools, backup/logical acquisition, advanced file system acquisition and physical acquisition. This is rather cumbersome and only looks at historic events. Such techniques cannot work with real-time information collection or a roving target. Ideally real-time data acquisition from a device that is associated with a 4G network is an optimal solution, in both extraction of data and the timing associated with such extraction. Additionally, the target is aware of the investigation with traditional extraction techniques.

To examine our hypotheses and to prove that data can be captured, we have developed experiments with a test network. In our experiments Skype video calls, chats and messaging was also tested. Additionally, popular smartphone applications such as eBay and the Leap card were also tested in our experiments for functionality on the network and to also determine if personal data could be extracted over the test network.

The rest of paper is organized as follows: We discuss on related work of this research context in Sect. 2. We present challenges and research questions in Sect. 3. We propose our approach in Sect. 4. Experiments and discussion will be described in Sects. 5 and 6 respectively. Finally, we conclude and discuss on future work in Sect. 7.

## 2    Related Work

In [4], authors outline an experiment involved the usage of a BladeRF and a Raspberry Pi2 amongst other components. The experiment successfully demonstrated that a Software Defined Radio (SDR) solution utilising a BladeRF and Raspberry Pi can establish links at more than 90 km distance.

The security uses of Universal Software Radio Peripheral (USRP) is examined by [5] on how SDR can be used to perform accurate radio frequency fingerprinting and to identify spoofing attacks in critical 802.15.4 based infrastructure networks. They stated that a recent analysis of WPANs in 10 US cities showed that healthcare and utility control networks operated with faulty or no security. Attacker can readily extract encryption keys from inexpensive hardware when security is not given design priority.

The security concerns associated with data being transmitted can be addressed with ciphering or encryption. From a forensic perspective this can prove problematic as the more secure the data the less likely it is for law enforcement to extract that data. In [6], authors recognised the requirement for privacy in relation to data but also the need for technologies that will forensically defeat data encryption at the same time respecting such privacy. They also recognised that the examination and analysis phases of a digital forensics investigation are impacted the most with the usage of encryption technology.

The vulnerability of the GSM network makes it susceptible to third party hacking. Authors in [7] explained how Man in the Middles attacks exploit the GSM standard where A5/0 encryption is switched off and a rogue base station can then log all the International Mobile Subscriber Identifier (IMSI)/International Mobile Equipment Identity (IMEI) Information for a device that connects to this device on a triggered re-authentication. By acquiring all this data about a user device the hacker then has access to voice and data traffic. With this data it is also possible to clone Subscriber Identity Module (SIM) card the author outlines [5], which can then lead to a cloned handset where the hacker has the proper IMSI/IMEI combination as well as a Temporary Mobile Subscriber Identity (TMSI) that is associated with the correct data allowing usage of a device that the genuine network provider will be fooled into believing has the correct authentication and credit to associate on the network and generate calls.

It was stated in one paper [8] that GSM speech service is secure up to the point where speech enters the core network and for true security it should be the end user and not the network provider that implements encryption. The authors concluded that by using particular tools, the voice traffic could be encrypted, transmitted and then decrypted and understood.

One paper [9] explained that the security mechanisms associated with GSM are comprised of the SIM, the end-user device and the provider's GSM Network.

Authors in [10] described an approach to build an unsafe (fake) LTE network to show the potential of attackers could launch DoS (denied of service) attacks, malicious calls and all the traffic (voice and data) can be eavesdropped. However, it is only a prototype and they did not show how it works in real-world applications such as using Skype, e-banking, etc.

Authors in [11] aimed at modelling all possible adversarial attacks for 4G LTE networks including detach/downgrading ones. Authors showed that they can make the malicious injection to the network and with the IMSI detach command, the victim's devices did not detach from the 4G LTE network. There is no real-world application was tested either.

In another paper [12], authors also presented adversarial attacks for 4G LTE. They also described the distance covered by attacks as well as showed experiments with social media apps. However, it is not clear which mobile OS they were testing. In addition, no experiments with e-commerce apps.

A survey on existing authentication and privacy-preserving schemes of 4G and 5G cellular networks was carried out [13]. It also indicated that the topic of network security and telecom networks are of interest to the academic community. The concerns of GSM vulnerabilities are outlined.

## 3 Problem Statement and Challenges

An LTE network utilises security based on 3GPP standards [33.401] which can make it extremely complex to lawfully intercept data on the end user device belonging to a subject of an investigation that is associated with an LTE network.

A solution has to be developed whereby the end user device is tricked into migrating off the LTE network and onto a network that is controlled by the investigator. The new network will be based on GSM, which will use 2G for voice and short messaging and 2.5G for data services using General Packet Data Service (GPRS). The GSM network can be more easily controlled by an investigator as the normal cipher streams that are used in 2G networks—A5/0, A5/1, A5/2 etc. can be deactivated for the interception network. This means that traffic that is transmitted on the network can be viewed.

Mobile networks, in particular GSM networks were developed to have the same security protocols as fixed networks had at that time. This means that such networks are vulnerable to third party access, unknown to the network operator.

This presents an opportunity for law enforcement to use such networks to gather evidence that may be hitherto difficult to obtain. Two threats are of use to law enforcement when discussing wireless networks [14].

Masquerading is when one entity pretends to be another entity—usually a rogue Base Station pretending to be a genuine network BS. In doing so the rogue entity may be able to obtain data that it would not normally have access to. Unauthorised Denial of Service is when a rogue entity actively prevents resources being made available to a normally authorised user. In such cases network access can be denied to a user and thereby force them to seek a different and stronger network signal to connect to. Such attacks are a genuine threat to the reliability of cellular networks [15].

These two vulnerabilities lead to the situation where eavesdropping is a realistic concern for cellular networks. By inserting themselves into the network cell an intruder can access any data that is transmitted through the air in that cell. Also an intruder can overload the network with network retransmissions and thereby lead to denial of service for that cell.

From the law enforcement perspective that presents opportunities for data acquisition, as any data that is sent across the physical layer of a cellular network in a compromised cell or rogue cell would be available to the data transmitted and routing information used including the location of the target user. For criminals such methodologies can be used to generate data for SIM cloning that can be used to steal a user's identity on the cellular network.

Device discovery is an intrinsic element of the cellular network. It is also a key prerequisite for the introduction of proximity-aware services in an LTE network [16]. Device discovery provides an augmented sense of the surrounding network that allows for meeting the communication peers in time, frequency and space prior to the establishment of an actual direct connection [17].

# 4 Proposed Approach

## 4.1 IMSI Catching

Law Enforcement has long used IMSI catchers to gather intelligence about suspects [14] as such tools are able to determine the IMSI of devices that are within its range. This allows these devices to target specific handsets and to see what traffic is flowing from the captured devices. These devices actively interfere with the communications between handsets and base stations by acting as a transceiver. This is more commonly known as a "man-in-the-middle" attack.

There are a number of academic papers in relation to this topic—[18, 19] which indicate the various stages of an IMSI catcher summarising the four steps using [14] as a basis (i) Obtain the IMSI; (ii) Complete the Mobile Station connection; (iii) Complete the network connection; (iv) Cell Imprisonment.

## 4.2 Proposed Approach

Sophisticated IMSI catchers are able to do more than just capture an IMSI. They can eavesdrop on telephony calls, short message services and data services. Such devices can jam normal network service and forcibly migrate a User Equipment (UE) from a 4G/LTE network to a 2G/GSM network where encryption is less or non-existent. This is something that we will attempt to achieve in the laboratory tests and associated field trials.

The test network will capture the IMSI of any device that connects to and is authenticated by the network. The SDR software will use part of the IMSI to create a specific phone number for a user to use on the test network. In a real-time environment the IMSI would be used to identify the targeted device and the user's own phone number would be used. Additionally, the SDR solution that we used can block users with IMSI that are not of interest from accessing the test network accidently.

The test network was based on the captured IMSI of the SIM in the mobile device. The test network recorded the IMSI the first time the SIM connected to the test network. The details were recorded in the Home Location Register (HLR) of the test network and a test network Command Line Interface (CLI) assigned to the IMSI. The test network could whitelist access based on IMSI details. This meant that only authenticated IMSI could connect to and roam on the test network. Such an approach prevented non-test devices from being mistakenly connected to the test network.

## 5   Experiments and Findings

### 5.1   Testing Platforms

In order to validate our proposed approach, we set up a testing platform as showed in Fig. 1. In our experiment, we built a hardware and software solution for developing a GSM Base-station. This station includes a 2G and a 2.5G mobile networks to intercept traffic from a mobile device that was roaming on a 4G/LTE network connection by

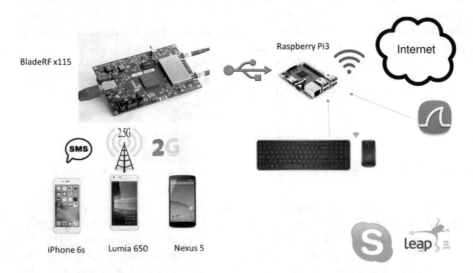

**Fig. 1**   The Testing platform

emitting a stronger signal near the target device. This would make the device move off-net and onto the test network, where security and encryption was non-existent.

The Testing platform used the following main devices: Raspberry Pi3 with "Jessie" image, Nuand BladeRF x115, 4 handsets (Apple iPhone 6 s with iOS 10.1 and 10.2, Apple iPhone 5 s with iOS 10.2, Nokia Lumia 650 with Microsoft 10 Mobile OS v.1607. OS Build 10.0.14393.693, Google Nexus 5 with Android OS v6.0.1 build number M4B30Z).

A YateBTS operating system is used in our experiments. It relies on Software Defined Radio [4] for transmission and reception of signals on the GSM band that we choose to work with (900 MHz). It completely mimics the entire GSM functionality as combining YateBTS with a BladeRF means that all the hardware and software components for a functioning GSM network are contained within a small footprint that is significantly less complex than a normal GSM setup. The BladeRF acts as a base transceiver station in YateBTS, where it transmits and receives GSM signals and all signal processing requirements. The system allows for calling traffic, short message service traffic and when configured, data traffic.

**Experiment Configuration**

After successfully setting up the BladeRF, it was connected to the Raspberry Pi3 with the relevant configuration. Next, a Graphical User Interface (GUI) is installed to allow a user to modify and edit parameters in the Base Transceiver Station (BTS) GSM and GPRS configurations. This utility is called "Network in a Box" or NIB.

The next step is to configure the BladeRF for GSM functionality. The Radio.Band is set at 900 MHz, with a Radio.C0 of 75 MHz downlink. The identity of the Mobile Country Codes (MCC) and Mobile Network Codes (MNC) were set at 1 for each as this was a test network. The national MCC for Ireland is 272 and the MNC value that could be chosen is 15 (Irish Eir telecom company has a MNC value of 7 for example). For some tests to ensure functionality we did rename the network 272 15 (as no operator has a MNC of 15 in Ireland). Therefore, the location area code (LAC) was set as 1000 as the MCC was set at 1 and the MNC as 01 (these are standard values for test cellular networks). When presented on the mobile device the network name will be Test Public Land Mobile Network (PLMN) 1-1. This will be the name to look for when connecting to the test network. In this experimental network we wanted to let anyone register and connect, so any handset could register onto the test network.

At this stage the GSM setup was complete. We then disabled the firewall to allow GPRS to work and allowed 5 concurrent channels to operate (one per handset) and set the MS IP Base at 192.168.2.120. Then GPRS was established. Initially it operated at slow speeds, but by having a high throughput broadband connection to the internet we were able to achieve significantly faster speeds to support on-line applications such as Skype video calls and Leap card top-ups. The test network was established with the following end user devices (UE) of handsets as described above.

All devices connect to the test network and would automatically re-connect to the network when the phone was brought back into the room where the network was located.

To investigate these theories in a qualitative manner we set six testing scenarios on multiple UE devices, all with the objective of proving that it is possible to degrade a 4G connected UE and migrate it to a less secure 2G network connection.

## 5.2   Scenario 1: Unencrypted GSM Calls

With three devices now registered on the network we made a voice call between the devices that were on-net. All calls connected, initially in a slower manner than a conventional mobile network, but subsequent connectivity was as quick as a conventional network. The voice quality of the call was acceptable as a normal conversation could be held without any noticeable loss in quality.

Real-time calls were generated and connected in a similar timeframe to what would be expected in a normal LTE connection. It was observed that the speed didn't vary on different makes of mobile device. The network analysis shows the calling parties using the Apple iPhone 6S and the Nokia Lumia 650 thus enabling an investigator to note who is being called and when.

The test network is setup with no hops and a single ARFCN. This means that the voice traffic part of the intercepted traffic can be identified. However, it will take further manipulation to actually render this captured data useful to listen to the actual intercepted calls. Given the way the test network is setup this may not be a significant burden for an investigator with the right skills. The use of other tools such as an Airgap and a Hexadecimal reader may allow an investigator to decrypt the captured data and listen to the voice traffic.

## 5.3   Scenario 2: Unencrypted SMS Text Messaging

Once connected to the test network—Test PLMN 1-1, we interacted with a second Bot called Eliza. This was to test SMS messaging functionality for all devices. This Bot's contact CLI is 35,492 which is only accessible when connected to the test network. Both this Bot and the David Bot are test elements for YateBTS to test for functionality.

Following the voice calls a standard SMS text was sent between the end user devices. As explained earlier GSM SMS "texts" are normally sent with a cipher—A5/3 or A5/1 etc. This means that unless an investigator knows the cipher it is extremely difficult to read the contents of the text.

As part of the YateBTS architecture, ciphers can be switched off, meaning that in theory it is possible to intercept and read the texts sent from an intercepted handset.

In the laboratory tests that were carried out, multiple GSM SMS messages were sent to and from phones on the test network. By connecting the handsets to a network with no ciphering enabled it was demonstrated to be possible to intercept text messages and read them in clear text.

Figures 2 and 3 show screen shot samples from Wireshark of how it is possible to read clear text GSM SMS that is sent on the test network from an intercepted UE.

As can be seen from the screen shots, there are three parts to each GSM SMS. The first part is when the originating MS sends the message to the Network. The second part is when the Network sends the data to the recipient MS and the third part is the recipient MS acknowledging receipt of data (the SMS text message). Of most relevance to an investigator is the second part, in particular the section entitled "GSM SMS TDPU". By examining this section of the GSM SMS package the wording of the text is visible in the sub-section TP-User Data.

Figures 2 and 3 also show screenshots of the texts as sent to and from the mobile handsets. These messages correspond with the information that is presented in Wireshark's analysis.

These tests were run on multiple occasions to demonstrate that the interception of GSM SMS texts is achievable when hijacking a handset to the test network.

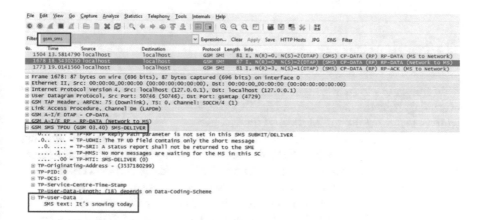

**Fig. 2**  Wireshark capture of first SMS

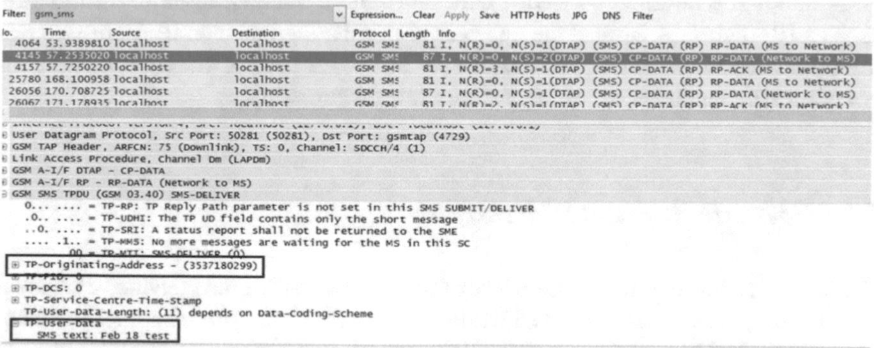

**Fig. 3**  Wireshark capture of second SMS intercepted

**Fig. 4** Chat with the Eliza bot

Of further use to an investigator is the identification of the parties involved in the messaging. Network analysis shows the BCD (terminating party) in TP-Originating-Address section of the "GSM SMS TDPU".

Figure 4 shows a network analysis and the screenshot of a chat session with the Eliza bot. This also shows how SMS messages can be intercepted and read in clear text when transmitted across the test network. This indicates that GSM based texts can be de-ciphered and read in clear text once they are sent via the test network.

As mentioned above, the next four scenarios focus on the data and internet functionality. Fajkus [20] cast doubt on the real-world application of SDR with a USRP N210. We also wanted to prove that it is possible to have a functioning General Packet Radio Service (GPRS) based on the test network. The data connection will use the APN that is pre-installed on a SIM card and ignore operator specific information such as IP Addresses. Instead the test network will assign its own IP Address range to a target device, in the range of 192.168.99.x. The range 99.x was chosen as it does not clash with any other known IP Addressing functionality.

## 5.4   Scenario 3: Basic Web Surfing

The key to having an effective GPRS connection, one that would convince a targeted user that they are still roaming on LTE is to have a sufficiently strong internet backhaul connection. Just like a traditional cellular network its backhaul needs to be of high bandwidth to allow high data throughput in a manner that would be perceived by a user when using a mobile data connection. Figure 5 shows that GPRS was activated and working on the test network. Figure 6 shows time-lapse of a download on the GPRS connection as well as the ability to open and use a high data usage iOS application, in this case the Sky News app. In the example the download speeds were not noticeably different from a normal 3G/4G or Wi-fi network download.

In all the tests that were carried out the speed provided by the GPRS connection was fast enough to allow an unaware user to have the perception that they were still using their own mobile provider's data connection. It should be noted that iMessages were sent between Apple devices that were connected to the test network and to traditional networks and all messages went through as blue, indicating that the messages were sent as iMessages.

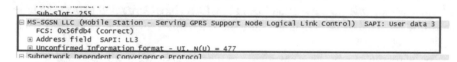

**Fig. 5** GPRS functionality

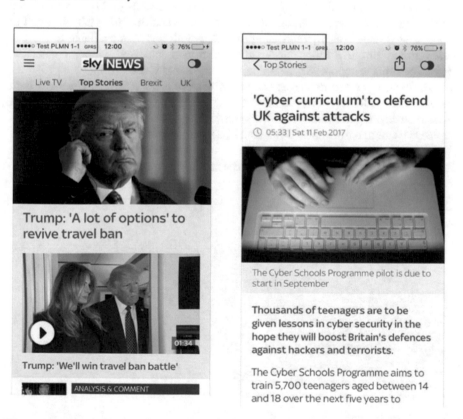

**Fig. 6** App functionality over GPRS (time-lapse view)

## 5.5    Scenario 4: Skype

This scenario and the next two were tested for functionality on the GPRS network, to see if the network could support application functions and to also investigate whether it was possible to extract user data that was passed over the air within the test network.

For Skype, the tests began to look at other communications that could be "hosted" and intercepted on the test network. Skype and other social media applications that utilise voice services require a high throughput bandwidth to function. A concern with hijacking a handset to the test network is that the quality of the bandwidth of a GPRS network will be much less than that of a typical LTE or wireless access network. Initial testing was conducted on a standard laboratory network with multiple devices connected and at a bust time of day for internet access—Monday morning at 11 a.m. for example. In these situations, the GPRS signal struggled to host the high overhead required by Skype, to not only place the user on-line but to support voice calls between parties.

By increasing the backhaul available to the test network it was possible to have Skype connect over the GPRS network. GPRS at the outermost speeds can deliver a bandwidth of 53.6 kb/s download, but can achieve speeds of 114 kbps with the right conditions. Skype requires throughput of between 30–100 Kb/s for a call to be activated. However, by increasing the bandwidth that was available to the test network to allow an 8 Mbps download speed from a dedicated Internet connection it was possible to support Skype for text messages and video calling.

The sending of text messages on and off-net in clear text was achievable—both on the same GPRS network and between the GPRS and off network handsets. This meant an investigator could see SMS traffic from a subject device. Additionally, voice calls, both on-net and off-net could be monitored and recorded in real-time.

Wireshark analysis of the messages and the video call is presented in Fig. 7. What is noticeable however is that even though the traffic is being carried on the test network

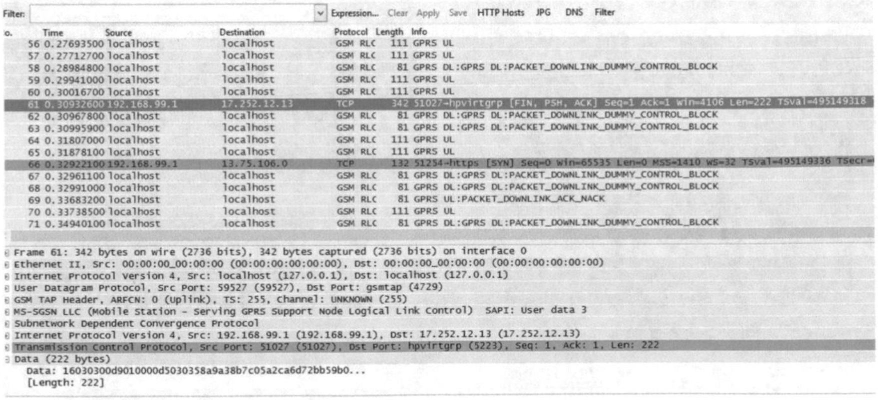

**Fig. 7**  Wireshark analysis of Skype call

which is unencrypted with no ciphering, Skype implements its own encryption using the Diffiie Hellman algorithm [21]. This means that although an investigator can provide sufficient bandwidth to allow the Skype service to operate it is not that easy to analyse the resulting network output. The investigator will have to know what traffic to capture, and then decrypt the traffic with the correct key and reassemble it in the correct order to have any chance of understanding the information.

Despite this hurdle, it would be possible, given time and skills, to do this. There is information however that it is available from the network analysis that is of use without the need for decryption.

## 5.6 Scenario 5: Leap Card Android Application

In Ireland, public transport can be paid for by cashless transactions involving a type of Debit card, akin to the Oyster card in London or Hong Kong. The Irish equivalent is known as a Leap Card. Users can link their Leap card to a credit card for updating their travel credits. Additionally there is an Android app available for users to check their credit balance and to purchase additional credit.

The Leap card requires a strong data connection to function. The test network's GPRS function was able to support the application and allow a user to view a card's details—credit balance and also enter the process of topping-up the credit on the card. Given that this is only an Android based application we used the Google Nexus 5 UE with Near Field Communication (NFC) enabled.

The NFC Functionality worked without issue and Fig. 8 demonstrates the flow of screen information that was received. This testing utilised a pre-production Android

**Fig. 8** Leap card application screenshots over a GPRS connection

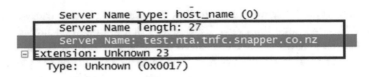

Server Name Type: host_name (0)
Server Name length: 27
Server Name: test.nta.tnfc.snapper.co.nz
☐ Extension: Unknown 23
Type: Unknown (0x0017)

**Fig. 9** Leap card application developer identified

Application that featured production security measures but still used the developer's own details.

From examining network analysis we were able to determine the following:

1. The developer of the application is a company called Snapper from New Zealand (Fig. 9)
2. The application uses godaddy.com certificate information (Fig. 10)
3. Payment information is encrypted once you access the payment section of the application (Fig. 11).

```
.......U ..1.U...
U....US1 .O...U..
..Arizon a1.0...U
....Scot tsdale1.
0...U... .GoDaddy
.com, In c.1-0+..
U...$htt p://cert
s.godadd y.com/re
pository /1301..U
...*Go D addy Sec
ure Cert ificate
Authorit y - G20.
..160/08 005/382.
.1707220 35338ZOI
```

**Fig. 10** Leap card application certificate authority identified

Session ID Length: 32
Session ID: e6d3b7823f854f669568f67bcce7e188b7e399899761fe9e...
Cipher Suite: TLS_ECDHE_RSA_WITH_AES_128_GCM_SHA256 (0xc02f)
Compression Method: null (0)

| | | | | |
|---|---|---|---|---|
| 3025 19.8309040 localhost | localhost | SNDCP | 555 SN-UNITDATA N-PDU 2638 (segme |
| 3026 19.8309940 test.tnfc.leaptopup | 192.168.99.1 | TLSv1.2 | 541 Certificate |
| 3027 19.8318220 test.tnfc.leaptopup | 192.168.99.1 | TLSv1.2 | 395 Server Key Exchange |
| 3028 19.8340400 localhost | localhost | GSM RLC | 01 GPRS DL.GPRS DL.PACKET_DOWNL] |

**Fig. 11** Leap card application cipher identified

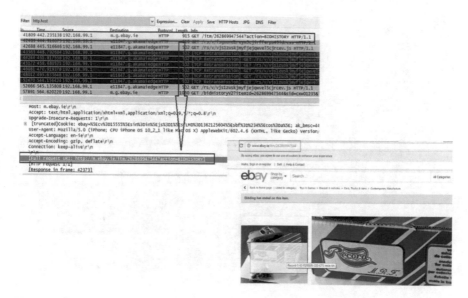

**Fig. 12** eBay information—user history

## 5.7 Scenario 6: Ebay on IOS 10.2

When the user iPhone6S was connected to the network and GPRS enabled we connected to the website of eBay.com using the Safari web browser. The user data that
was inputted such as the eBay member name was visible in network analysis, see
Figs. 12 and 13. This included the user's EBay login details, originating IP Address
and provided sufficient detail to determine who the user could possibly be.

Additionally, the URL of the pages visited on the eBay site were also retrieved.
The specific urls could be copied and the webpages re-built in real-time for example to
determine what site, or object the user was interested in. The user's history was readily
available from the network analysis, indicating that intercepted device could provide
real-time browsing habits of a targeted user. Furthermore, the webpage layouts were
also retrieved from network analysis. This indicated that it is possible to track a user's
personal data, purchasing and browsing habits on an intercepted device.

## 6 Discussion

The objective of this research is to demonstrate that a 4G connected end user device
could be forced off its home network and migrated to a less secure 2G and 2.5G
network in a manner that a user would be unaware of. Once migrated to a network that
was controlled by law enforcement for example, any data that was being sent from the
targeted devices could be retrieved in real-time. In the cases of unencrypted SMS and

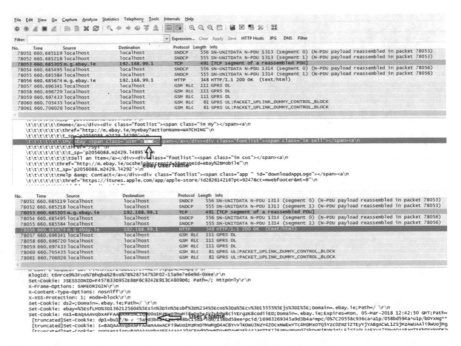

**Fig. 13** eBay userID and user's name (redacted for privacy)

HTTP requests this information would be provided in clear text. Such SMS are sent by many organisations as part of two factor authentication process. As demonstrated, unless the ciphers are robust, this sensitive information can be extracted by a third party with relative ease. For law enforcement the tests demonstrated that it is possible to develop an effective and low cost solution that captures data in real-time from a 4G associated mobile device (for a cost of around €1000 at the time of writing this paper). It is not necessary to require the seizure of a device in order to extract sensitive data. On the other hand, a hacker can also potentially use this technology for their benefit. The interception of private data that is being transferred from device to network and on and off-network can provide ample opportunity for data theft at the very least. At the very worst a hacker can track a device's geo-location and by inference track the user's movements. Financial information, social media personal information [22], on-line passwords and internet history can all be removed, without detection, from the targeted devices.

Using tools such as Airgap would allow an investigator to decode the voice package of a GSM packet and listen to a recorded conversation that has occurred in real-time. Currently such conversations would have to be isolated in packets as part of a network packet analysis and then saved in a specific format, decoded, reassembled and played as a .wav file for example.

Additionally, the research question was posed as to how would GSM ciphering streams work or be disabled in such a solution? This was easily dealt with by simply removing any ciphering across the network. There was no indication provided to the user that the data they were sending (SMS, urls etc.) was in any way unsecure. Furthermore the choice of using a Regular Expression or the whitelisting of IMSI means that an investigator can guarantee the security of the interception baste station as no "unauthorised" devices will be able to connect to it. As part of the IMSI catching functionality of the test network any device that connected to and roamed on the test network had its IMSI recorded. In this way any activity that device carried out on the network was recorded. Furthermore, the ability to only allow specific IMSI's to connect to the network illustrated that the targeted element of the test network could be used in a more focused manner by law enforcement—by only allowing a targeted device to discover and connect to the network.

In the case of proprietary ciphers such as Skype, such data would still be encrypted and would require rainbow tables to break the code or assistance from the service provider. It is possible with a large enough hard drive (c. 1 TB) containing the full rainbow table connected to the test network to decrypt the Skype cipher. However, as Skype use SQL databases, it may be more practical to seize the devices that are targeted to retrieve information in the short term. As part of this test network the functionality would have to be more or less seamless and in particular, given the preponderance of smart devices, data connectivity was a point that was focused upon. If a user can still have reasonable download speeds, would their user experience be in any way diminished if migrated to the test network? The answer we found was "No". What the Skype, eBay and Leap Card testing did demonstrate is, that with a sufficiently fast internet backhaul connection working with the Pi3/BladeRF combination that a fast data connection can be provided over GPRS that supports LTE functionality. This real world experience does contradict what other researchers [23] experienced with a USRP N210. The modern BladeRF is able to support concurrent calls, GPRS and SMS on at least 4 devices from the experiment. The tests illustrated that Skype chats could be carried out with no noticeable degradation in call quality, messages were sent and received in real-time and users were able to send iMessages on Apple devices without issue. We would suggest that the internet backhaul connection is a crucial aspect of this functionality as we used a high-speed line to run tests which provided for fast GPRS speeds.

The Leap card application demonstrated that it is possible to use an application that connects to a payment back-end and process credit updates while using the test network. For a target device that is roaming on the test network there was no noticeable degradation in the ability to have full application functionality.

A target device's internet search history and EBay history could be extracted in real-time from a device that was connected to the test network. This demonstrated that law enforcement can view on-line purchases and search history on a device, even when the device is not physically in the investigator's possession.

# 7    Conclusion and Future Work

This paper illustrates that it is possible to build a relatively inexpensive solution for extracting data from a 4G connected mobile device. The use of 2G and 2.5G in the test network and the control that law enforcement gains with having its own solution to extract data in real-time from a 4G device. The ability to run smartphone applications over GPRS for three different mobile operating systems indicates that the user experience would not be unduly affected or interfered with to an extent that the user would become aware of a change in their network connectivity.

Further work could be carried out on decoding the ciphers used in GSM voice calls, in particular, so that this functionality could be added to Wireshark or a Bash script where the hex code of the recorded call can be recorded, downloaded, decrypted and saved as a .wav file in real-time (or as near as practical) to allow an investigator to know what a user is saying on a targeted device. In tandem with the clear SMS texts, HTTP requests and URL of browsing history can provide real-time intelligence. Such a solution could be portable and effectively provide law enforcement with a mobile interception tool that can scan, capture and decode any information emanating from a device that has connected to the test network. We also look at the extension of our approach to examine the security of two-way radio communication services [24] as well as of the 5G networks.

# References

1. Bhandari, S., et al.: Android inter-app communication threats and detection techniques. Comput. Secur. **70**, 392–421 (2017)
2. Sgaras, C., Kechadi, T., Le-Khac, N.-A.: Forensics acquisition and analysis of instant messaging and VoIP applications. Lect. Notes Comput. Sci. **8915**, 188–199 (2015). https://doi.org/10.1007/978-3-319-20125-2_16
3. Faheem, M., Le-Khac, N.-A., Kechadi, T.: Smartphone forensics analysis: a case study for obtaining root access of an android samsung S3 device and analyse the image without an expensive commercial tool. J. Inf. Secur. **5**(3), 83–90 (2014). https://doi.org/10.4236/jis.2014.53009
4. Akhtyamov, et al.: An implementation of software defined radios for federated aerospace networks. Acta Astronaut. **123**, 470–478 (2016)
5. Ramsey, B.W., et al.: Wireless infrastructure protection using low-cost radio frequency fingerprinting receivers. Int. J. Crit. Infrastruct. Prot. **8**, 27–39 (2015)
6. Balogun, A., Zhu, S.Y.: Privacy impacts of data encryption on the efficiency of digital forensics technology. Int. J. Adv. Comput. Sci. Appl. (IJACSA) **4**(5), 36–40 (2013)
7. Gold, S.: Cracking wireless networks. Netw. Secur. Mag. (November 2011)
8. Islam, S., Ajmal, F.: Developing and implementing encryption algorithm for addressing GSM security issues. In: IEEE International Conference on Emerging Technologies. Islamabad, Pakistan (October 2009)
9. Hossain, A., et al.: A proposal for enhancing the security system of short message service in GSM. In: 2nd International Conference on Anti-counterfeiting, Security and Identification. Guiyang, China (November 2008)
10. Shan, H., Zhang, W.: Forcing a targeted LTE cellphone into unsafe network, DEFCON 24. https://www.defcon.org/html/defcon-24/dc-24-speakers.html#Shan

11. Hussain, S., et al.: LTEInspector: a systematic approach for adversarial testing of 4G LTE. In: Network and Distributed System Security Symposium. CA, USA (February 2018)
12. Shaik, A., et al.: Practical attacks against privacy and availability in 4G/LTE mobile communication systems. arXiv preprint arXiv:1510.07563 (2015)
13. Amine, F., et al.: Security for 4G and 5G cellular networks: a survey of existing authentication and privacy-preserving schemes. J. Netw. Comput. Appl. 101(1), 55–82 (2018)
14. Rahmann, M.G., Imai, H.: Security in wireless communication. Wireless Pers. Commun. 22, 213–228 (2002)
15. Tas, I.M., et al.: Novel session initiation protocol-based distributed denial-of-service attacks and effective defense strategies. Comput. Secur. 63, 29–44 (2016)
16. Tsolkas, D., Passas, N., Merakos, L.: Device discovery in LTE networks: a radio access perspective. Comput. Netw. 106, 245–259 (2016)
17. Doppler, K., Ribeiro, C.B., Kneckt, J.: Advances in D2D communications: energy efficient service and device discovery radio. In: 2nd International Wireless conference (Wireless VITAE). Chennai, India (February 2011)
18. Xenakis, C., Ntantogian, C.: An advanced persistent threat in 3G networks: attacking the home network from roaming networks. Comput. Secur. 40, 84–94 (2014)
19. Dabrowski, A., et al.: IMSI-catch me if you can: IMSI-catcher-catchers. In: 30th Annual Computer Security Applications Conference. LA, USA (December 2014)
20. Fajkus, M., et al.: Speech quality measurement of GSM infrastructure built on USRP N210 and OpenBTS project. Infor. Commun. Technol. Serv. 12(4) (2014)
21. Chow, S.S.M.: Removing escrow from identity-based encryption, public key cryptography–PKC 2009, pp. 256–276. Springer, Berlin/Heidelberg (2009)
22. Thantilage, R., Le Khac, N.A.: Framework for the retrieval of social media and instant messaging evidence from volatile memory. In: 18th IEEE International Conference on Trust, Security and Privacy in Computing and Communications (Trustcom 2019) (2019). https://doi.org/10.1109/trustcom/bigdatase.2019.00070
23. Chen, G., Rahman, F.: Analyzing privacy designs of mobile social networking applications. In: Proceedings of 2008 IEEE/IFIP International Conference on Embedded and Ubiquitous Computing. Washington, DC, USA (2008)
24. Kouwen, A., Scanlon, M., Choo, K.-K.R., Le-Khac, N.-A.: Digital forensic investigation of two-way radio communication equipment and services. Digit. Invest. 26(1), S77–S86 (2018). https://doi.org/10.1016/j.diin.2018.04.007

**Neil Redmond** received MSc in Forensics and Cybercrime Investigation from University College Dublin 2017. He also received a Masters in Business Administration in 2004 and a Bachelor of Engineering in Electronic Engineering in 1997, both from Dublin City University. He/ is currently a Senior Manager for Cyber Security at the Deloitte Ireland LLP. His research interests include telecom and network vulnerabilities and security strategy for IT and telephony networks.

**Le-Nam Tran** received the B.S. degree in electrical engineering from Ho Chi Minh City University of Technology, Ho Chi Minh City, Vietnam, in 2003 and the M.S. and Ph.D. degrees in radio engineering from Kyung Hee University, Seoul, Korea, in 2006 and 2009, respectively. He is currently a Lecturer/Assistant Professor with the School of Electrical and Electronic Engineering, University College Dublin, Ireland. His research interests are primarily on applications of optimization techniques on wireless communications design. Some recent particular topics include energy-efficient communications, cloud radio access networks, massive MIMO, and full-duplex transmission. He has authored or co-authored in some 80 papers published in international journals and conference proceedings. Dr. Tran is an Associate Editor of EURASIP Journal on Wireless Communications and Net-working. He was Symposium Co-Chair of Cognitive Computing and Networking Symposium of International Conference on Computing, Networking and Communication (ICNC 2016). He is a reviewer for many top-tier international journals on signal processing and wireless communications, and has also been a Technical Programme Committee Member for several flag-ship international conferences in the related fields.

**Kim-Kwang Raymond Choo** received the Ph.D. in Information Security in 2006 from the Queensland University of Technology, Australia. He currently holds the Cloud Technology Endowed Professorship at the University of Texas at San Antonio (UTSA). In 2015, he and his team won the Digital Forensics Research Challenge organized by the Germany's University of Erlangen-Nuremberg. He is the recipient of the 2019 IEEE Technical Committee on Scalable Computing (TCSC) Award for Excellence in Scalable Computing (Middle Career Researcher), 2018 UTSA College of Business Col. Jean Piccione and Lt. Col. Philip Piccione Endowed Research Award for Tenured Faculty, British Computer Society's 2019 Wilkes Award Runner-up, 2019 EURASIP Journal on Wireless Communications and Networking (JWCN) Best Paper Award, Korea Information Processing Society's Journal of Information Processing Systems (JIPS) Survey Paper Award (Gold) 2019, IEEE Blockchain 2019 Outstanding Paper Award, Inscrypt 2019 Best Student Paper Award, IEEE TrustCom 2018 Best Paper Award, ESORICS 2015 Best Research Paper Award, 2014 Highly Commended Award by the Australia New Zealand Policing Advisory Agency, Fulbright Scholarship in 2009, 2008 Australia Day Achievement Medallion, and British Computer Society's Wilkes Award in 2008. He is also a fellow of the Australian Computer Society, an IEEE senior member, and co-chair of IEEE Multimedia Communications Technical Committee's Digital Rights Management for Multimedia Interest Group.

**Nhien-An Le-Khac** is a lecturer at the School of Computer Science (CS), University College Dublin (UCD), Ireland. He is currently the program director of MSc program in Forensic Computing and Cybercrime Investigation (FCCI)—an international program for the law enforcement officers specializing in cybercrime investigations. He is also the co-founder of UCD-GNECB Postgraduate Certificate in fraud and e-crime investigation. Since 2008, he is a research fellow in Citibank, Ireland (Citi). He obtained his Ph.D. in Computer Science in 2006 at the Institut National Polytechnique de Grenoble (INPG), France. His research interest spans the area of Cybersecurity and Digital Forensics, Data Mining/Distributed Data Mining for Security, and Fraud and Criminal Detection. Since 2013, he has collaborated on many research projects as a principal/co-PI/funded investigator. He has published more than 150 scientific papers in peer-reviewed journal and conferences in related research fields, and his recent edited book has been listed the Best New Digital Forensics Book according to the Book Authority.

# Towards an Automated Process to Categorise Tor's Hidden Services

Andrew Kinder, Kim-Kwang Raymond Choo, and Nhien-An Le-Khac

**Abstract** It has been argued that the anonymity the dark web offers has allowed criminals to use it to run a range of criminal enterprises, acting with impunity and beyond the reach of law enforcement. By designing a process that can identify sites based on their criminality, law enforcement officers can devote their resources to finding the people behind the sites, rather than having to spend time identifying the sites themselves. The scope of the study in this chapter is focused solely on Tor's hidden services. The research problem was to identify what percentage of hidden services are accessible and how many of these are connected to criminal/illicit activities. Additionally, our research also aims to determine if it is possible to automate a system to identify sites of interest for law enforcement by categorising them based on the prevalent crime type of the hidden service. In this chapter, we look at how hidden services are set up. To facilitate this, an experiment was conducted where a hidden service was set up and hosted on the Tor network. It is connected to the Tor network and obtained an un-attributable IP address, identified over 12,800 .onion addresses from which it scraped the HTML from the home page, before checking this against a pre-determined list of keywords to identify illicit sites and categorise each of these dependant on their type of criminality. Our approach successfully identified criminal sites without the need for human interaction making it a very useful triage solution. Whilst further work is required before its categorisation process is sufficiently robust enough to provide an accurate, unquestionable strategic overview of hidden services, the tool in essence, works very well in achieving its primary function; to identify criminal sites across the dark web.

**Keywords** Tor network · Dark web investigation · Forensic process

A. Kinder (✉)
National Crime Agency, London, UK
e-mail: andrewkinder@hotmail.com

K.-K. R. Choo
University of Texas at San Antonio, San Antonio, USA

N.-A. Le-Khac
University College Dublin, Dublin, Ireland

© The Editor(s) (if applicable) and The Author(s), under exclusive license to Springer Nature Switzerland AG 2020
N.-A. Le-Khac and K.-K. R. Choo (eds.), *Cyber and Digital Forensic Investigations*, Studies in Big Data 74, https://doi.org/10.1007/978-3-030-47131-6_10

221

# 1  Introduction

By its very name, the dark web invokes images of a shadowy under world, lurking on mysterious servers, hidden from law enforcement. Whilst it is argued that the actual main goal of networks like The Onion Router (Tor) was to enable freedom of speech in countries that employ political censorship [1], it is believed to have degenerated into a rough, virtual neighbourhood in need of extra police patrols, where drugs, guns, indecent images of children (IIOC), money laundering and other illicit services are traded. In 2015 the Australian television news program, 60 min, described the dark web as a "vast, secret, cyber underworld that accounts for 90% of the internet" [2] and some recent reports have stated that its notoriety and popularity is such, that drug dealing is moving away from the streets and onto the dark web [3].

Whilst the '90% of the internet' claim is thought to be wildly inaccurate, (other studies place the figure at, around 30,000 live sites at any one time [4]) it is the case that monitoring the dark web can be an arduous task. No central entity or authority stores the entire list of onion addresses and as hidden services rarely link to one another, traditional web crawling has limited use [5]. As such, there are many sites that have to be identified, and then trawled through in order to establish which ones are worthy of further investigation, as opposed to ones advocating free speech, the promotion of human rights, whistle blowing, or simply pictures of cats.

In the context of digital forensics, one of the research challenges today is how to quickly scrap across a large number of published hidden services, collect and store the html from the site and then automatically assess the sites against a list of key words to place them into certain categories dependent on the type of criminality the sites are involved in. The idea was not to spend months or even weeks analysing data, but to design it so that it can be each week the results available to be examined in a short space of time.

Hence, in this chapter we present an approach that can be used to automatically collect (scrape) data from dark web sites, categorise them into different criminal categories and identify those sites that may potentially be of interest to law enforcement. It will also go some way to estimating the true size of the dark web, as well as its make-up, to determine what percentage of the dark web is associated with criminal activity. In addition, an experiment was conducted where a hidden service was set up and hosted on the Tor network.

# 2  Literature Survey

Whilst entire books have been devoted to writing scripts to scrape websites on the clear web [6], there is limited work on similar techniques for collecting data from the dark web. To date, published research on the dark web has primarily focused on deanonymising users of Tor as opposed to carrying out content analysis of the hidden services themselves.

A number of studies focused on analysing network traffic by setting up Tor nodes and examining the requests that pass through a modified router to determine, amongst other things, which ports Tor uses and the geographic makeup that the Tor network utilises [7]. No effort was made to analyse the content of hidden services or even identify them from the network analysis.

A similar study [8] also analysed traffic over the Tor network. They found that the most popular use of Tor was to download films and TV programmes using the BitTorrent protocol. Again, no effort was made to investigate hidden services themselves.

One study that did look at hidden services, or at least one of them, focused on the infamous "dark market" Silk Road [9]. This study carried out daily scrapes of the market place and analysed the results. By incorporating the authentication cookie from a valid log on into the scraping tool, the collection process could be automated. The site was then 'crawled' using the software HTTrack which took around 48 h to download 244 MB of data. (Further crawls took on average 14 h to complete). The benefits of this method of collection are that it generates a rich data source, at least in relation to the Silk Road marketplace. The downside is that it does take a long time, the activity may draw suspicions from the Silk Road administrators and that despite many hours of crawling, data from only one hidden service has been obtained. The analysis carried out by the researchers was interesting in that despite offering a wide range of illicit services, illegal narcotics were by far the most popular commodity traded with an estimated USD 1.2 million a month across all sellers. It will be interesting to see if this trend is repeated across illicit content across the dark web itself.

One study that purported to identify hidden services was conducted by Zander and Murdoch [10] and built on earlier work completed by [11]. This focused on revealing hidden services by correlating clock skew changes with times of increased load (and therefore temperature). This study didn't focus on identifying unknown hidden services, rather the servers on which known hidden services were run. Whilst undoubtedly a novel and useful approach that has potential in identifying the people behind certain hidden services; it did not focus on finding new services or categorising them.

Whilst it can be argued that there are not many studies relating to categorising the content of hidden services, there are certainly some. One study [12] looked at carrying out attacks against a vulnerability they had discovered, which they then used to determine how much traffic went to hidden services. They used this to rate the sites popularity, explore opportunities to potentially take the site down and to deanonymise them, i.e. locate the servers that hosted them. They did this by setting up a number of Tor nodes to act as Hidden Service directories, and analysing the traffic that passed through them. Additionally, they also managed to obtain the Hidden Service descriptors for all hidden services. At the time of the study this was around 1200. Whilst this is useful in that the .onion address can be obtained from the hidden service descriptors, the content relating to the sites cannot.

Further work building on the above study was carried out by [5]. In addition to gauging the popularity of individual hidden services, this study looked to analyse the content of the sites themselves. Using the techniques described in their previous study [12] identified 39,824 unique hidden services. This, they state, was a great deal more than the 1657 sites advertised at the time on the hidden wiki and ahima.fi indexing sites, adding strength to the argument that the advertised sites do not accurately portray the true size of the dark web.

With their 39,824 onion addresses, [5] carried out research to determine the makeup of the sites. Hidden service descriptors were identified for 24,511 of the sites, of which only 22,007 had open ports. The majority of these (13,854) were linked to a botnet (Skynet) running on port 55,020, leaving 8153 hidden services that could potentially be connected to. From these, successful connections were made to 6579 sites. This is only 16% of the headline figure of 39,824 but still more than advertised on the hidden wiki and ahima.fi combined.

After an element of data manipulation and processing, including removing all non-English language sites, only 1813 of the 6579 sites contained enough data to enable them to be categorised. Categorisation was done using the software "Mallet" and the web service "uClassify". These are freely available tools that allow for the automated classification of text (provided a degree of pre-processing has been carried out beforehand). The study identified eighteen separate categories, the two most popular being 'adult' and 'drugs.' The research stated that those sites devoted to drugs, counterfeit products, weapons or adult content accounted for 44% of the hidden services they were able to categorise, the other 56% cover a range of topics. Interestingly, one of these topics, 'services', accounted for 4% of the total sites and included services such as legitimate escrow services as well as less legitimate services such as hiring a hitman or money launderer. It is unclear why these criminal and non-criminal services were grouped together in this way.

Whilst the study is valuable in that it is one of the first to attempt to classify the content of hidden services, it does have a number of limitations. Firstly, the number of hidden services that the study was able to categorise appears a little low. Secondly, there is no mention of exactly how the categorisation process took place. From research carried out, the classification tools used require the raw data (the scraped HTML) to undergo a thorough element of pre-processing before it can automatically start to identify topics within the data. This increases the time the entire process takes to complete. Whilst this may be sufficient for the research project it was designed for, it may not be suitable for carrying out weekly or more frequent scrapes that then categorise the data 'on the fly.' Finally, it is perplexing that IIOC does not appear to get a mention at all with none of the sites appearing to contain any.

A similar study also carried out 'attacks' on the Tor network to identify hidden services and then to categorise them [13]. This study obtained a copy of Tors source code (being an open source project, it is publically available) and, after setting up a Tor node, adapted their Tor client to log all publication requests from hidden services, as well as requests from clients for a copy of the descriptor. By matching up the two, the researchers were able to determine the .onion addresses of the requested hidden services and the level of popularity on any given day. Over the course of the study (six months), this method identified around 80,000 hidden services. Although as authors claim, many of these were short lived and only seen briefly.

As well as identifying the .onion addresses of hidden services, the study also went some way to categorise the content found within them. The study spent a month crawling across a small dataset (nfd) of hidden services and scraped the html content from them. This was then stored in a database to be reviewed and manually classified. It was thought that automatic classifiers would be, "insufficient due to difficulty in interpreting context and meaning". Savage and Owen [13] additionally, they found that it was difficult to differentiate between legal and illegal content due to intricacies and differences between legal jurisdictions. That aside, they did find that the majority of sites were, "of questionable morality/legality". They found that although there were more drugs sites than any other (followed by market and fraud sites) it was the abuse sites that were the most popular and received the most traffic. They also discovered that in the vast majority of cases, no effort was made to hide child abuse sites. They argue that the anonymity Tor offers gave web-masters the confidence to openly advertise sites containing Child Sexual Exploitation and Abuse (CSEA) material.

The main limitations with this study and its methodology lie with the manual classification. Whilst it may be argued that manual classification is more accurate than automatic methods, it is simply not feasible to carry this out on a large data set with meaningful timescales. To carry out such work on a regular basis will require some form of automation to ensure the results produced are an accurate representation at that point in time and not as they were weeks or months before.

Other researchers have noticed that there has been little in the way of previous studies designed to analyse the content of hidden services. Guitton [14] carried out a study to do just that. He collected and categorised .onion addresses from three indexing sites (Hidden Wiki, Snapp BBS and Ahmia.fi) in order to test two hypotheses; "Content on hidden services relates to challenging the state authority in order to establish a democratic order" or "Content on hidden services challenges ethics". Categorising the sites was done solely by identifying the description of the site on the hidden service indexing page and then manually checking (in all but the CSEA sites). As mentioned above, this method will not be feasible for large data sets. In this instance, there were only 1171 individual entries found across the three sites. The results of the study found that the 1171 sites could be categorised across 23 categories, both "ethical" and "unethical". It was found that 45% of all listed services were what the researcher deemed to be in the unethical category, with 206 CSEA sites making up the highest percentage (18%). Surprisingly only 50 (4%) drugs sites were found, although there were 70 (6%) 'Black Market Sites' which may also be linked to drugs. The author felt that whilst 55% of hidden services are what he deems "ethical services," the unethical services are so unsettling "efforts should be made to stop the development of hidden services".

Leaving the ethical question aside, the study does utilise a simple way of obtaining lists of onion addresses, that being to use the data already available in the form of indexing sites. However, the method of categorisation is again labour intensive and impractical on a medium to large scale.

One study did use automated processes to categorise known hidden service sites as well as carry out linguistic analysis on them. The goal was to provide, "an up-to-date model of (Tor's) thematic organisation and linguistic diversity" [1]. A crawler was used that downloaded the html from hidden services webpages. It was thought that it was of greater benefit to get a little bit of data from a large number of hidden services as opposed to a lot of data from a few services. There is no discussion as to where they obtained the list of onion addresses from.

Prior to analysing the content of the hidden services, an amount of pre-processing took place including extracting the text from the html code. The subsequent analysis found that 83.27% of the content was in the English language. It was noted that this is similar to the percentage share the English language had in relation to clear websites in the late 1990s, prior to the world-wide popularisation of the internet [1]. Of interest is that although over 7000 sites were identified, the majority of these were unreachable with less than 1500 online at any one time.

Of the sites that they were able to access, a combination of an automated and manual categorisation approach was adopted. This used the Mallet software as used previously by [5]. Mallet identified topics based on common used words which the researchers then manually classified into categories. They found that the majority of hidden services "exhibit illegal or at least controversial content" with 60% of hidden services devoted to trading in drugs, weapons, counterfeit money/documents, stolen credit cards and hacked accounts and around 8% that contained CSEA material. They concede that there were some hidden services with 'noble intentions' but it was very difficult to divide the hidden services they found into good and bad.

As with other studies mentioned previously, the classification technique may not be suitable for quick time analysis of large data sets. Additionally, it has yet to be used on un-seen data to test it out. However, scraping only a small amount of data (the html) from the hidden services lends itself well to being carried out across a large number of sites.

A number of commercial companies have carried out their own dark web research and published their findings. One such study [4] carried out automated categorisation of 29,532 hidden services. They manually classified 1000 sites and then used this data to train their automated process. Their results showed that of the sites they discovered, more than half (54%) were inaccessible. Of those that were inaccessible 48% were classified as illegal under UK and US law. The two biggest categories were file sharing (29%) and leaked data (28%). Drug sites made up only 4% of the sites with weapons a mere 0.3%.

Another study offered to "look inside the internet's massive black box" [15]. However, this was simply a bland report on how to access Tor with no effort made to categorise the content within it. A comprehensive report carried out by Trend Micro makes efforts to determine "what bad stuff goes on in the deep web [16]. Whilst they identify a number of illicit services, such as assassination, drugs, malware and money laundering, there is no real breakdown on the number and therefore makeup of these sites, nor the methodology deployed to obtain the data. That aside, it is a good introduction as to how the dark web works and what sort of content is on it.

In the context of privacy-preserving web browsers, authors in [17, 18] researched particular Web browsers found out that they still leave evidence in RAM or on the local drive about the user's browsing activities, like keyword searches, websites visited, and viewed pictures. This research is not in the context of dark web but it could lead to the development of forensic acquisition techniques of artefacts from dark web.

# 3 Problem Statement

As can be seen in Sect. 2, there have been a small number of studies carried out devoted to understanding the actual content of hidden services. This study looks to build on these and develop an automated process that goes someway to identifying and categorising criminal sites. The aim is to help provide a strategic overview of the type of criminality that exists on the dark web, but also to act as a form of automatic triage service in order to identify the sites that may require further investigation by law enforcement.

The problems/questions that need to be overcome/answered in order to do that are as follows:

1. What percentage of hidden services are accessible and how many of these are connected to criminal/illicit activities?
2. Is it possible to automate a system to identify sites of interest for law enforcement based on the prevalent crime type of the hidden service?

The answer to the first question can be used to gather a strategic oversight of the content on the dark web as it will provide valuable data as to the makeup of it. In order to determine whether a site is criminal or illicit, it will first have to be assessed to see if it falls into one of a number of crime types. Therefore, as well as showing how many sites of interest there are, it will also be possible to break down the 'bad' sites into crime categories and determine which are the most prevalent on the net. This strategic data can then be further interrogated in order to answer the second question which focuses on a more tactical use of the tool.

The process, which has been developed as a forensic tool, has been designed to identify all illicit/criminal hidden services. Depending on the strategy of the Law Enforcement Agency using the tool, any particular crime category can be focused on, with all the sites containing 'hits' analysed further. By automating this service, the initial analysis can be run during on a regular basis with the results immediately available for review once complete. These can then be used by a law enforcement unit to help prioritise their work.

There are other benefits to carrying out the research this way. Over time, as each weekly scrape is stored, a historic database of hidden services content will be gathered. This will be a valuable, rich source of data to analyse when looking at the evolution of any particular site. Additionally, by saving the whole HTML, and not parsing the keywords out when initially carrying out the scrape means that custom key word searches can be ran across the database. This can then be used to react to any given situation that requires searching across dark websites. Examples include looking for discussions about recent data leaks, or trying to identify a vulnerable child. As such, it is believed that this research offers a unique insight into the dark web and provides investigators with valuable data to assist in the policing of it.

# 4   Adopted Approach

In order to answer the research problem posed above, our research was first carried out on how hidden services are set up. To facilitate this, an experiment was conducted where a hidden service was set up and hosted on the Tor network. Initially, no efforts were made to advertise, publicise, or share the .onion address. After a period of time, the .onion address was posted onto a paste-bin site and onto the indexing site, 'ahmia.fi.' The access log was then examined to gauge the level of visitors both before and after the site was 'made public'.

A process, which was then developed as a forensic tool, has the following steps:

1. Connected to the Tor network and obtained an IP address.
2. Navigated to three .onion sites that list hidden services, normalised these, removed the duplicates then placed the list of addresses (over 12,800) into a SQLite database.
3. Iterated through each entry in the database in order to scrape the html from the home page of each .onion address. Again, all this was done through Tor.
4. Checked the html data from each site against a pre-determined list of keywords to identify sites depending on type of criminality.

## 4.1   Setting up the Hidden Service

In order to establish how .onion sites operate, a hidden service was set up. A benign, non-criminal theme was chosen that may pique some interest, but remain firmly on the right side of the law, if not the force! In order to set this up, an apache web server was installed on a virtual machine running Ubuntu.

A very simple page (index.html) was designed in HTML and this along with the media associated with it (a jpeg) was placed in the /var/www/html folder. By navigating to via a browser to the local home page (127.0.0.1) the web page was observed to be displaying correctly

Tor was installed onto the machine and the configuration file (Torc) altered to contain the following lines [19]:

```
HiddenServiceDir/home/hs/hidden_service/
HiddenServicePort 80 127.0.0.1:80.
```

By restarting Tor via the command line, two files were created in the folder mentioned above. One of these files, (private_key) was the private key associated to the .onion site) whilst the other (hostname) contained the .onion address itself; http:// ikwbuu3dlw4midsb.onion.

To test the site, a number of visits were made on a separate, windows computer running the Tor browser, the last of these visits was on 09/11/2016. A screenshot of the site accessed through Tor can be seen in Fig. 1.

The site was left up for five days with no advertisement or sharing of the .onion address. After this time, the apache access logs were examined. At this time, these showed that the last recorded visit to the site took place at 0753 h on 09/11/2016. This was the visit that was carried out as part of this study to determine if the site was up and running as mentioned above. There were no other visitors to the site in these five days. At 08:23 h on 14th November, a message was posted on the sharing site, Pastebin, advertising the hidden service as a new source of Star Wars Episode VIII gossip, news and leaks (see Fig. 2). The onion address was also added to the ahmia.fi indexing site using their add service function.

The access log files of the apache server running the hidden service were examined again. These showed that around 20 min after the site was advertised, the first visitors arrived. Within 60 min a handful had a look at the site, some improvement on the

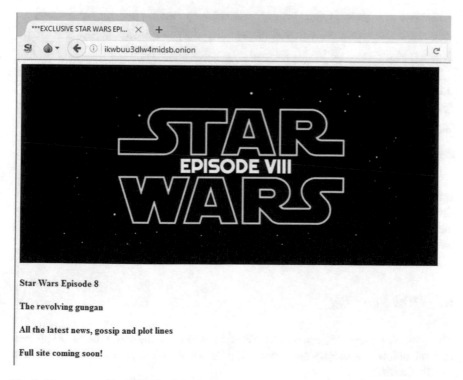

**Fig. 1** Home page of http://ikwbuu3dlw4midsb.onion

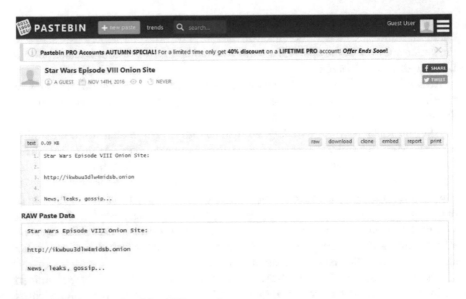

**Fig. 2** Pastebin post advertising hidden service

```
127.0.0.1 - - [09/Nov/2016:07:53:52 +0000] "GET /favicon.ico HTTP/1.1"
404  514  "-"  "Mozilla/5.0  (Windows  NT  6.1;  rv:45.0)  Gecko/20100101
Firefox/45.0"

127.0.0.1 - - [14/Nov/2016:08:40:23 +0000] "GET / HTTP/1.1" 200 604 "-"
"Mozilla/5.0 (compatible, MSIE 11, Windows NT 6.3; Trident/7.0; rv:11.0)
like Gecko"
```

**Fig. 3** Apache access log for hidden service

zero visitors the site had prior to the address being shared with others. A snap shot showing the 5 day gap can be seen in Fig. 3. This experiment strongly supports the argument that without advertising the site, hidden services remain hidden, thus making the indexing and trawling of them all the more difficult.

## 4.2 Web Scraping

There were a number of component parts and functions to the script that are explained in detail below. The script was written using Python 2.7 on a machine running Ubuntu v16.04 (xenial).

### 4.2.1 Connecting to Tor

In order to connect to Tor, the configuration file was amended to tell the ControlPort listener to listen on port 9051 to any communications to the Tor controller. It is used to connect to Tor and obtain an IP address, in order to access the .onion sites, the script also needs to ensure that all the requests it makes are routed through Tor via a proxy. Tor itself does not act as a proxy so in addition to installing Tor, Privoxy was also installed. The Privoxy configuration file was manipulated to route all traffic through the SOCKS servers at localhost port 9050. This ensures that any time this proxy is used, it will use the Tor network [20]. In order to do this, a function called 'request' was written (see Fig. 4).

The first element of this function is to add http://to the start of the URL if it wasn't already there. By default, Privoxy listens on port 8118, and then forwards the traffic to port 9050, port 9050 being the port the Tor socks listens on [20]. By using the proxy handler function of the URLlib2 python library to send traffic by the local host on port 8118, all traffic is subsequently routed via Privoxy onto the Tor network. This ensures that every time this function is called, the URL that forms part of the argument is visited over the Tor network.

In order to optimise the tool, a timeout of 20 s was added. This increased the performance of the script dramatically. Hidden services are notorious for being unstable and as the studies above have shown, a large number of them can be down at any one time. As such, without a specified timeout, the script can wait for an answer from the hidden service for a long amount of time. Research carried out on a sample size of 1000 estimated that to go through 12,000 .onion addresses would take almost a week, this is clearly not acceptable for such a tool, so the time out was added.

```
# request a URL
def request(url):
    #prefix with http:// if not already.
    if not "http://" in url:
        url = "http://" + url
    # communicate with TOR via a local proxy (privoxy)
    def _set_urlproxy():
        proxy_support = urllib2.ProxyHandler({"http" : "127.0.0.1:8118"})
        opener = urllib2.build_opener(proxy_support)
        urllib2.install_opener(opener)

    # request a URL via the proxy
    _set_urlproxy()
    request=urllib2.Request(url, None, headers)
    return urllib2.urlopen((request),timeout=20).read()
```

**Fig. 4** 'request' function

### 4.2.2 Obtaining Onion Addresses

As mentioned above, there is no central DNS repository, nor a google service that indexes and creates directories of .onion addresses. This makes detailing them difficult. That aside, there are a number of sites on the dark and clear web that serve up lists of onion addresses. These include:

Hidden Wiki  http://gxamjbnu7uknahng.onion
VisiTor      http://visitorfi5kl7q7i.onion
Harry71      http://skunksworkedp2cg.onion
Ahima       http://msydqstlz2kzerdg.onion

Additionally, links and lists of onion addresses are often posted on PasteBin, as indeed was the URL for the hidden service created as part of this research. Of the indexing sites listed above, VisiTor, Ahima and Harry71 all have a single page where all the .onion addresses are listed, the Hidden Wiki does not. As such the research focused on the sites with the single index page as the source of its onions.

In order to facilitate extracting the .onion addresses from the sites, a regular expression for an onion address was created. This can be seen in Fig. 5. Onion addresses are 16 character alphanumeric strings containing any lowercase letter a–z and any digit between 2 and 7 that represents the first eighty bits of a SHA1 hash of the identity key for the hidden service, encoded in base32 [21].

The reason a regular expression was used was down to the inconsistencies with the way the indexing sites displayed their lists of onions. Some prefixed with the http:// whilst others did not. Some placed a/at the end of the URL whist others did not. By just grabbing the <address>.onion from all three sites, identifying unique (and conversely repeat) URLs was much more straightforward.

Figure 6 shows the code that was written to obtain the list of onion addresses. It visited each site and returned the content of each site (in this case a text list of onion addresses). Error handling was built into handle a situation where a site may be down. If so, a string of 'No Onions' was returned and a message printed to the screen informing the user that the site was not reachable. The onion_regex as defined in Fig. 5 was then used to extract all .onion addresses into individual lists. These were then combined using the .extend command and the set command was used to produce one definitive list of unique .onion addresses. This list was then simply written to a temporary text file called index.txt.

```
#Define the regex for an onion address for later useage
onion_regex = "([a-z2-7]{16}.[onion]{5})"
```

**Fig. 5** Regular expression for a.onion URL

```
#Visit index sites and download onion addresses:
print ("Obtaining Onion Addresses from indexing sites...")
try:
    Raw_Onion_List_visitor = request("http://visitorfi5kl7q7i.onion/onions")
except Exception as e:
    print "Cannot connect to Visitor indexing site"
    Raw_Onion_List_visitor = str("No onions")
try:
    Raw_Onion_List_Harry71 = request("http://skunksworkedp2cg.onion/sites.txt")
except Exception as e:
    print "Cannot connect to Harry 71 indexing site"
    Raw_Onion_List_Harry71 = str("No onions")
try:
    Raw_Onion_List_ahmia = request("http://msydqstlz2kzerdg.onion/onions/")
except Exception as e:
    print "Cannot connect to Ahmia indexing site"
    Raw_Onion_List_ahmia = str("No onions")

Raw_Onion_List = re.findall(onion_regex, Raw_Onion_List_visitor)
Raw_Onion_List_2 = re.findall(onion_regex, Raw_Onion_List_Harry71)
Raw_Onion_List_3 = re.findall(onion_regex, Raw_Onion_List_ahmia)

Raw_Onion_List.extend(Raw_Onion_List_2)
Raw_Onion_List.extend(Raw_Onion_List_3)
Raw_Onion_List = list(set(Raw_Onion_List))
```

**Fig. 6** Gathering onions

### 4.2.3 Database Design

Once the list of .onion addresses was obtained, the next step was to determine where they were to be stored. The storage medium had to have connectivity with the python script and be robust enough to store and retrieve large amount of data. For those reasons and the ease of use it afforded, SQLite was chosen. In order for the script to carry out this function automatically, a data frame was created with just one column, named 'url.' Using the 'read_csv' function of the pandas library, index.txt was read and the .onion addresses appended to the dataframe in the column, 'url.' An empty SQLite database with one table (called onions) was created with the following column headings: index (To keep track of each entry), URL (To store the actual URL), timestamp (Time the URL was scraped), content (To store the content from the site), and <categories>—A series of 8 columns each named after a particular category of hidden service. Namely: marketplace, drugs, weapons, csea, fraud, cyber, pornography and extremism.

The data frame was then loaded to the empty SQL database with the list of URLs populating the URL column in the table. This process was carried out by the function 'loader' as seen in Fig. 7.

```
#Load rows into an SQLite Database from a CSV Data Dump

def loader():

    # Ingest the CSV data file
    onion_df_header_row = ['url']
    onion_df = pd.read_csv('index.txt', engine='python', sep='|', header=None,
                           skip_blank_lines=True,warn_bad_lines=True,
                           names=onion_df_header_row, skiprows=0, encoding='ISO-8859-1')

    # Setup the empty SQLite Database
    conn = sqlite3.connect("index.db")
    curs = conn.cursor()
    curs.execute('CREATE TABLE IF NOT EXISTS onions (\
                     [index]             INTEGER,\
                     url                 TEXT,\
                     timestamp           VARCHAR,\
                     content             TEXT,\
                     marketplace         INTEGER,\
                     drugs               INTEGER,\
                     weapons             INTEGER,\
                     csea                INTEGER,\
                     fraud               INTEGER,\
                     cyber               INTEGER,\
                     pornography         INTEGER,\
                     extremism           INTEGER);')
    conn.commit()

    onion_df.to_sql(con=conn, name='onions', if_exists='append', flavor='sqlite')

    conn.close()
```

**Fig. 7** 'loader' function

### 4.2.4  Scraping the Sites

**Data Collection**

With the database set up and filled with a list of URL's, the next step was to populate it with the content of the sites. In order to facilitate this, a function called 'scrape' was defined. This took one argument, a URL, fed this into the 'request' function then fed the result of that function into 'Beautiful Soup' which parsed the html and returned it in an easy to read format. Beautiful Soup is a python library designed to easily parse data from web scrapes [22]. In this instance only a basic element of the library is being used (prettify) but it has been included and to easily incorporate future modifications/development (Fig. 8).

```
def scrape(url):
    try:
        # prefix with http:// if not already.
        if not "http://" in url:
            url = "http://" + url
        html = request(url)
        soup = BeautifulSoup(html, "lxml")
        return soup.prettify()
```

**Fig. 8** 'scrape' function

```
# Handle errors
except socket.timeout:
    errorlogger('HTTPTimeout: ', url)
    return "Timeout: No Response"
except urllib2.HTTPError, e:
    errorlogger('HTTPError = ' + str(e.code), url)
except urllib2.URLError, e:
    errorlogger('URLError = ' + str(e.reason), url)
except httplib.HTTPException, e:
    errorlogger('HTTPException = ' +str(e), url)
except AttributeError as e:
    return "No attribute found"
except Exception:
    errorlogger('generic exception!', url)
```

**Fig. 9** Error handling in the function **'scrape'**

### Error Handling

As the internet, and the dark web in particular, can be relatively fickle when it comes to connectivity, surfing it can often lead to errors which, left undealt with, could cause the program to crash. Figure 9 details some error handling that was incorporated into the code to deal with these eventualities. This ensures that any foreseeable error is dealt with and the script can continue onto the next URL.

### Harvesting Onions

The next step was to iterate through each .onion address in the database and scrape the html from the home page. Only the home page was chosen as it was thought, as argued by [1] that a little bit of data about as many sites as possible was preferable to a lot of data focused on only a few sites. This 'mile wide, inch deep' approach was determined to be suitable, as once the sites that require further attention have been identified, a more thorough scrape could be carried out.

A for loop was written into the script that ran through each entry in the database and used this URL as the argument for the 'scrape' function as detailed above. This returned the entire HTML of the home page which was stored in the database under the column entitled 'content.' Additionally, a time stamp was incorporated into the loop to ensure that this was an accurate representation of the actual time the data was scraped. As the processes took a number of hours to run, each site was visited at a slightly different time.

## 4.3 Categorisation

### 4.3.1 Key Words

The key to ensuring the data was analysed sufficiently to answer the problem statements accurately, was to ensure the list of keywords were such that they maximised the amount of hits while minimising the number of false positives. As previous studies found that the anonymity hidden services offer means that little or usually no effort is made to obfuscate the content of the site [13], the list of key words can be kept relatively simple. For example, whilst CSEA sites on the clear web may contain obfuscated words such as P.'E.D.'O, on the dark web it is usually just written as 'pedo'. As such, far fewer words were needed to categorise sites. With fewer, less ambiguous words, focused solely on the category, it was thought that the number of false positive results would be reduced.

### 4.3.2  Data Mining

In order to check the key words against the HTML content scraped from the hidden services, two functions were written, 'mine' and 'keyword_scan'.

The 'mine' function makes a soup by parsing the html supplied in the first argument through beautiful soup using lxml. It then uses the find_all function of beautiful soup to look for the string passed to it as the second argument, 'keywords.' By selecting IGNORECASE, it doesn't matter if the text is in upper or lower case. The number of hits are returned and if there are any exceptions, a return of 0 is made. For example if 25 keywords were found, then the return or the sites 'score' for that category would be 25.

The function 'keyword_scan' iterates through each entry in the database and runs the 'mine' function across the previously collected HTML stored in the 'content' column. The three arguments it takes are curs (to identify which database connection it is), keyword, to identify which keyword lists to use and table header, to determine which column to place the results in. The 'keyword_scan' function is then called for each crime category.

## 4.4  Self-regulation

Once the collection and categorisation has completed, the final process the tool does is to create a subfolder in the 'Scripts' folder with the current date and time as the name. It then moves the entire database and the 'errorLog.txt' into this folder. This means that every time the script is run, the results are automatically stored in their own, easily identifiable folder.

To ensure the tool ran on a regular basis, the Linux system tool, 'cron' was used to schedule a regular running of the script. This schedules the script to run each Friday (the 5th day) at 1400 h.

Each hidden service was given a score based on the number of key words that were found in the scraped html. The higher the score, the more key words were found. Rather than pigeon hole sites into distinct categories, the focus was on how many sites had positive scores in any category. The result of this is that any one site may have scores of one or more in all categories and be counted in that categories total.

**Table 1** Performance statistics for onion soup

|  | Scraping | Categorising | Complete process |
|---|---|---|---|
| Total sites: 12,882 | 2 days, 2 h, 2 min | 3 h 49 min | 2 days, 5 h and 51 min |
| Average per site | 14 s | 1 s | 15 s |

The benefit of this is that it potentially gives a more accurate picture of the nature of the hidden services as some of the sites may contain content relating to a range of criminal activities, an example being an online shop that sells drugs, guns and CSEA material.

## 5 Evaluation

### 5.1 Performance

The script started running at 13:10 h on Nov 18th. The last website was scraped at 15:12 h on Nov 20th. The categorisation process finished at 19:01 h the same day. The full metrics can be seen in Table 1.

This goes some way to prove that the tool can run automatically with the results ready for analysis in a matter of days. This answers part of the problem statement regarding the speed of the process. If the categorisation process is a success, then this tool has the potential to be used regularly without the need for weeks of data collection followed by processing and then analysis.

### 5.2 Analysis of Hidden Services

#### 5.2.1 Number of Onions

The tool found 12,882 unique .onion addresses that it tried to scrape the html from. It did this successfully for 4532 sites (35%) as the rest could either not be connected to (7257, 56%) or contained no data (1093, 9%). These statistics can be seen in Fig. 10. This is consistent to some extent with the studies as detailed in Sect. 2, but the figure is considerably more than 1450 which was the "highest number of concurrently online hidden services," as one study found. Spitters et al. [1] Conversely, it is certainly less than the "90% of the internet" as one reporter commented [2].

**Fig. 10** Accessible hidden services

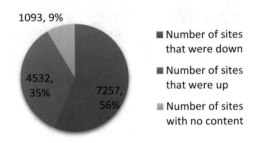

### 5.2.2 Categorisation

**State of Hidden Services**

By using a simple SQL statement, the database was queried to determine how many of the sites had 1 or more, 2 or more and then 3 or more key words associated to each category. The results of this can be seen in Fig. 11.

As can be seen, sites with no key words was the most common, followed by Market Places, Drugs, Fraud, Cyber-Crime, Pornography, Weapons, Extremism and lastly CSEA. These results are similar to the work carried out by Savage and Owen [13] whose top three were Drugs, Markets and Fraud.

As [13] state, it is very difficult to categorise criminal and non-criminal sites. However, in an attempt to answer question one in Sect. 3 a crude comparison between illicit and non-illicit sites was made and can be seen in Fig. 12.

As the above shows, over half of the services where data was obtained fall into the non-illicit category, whilst under half fall into the illicit category. This is common with the majority of studies, such as one that had what they termed 'unethical services' accounting for 45% of hidden services [14]. However, it should be stated that this is a particularly crude assessment and that the figure of 42% should be treated with caution.

**Fig. 11** Hidden services by category

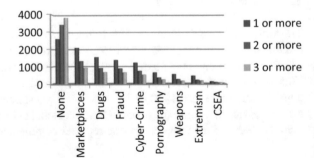

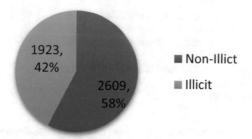

**Fig. 12** Illicit versus non-illicit hidden services

### 5.2.3 Identifying High Risk Sites

Analysis was also carried out to see if the tool could be used as a triage tool and identify high risk sites for further examination (as per the second research question). This was done by analysing the top 'scoring' sites in each section. Sites that scored high across all the categories were discounted, as it was found that these tended to be indexing sites such as the ones used as part of this research. Instead, sites which scored high in just one category were identified (with the focus on Drugs, Weapons, CSEA and Fraud).

**Drugs** A site that 'scored' 56 on the drugs section, yet relatively low on all the others was o3shuzjrnpzf2aiq.onion. If the tool works correctly, this should indicate that this is a site devoted to drugs. When the html is exported out of the database and into an empty HTML file, it can be opened in a browser to get some idea how it would display had you visited the site at the time the scrape was made. The results for this particular site can be seen in Fig. 13 (note that as only the HTML and no associated media was downloaded, the page does not display correctly as there are no images).

## · MDMA

3,4-methylenedioxy-N-methylamphetamine

202 649
Есть адреса
  ○ 1 gr: > 2 адресов
  ○ 2 gr: > 2 адресов
  ○ 4 gr: > 2 адресов

**Fig. 13** Excerpt from o3shuzjrnpzf2aiq.onion

armsmeo3z5kuf2dp.onion

armsmhmd4c3hb5xu.onion

gunsammoaahtpmqs.onion

**Fig. 14** Weapons.onion sites

**Fig. 15** Excerpt from gunsammoaahtpmqs.onion (BMG—Black Market Guns)

This is just one excerpt from the site. It appears to be a Russian Language online shop selling drugs, as following the MDMA entry is a long list of other drugs, displayed in a similar format.

## Weapons
Sites that scored high on the weapons category are as follows (Fig. 14).

An excerpt from one of the sites can be seen below. This is clearly a website devoted to selling weapons online. Its name is actually BMG—Black Market Guns (Fig. 15).

## CSEA
As only the HTML was obtained from the sites and no images, there was no concern about analysing CSEA sites as no Indecent Images of Children were collected as part of this research. However, sites which hosted such images were identified as detailed below. Once such site being found at babyixntjlabwkpi.onion. This scored 19 for CSEA and is the CSEA forum known as BabyHeart. An excerpt can be seen in Fig. 16 and it is clear from this that this is aimed at very young (0–5 year old) CSEA material.

## Fraud
One particular site that scored high in the fraud category was 35mdw76nu7azox7e.onion. This scored 45 in the fraud Category and appears to be a site devoted to selling pre-paid cards as can be seen in Fig. 17.

- Producers Zone (0-5yo)
    Topics
    Posts
    Last post

- Application
  Only CP producers
    1 *Topics*
    2 *Posts*
    *Last post* Re: Rules to join in the Prod...
    by Twinkle View the latest post
    Mon Sep 05, 2016 5:21 pm

- VIP Zone (0-5yo)
    Topics
    Posts
    Last post

- Application
  Only rare CP
    1 *Topics*
    1 *Posts*
    *Last post* Rules
    by Dragoslav View the latest post
    Wed Nov 25, 2015 4:51 am

**Fig. 16** Excerpt from babyixntjlabwkpi.onion

### 5.2.4    Finding Hard to Find Sites

The above all shows that the research has some degree of success when it comes to categorising/identifying sites that contain a lot of keywords. Analysis was also carried out on some of the sites that only contain one key word in a particular category.

**CSEA**
Analysis was carried out on sites that only scored 1 key word hit in the CSEA category. Examples of these can be seen in Table 2.

A screenshot of the text found in gso4vx44vwnlanwe.html can be found in Fig. 18. This shows this is clearly a CSEA site.

The above sites would not have been categorised as CSEA sites had the minimum 'score' been set higher than 1. This shows how the tool is useful in finding sites that may not be found with more stringent 'rules' for categorisation.

**Other**
The tool didn't perform as well in other crime categories. Dip sampling a number

## Pre-Paid Cards  _____

Cards that we've loaded with funds from various acc

You can buy money right here right now! We have l⟨

_____

16 Reviews

80 Orders

4.81 stars

## Magnetic cards [USD]

### 145$

0.2188415

- Custom emboss name
- Shipping included
- Balance : Min. 1500 USD
- Balance : Max. 3000 USD
- PIN-Code attached
- Bonus: Carding guide

**Fig. 17**  Excerpt from 35mdw76nu7azox7e.onion

**Table 2**  Examples of low scoring CSEA sites

| Name of site | CSEA score | Type of site |
|---|---|---|
| rupedoszuuqmn6pg.onion | 1 | Russian language CSEA site |
| minori6o257wqhca.onion | 1 | Korean language CSEA site |
| gso4vx44vwnlanwe.onion | 1 | Placeholder for private CSEA site |

## Welcome to the Ultimate Pedo Archives

The address has changed and your account must be validated.

Send an e-mail to the known torbox-address with the three first
characters of your username and the three first characters of your password.

Then I send you the new address.

**Fig. 18**  Excerpt from gso4xv44nlanwe.onion

**Table 3** Examples of low scoring drugs sites

| Name of site | Drugs score | Type of site |
|---|---|---|
| gso4vx44vwnlanwe.onion | 1 | Download page for PC game |
| tetatl6umgbmtv27.onion | 1 | Chat with John Doe site |
| fhostingesps6bly.onion | 1 | Freedom Hosting 2–contains drugs refs |

**Table 4** Examples of low scoring weapons sites

| Name of site | Weapons score | Type of site |
|---|---|---|
| luxdezjnop7f2udi.onion | 1 | Russian language online shop (unsure what selling) |
| weapon5cd6o72mny.onion | 1 | Black market site selling weapons |
| telavivguw3ey5wh.onion | 1 | Online store selling weapons |

**Table 5** Examples of low scoring fraud sites

| Name of site | Fraud score | Type of site |
|---|---|---|
| msydqjihosw2fsu3.onion | 1 | Fake version of Ahima indexing site |
| blackshopd7kvadt.onion | 1 | Possible scam site |
| rb564gvo6isyhayz.onion | 1 | Public Git repository |

of sites that only 'scored' 1 in drugs, weapons and fraud, produced mixed results as seen in the Tables 3, 4 and 5.

Whilst the tool still managed to correctly identify potential fraud, drugs and weapons sites despite them only receiving a score of 1, there were also some false positives. As such it may be that rather than a categorisation tool, this tool, in its current state, is more suited to triaging Hidden services, to identify potential criminal sites for further investigation.

# 6  Conclusion and Future Works

The key point to make about the tool in its present state is that it works. It successfully routes itself through Tor to collect a list of onion addresses, collects the html from the home page of each of them and then carries out some form of categorisation of each of the sites. As such, it is believed that this tool has successfully answered the second problem statement, that being, "Is it possible to automate a system to identify sites of interest for law enforcement based on the prevalent crime type of the hidden service?" By using Onion Soup, the answer to that question is 'Yes.'

However, more care needs to be taken when answering the question, "What percentage of hidden services are accessible and how many of these are connected to criminal/illicit activities?" It can be argued that the tool is robust enough to answer the first part of the question as if a site is not accessible; it records it as such in the database. Therefore, at least in respect of the indexed sites visited, the value of 35% as reported above can be considered accurate.

On an individual basis, the tool does appear to work well as a triage tool by using its simple categorisation process to identify sites that may be of interest to law enforcement. However, this relatively crude method is not perfect and does throw up the odd false positive. When the tool is being used for triage, these can simply be discounted by the investigating officer when they are examined more closely. However, when being used to determine the make-up of the dark web as a whole, it is unfeasible to check every site manually and as such, the false positives may skew the results, deeming them inaccurate.

However, that aside, when compared to other research that used manual classification, the results from this study were similar. As mentioned above, the top three categories found by [13] were the same top three categories found by this research; those being Drugs, marketplaces and fraud. Additionally, the figure of 42% of the dark web being devoted to illicit services from this study is similar to [14] figure of 45%.

However, it should also be noted that the results differ from the results found by other studies. This study found that the number of CSEA sites, (that is sites with a CSEA score of 1 or more) was 3%. This appears low when compared with other studies such as Guitton [14] which reported child abuse sites accounted for 18% and [1], which found the value to be 8%. It may be that the keywords used are not sufficient enough to identify all the CSEA sites, or that simply in the years since those studies, the number of CSEA sites on the dark web has fallen and other platforms are becoming more prevalent, such as live streaming services [23]. As suggested above, the tool may benefit from a more rigorous testing of the keywords used. By looking at known sites, keywords can be identified that are unique to a particular crime category. These can then be used as the keywords, with the more ambiguous words discarded. This will reduce the number of false positives and make the tool more useful when providing a strategic oversight of the make-up of the dark web. Sentiment analysis could also be used to improve the automated classification process. However, this is likely to be time consuming and may limit the tools effectiveness as a triage tool.

Another future development is to build on the triage element of the tool, using the forensic framework to identify local versus synced artefacts [24] and develop a deep-dive scraper tool that collects all the data from any site it is directed at. This will allow for more in depth analysis of any onion sites that are identified as part of the triage process, meaning that once complete, the Onion Soup suite of tools will be a valuable addition in the fight against criminality on the dark web.

# References

1. Spitters, M., Verbuggen, S., Ataalduinen, M.V.: Towards a comprehensive insight in the thematic organization of the tor hidden services. Perspective **5**, 6–9 (2014)
2. Hayes, L.: 60 Minutes: the Dark Web. Retrieved 26 Oct 2016, from nine.com.au. http://www.9jumpin.com.au/show/60minutes/stories/2014/september/the-dark-web/ (2015)
3. The Economist.: Shedding Light on the Dark Web—Buying Drugs Online. Retrieved 13 Oct 2016, from The Economist. http://www.economist.com/news/international/21702176-drug-trade-moving-street-online-cryptomarkets-forced-compete (2016)
4. Intelliagg.: Deeplight: Shining A Light On The Dark Web. Onyx, London (2016)
5. Biryukov, A., Pustogarov, I., Thill, F., Weinmann, R.-P.: Content and popularity analysis of Tor hidden services. Retrieved 9 Nov 2016, from arXiv.org. https://arxiv.org/abs/1308.6768 (2014)
6. Mitchell, R.: Web Scraping with Python: Collecting Data from the Modern Web. O'Reilly, Sebastopol (2015)
7. McCoy, D., Bauer, K., Grunwald, D., Kohno, T., Sicker, D.: Shining Light in Dark Places: Understanding the Tor Network. Privacy Enhancing Technologies, pp. 63–76. Springer, Cham (2008)
8. Chaabane, A., Manils, P., Kaafar, M.A.: Digging into anonymous traffic: a deep analysis of the tor anonymizing netowork. In: 4th International Conference on Network and System Security (pp. 167–174). IEEE Computer Society, Grenoble, France (2010)
9. Christin, N.: Traveling the Silk Road: a Measurement Analysis of a Large Anonymous Online Marketplace. Carnegie Mellon University, Pittsburgh, PA (2012)
10. Zander, S., Murdoch, S.: An improved Clock-Skew measurement technique for revealing hidden services. In: 17th USENIX Security Symposium (pp. 211–225) (2008)
11. Murdoch, S.J.: Hot or not: revealing hidden services by their Clock Skew. In: CCS '06 Proceedings of the 9th ACM Conference on Computer and Communications Security (pp. 27–36). ACM Press, Alexandria, VA, USA (2006)
12. Biryukov, A., Pustogarov, I., Weinmann, R.-P.: Trawling for tor hidden services: detection, measurement, deanonymization. In: IEEE Symposium on Security and Privacy, pp. 80–94 (2013)
13. Savage, N., Owen, G.: Emperical analysis of tor hidden services. Inst. Eng. Technol **10**(3), 113–118 (2015)
14. Guitton, C.: A review of the available content on Tor hidden servcies: the case against further development. Comput. Hum. Behav. 2805–2815 (2013)
15. Sui, D., Caverlee, J., Rudesill, D.: The Deep Web and the Darknet: A Look Inside the Internet's Massive Black Box. Wilson Center, Washington DC (2015)
16. Ciacaglini, D.V., Balduzzi, D.M., McArdle, R., Rosler, M.: Below the surface: exploring the deep web. Trend Micro (2015)
17. Warren, C., El-Sheikh, E., Le-Khac, N.-A.: Privacy preserving internet browsers—forensic analysis of browzar. In: Daimi, K., et al. (eds.) Computer and Network Security Essentials. Springer, New York. DOI: https://doi.org/10.1007/978-3-319-58424-9_21 (2017)
18. Reed, A., Scanlon, M., Le-Khac N-A.: Forensic analysis of epic privacy browser on windows operating systems. In: 16th European Conference on Cyber Warfare and Security, Dublin, Ireland, June 2017 (2017)
19. Tor Project.: Tor: Anonymity Online. Retrieved 31 Oct 2016, from The Tor Project. https://www.torproject.org/ (2016)
20. Acharya, S.: Crawling Anonymously with Tor in Python. Retrieved 15 Oct 2016, from www.sacharya.com. http://sacharya.com/crawling-anonymously-with-tor-in-python/ (2014, March 5)
21. Mathewson, N.: Special Hostnames in Tor. Retrieved 20 Nov 2016, from The Tor Project. https://spec.torproject.org/address-spec (2006)
22. Crummy.com.: Beautiful Soup. Retrieved 11 Oct 2016, from Crummy.com. https://www.crummy.com/software/BeautifulSoup/ (2016)

23. National Crime Agency: National Strategic Assessment of Serious and Organised Crime 2016. National Crime Agency, London (2016)
24. Boucher, J., Le-Khac, N-A.: Forensic framework to identify local vs synced artefacts. J. Dig. Inv. **24**(1), S68–S75. https://doi.org/10.1016/j.diin.2018.01.009

**Andrew Kinder** received his MSc in Forensic Computing and Cybercrime Investigation from the University of Dublin in 2017. He was a serving Police Officer for 16 years, and in 2015 set up and managed the Dark Web Intelligence Unit at the National Crime Agency in the UK. In 2019 the team received the award for Exceptional Development or Use of Internet Intelligence & Investigations at the International Digital Investigation and Intelligence awards. Andrew recently left law enforcement and now works in Information Security as the Head of Security Operations for the Co-op. His research interests include methods of intelligence gathering from the dark web and using data science techniques in an operational setting.

**Kim-Kwang Raymond Choo** received the Ph.D. in Information Security in 2006 from the Queensland University of Technology, Australia. He currently holds the Cloud Technology Endowed Professorship at the University of Texas at San Antonio (UTSA). In 2015, he and his team won the Digital Forensics Research Challenge organized by the Germany's University of Erlangen-Nuremberg. He is the recipient of the 2019 IEEE Technical Committee on Scalable Computing (TCSC) Award for Excellence in Scalable Computing (Middle Career Researcher), 2018 UTSA College of Business Col. Jean Piccione and Lt. Col. Philip Piccione Endowed Research Award for Tenured Faculty, British Computer Society's 2019 Wilkes Award Runner-up, 2019 EURASIP Journal on Wireless Communications and Networking (JWCN) Best Paper Award, Korea Information Processing Society's Journal of Information Processing Systems (JIPS) Survey Paper Award (Gold) 2019, IEEE Blockchain 2019 Outstanding Paper Award, Inscrypt 2019 Best Student Paper Award, IEEE TrustCom 2018 Best Paper Award, ESORICS 2015 Best Research Paper Award, 2014 Highly Commended Award by the Australia New Zealand Policing Advisory Agency, Fulbright Scholarship in 2009, 2008 Australia Day Achievement Medallion, and British Computer Society's Wilkes Award in 2008. He is also a fellow of the Australian Computer Society, an IEEE senior member, and co-chair of IEEE Multimedia Communications Technical Committee's Digital Rights Management for Multimedia Interest Group.

**Nhien-An Le-Khac** is a lecturer at the School of Computer Science (CS), University College Dublin (UCD), Ireland. He is currently the program director of MSc program in Forensic Computing and Cybercrime Investigation (FCCI)—an international program for the law enforcement officers specializing in cybercrime investigations. He is also the co-founder of UCD-GNECB Postgraduate Certificate in fraud and e-crime investigation. Since 2008, he is a research fellow in Citibank, Ireland (Citi). He obtained his Ph.D. in Computer Science in 2006 at the Institut National Polytechnique de Grenoble (INPG), France. His research interest spans the area of Cybersecurity and Digital Forensics, Data Mining/Distributed Data Mining for Security, and Fraud and Criminal Detection. Since 2013, he has collaborated on many research projects as a principal/co-PI/funded investigator. He has published more than 150 scientific papers in peer-reviewed journal and conferences in related research fields, and his recent edited book has been listed the Best New Digital Forensics Book according to the Book Authority.

# The Bitcoin-Network Protocol from a Forensic Perspective

Cornelis Leendert (Eelco) van Veldhuizen, Madhusanka Liyanage, Kim-Kwang Raymond Choo, and Nhien-An Le-Khac

**Abstract** Network forensics is challenging within most police investigation. Adding a cryptocurrency to network forensics makes it an even more complex challenge. One of the cryptocurrencies that can show up during network forensics is Bitcoin. Bitcoin gained popularity over the last years among criminals as an alternative to fiat currencies. Because of this increasing popularity, the use of bitcoins by criminals can be found in more and more police investigations. The bitcoin is a cryptocurrency that completely depends on its participating computers. These computers communicate with the bitcoin network protocol to make everyone aware of the latest changes. The bitcoin network protocol uses a message paradigm to send and receive information between participants. To be able to investigate a protocol like the bitcoin network protocol an investigator needs to have specific knowledge to gain investigative insights regarding the network information that was collected. While there are many (academic) papers written about the bitcoin ledger, very little information is available to investigator to acquire the knowledge to investigate the network protocol. This chapter focuses on the knowledge gap that a police investigator might have when encountering bitcoin network protocol. By conducting an experiment in which the network traffic of a bitcoin client, receiving a small amount bitcoins, the relevant information was investigated. After the experiment was completed the collected data is processed following the phases of the generic process model for network forensics. This chapter identified four bitcoin messages that were marked as possible messages that contain relevant information. After analysing three out the four messages turned out to be relevant. These messages will allow the investigator to identify the following information: (i) Identify the software that was used for communicating with the bitcoin network; (ii) The use of a Bloom filter enables an investigator to test bitcoin addresses to determine if a bitcoin client is interested in

C. L. van Veldhuizen
Team High Tech Crime, Dutch National Police, The Hague, The Netherlands

M. Liyanage · N.-A. Le-Khac (✉)
University College Dublin, Dublin, Ireland
e-mail: an.lekhac@ucd.ie

K.-K. R. Choo
University of Texas at San Antonio, San Antonio, TX, USA

© The Editor(s) (if applicable) and The Author(s), under exclusive license to Springer Nature Switzerland AG 2020
N.-A. Le-Khac and K.-K. R. Choo (eds.), *Cyber and Digital Forensic Investigations*, Studies in Big Data 74, https://doi.org/10.1007/978-3-030-47131-6_11

them; (iii) With the help of the Bloom filter and open source information the transaction message could be determined; (iv) From the messages of the sending party the unique transaction identifier was calculated, enabling the investigator to retrieve details from this transaction from the block chain. With the help of the Bloom filter and transaction messages it's possible to determine to a degree close to certainty the transaction sent and received by bitcoin clients. The experiment has a limited amount of messages that were investigated. There might be more information available in the messages that did not get the attention in this experiment, also the bitcoin client used had no history of previous payments which likely has resulted in less network information and less pollution of historic information within the messages. For future experiments or research challenges can be found in investigating heavily used bitcoin clients or the bitcoin messages that did not get the attention within the experiment.

**Keywords** Network forensics · Bitcoin investigation · Bitcoin-network · Cryptocurrency analysis

## 1   Introduction

Network forensics, the analysis of network traffic, is not an easy task. The introduction of encryption and the increase in usage of encrypted network protocols made analysis of network data sometimes nearly impossible.

A profession that struggles with the increase network of encryption protocols are the digital investigators working for law enforcement [1]. Law enforcement has, under certain circumstances, the legal means to wiretap a person of interest. Unfortunately the investigative value of a wiretap within an investigation has downgraded over the recent years because of the use of encryption while communicating over the network.

Besides, encryption does influence law enforcement cases when there is the usage of encryption in modern ways of communication. Instant Message (IM) apps like WhatsApp [2], Telegram [3], Viber [4] or one of the many others are most likely a good source for collecting evidence. Forensic acquisition and analysis of the IM apps are challenges [5]. On top of it, many of these apps, called over-the-top (OTT) services and applications, have adopted end-to-end encryption in their recent versions. The result of this adaptation is that the potential evidence became inaccessible to law enforcement [6, 7].

Because the amount of information that becomes more and more encrypted, law enforcement seeks new sources of information within the network data. For example; changing the focus from the content of a network package to metadata analysis, behavioural analysis or the analysis protocols that are designed to be unencrypted or implement security mechanism on different levels of the Open Systems Interconnection (OSI) model. In this way the digital investigator can maximise the evidential value of network forensics in a police case. This is vital since network forensics and wiretaps are used in a wide variety of police investigation strategies.

Besides, a police investigation can have many strategies to find the criminals involved with or conducting a crime. Two of those strategies that are applied often in police cases are: (i) Investigate financial traces left behind while committing the crime; (ii) Investigate the digital traces of the crime. With a financial strategy the police tries to find out how criminals makes profit, how much profit is made and how the money is brought back into the legal financial system (the process of money laundering). This can be a complex strategy because to the process of money laundering is sometimes been done by an international network of shell companies and money mules [8].

The digital strategy focuses on the digital devices that were involved to commit the crime, since the growth of digital devices in the everyday life of people, criminals start to use the same or specialised, services and devices. A digital investigation is likely to cross multiple borders, face cryptographic algorithms and other ways to cloak one's identity and hide information. Making it difficult and time-consuming to follow a criminal in the digital domain.

Both of the strategies have to join forces within investigations where a digital currency is involved such as Bitcoin [9].

From a law enforcement view, that is already facing difficulties in its investigative strategies because of encryption, digital currencies and no relationship between the money and its owner urges the need to get a good understanding of the bitcoin on every aspect. Especially since there is a growth in police cases where bitcoin was involved. The value of bitcoin, the cut between the identity and the bitcoin address, the ability to transfer money all over the world in a quick, cheap and arguably private way did not go unnoticed by individuals with less good intent. The result of this interest also raised the number of police cases in which individuals were arrested due to their involvement within bitcoin.

Because law enforcement agencies are always trying to gather as much information from a wiretap as possible and there are likely opportunities within the network protocol that may be of benefit to the police. Having a good understanding of the bitcoin network protocol and the opportunities within the protocol are of great value in the current world of fighting cybercrime.

The determination of the presence of the bitcoin protocol in a wiretap is not hard to do but without the right domain knowledge of the bitcoin network protocol deeper investigative insights stays out.

This chapter is an attempt to explore the possibilities for prosecution within the bitcoin network protocol from a law enforcement perspective. Since the bitcoin protocol is pretty extensive, a research question has been developed to limit the scope of this chapter but still tries to answer a typical investigators research objective: "Based on recorded bitcoin network traffic, determine the transactions sent and/or received by the client".

As will be shown in Sect. 2 there are no other research papers, known by the author, who tried a similar approach. Therefore there is a knowledge gap within the law enforcement regarding bitcoin network traffic analyses. The following knowledge gaps will be fulfilled in this research: (i) How does the bitcoin client communicate with its participants; (ii) How valuable is the analysis of bitcoin network data for a police investigation? By fulfilling to the gaps mentioned above a foundation for answering the research question can be created. In accordance with these knowledge gaps, this research will also develop a set of deliverables that will create practical and useful solution for police investigations.

## 2 Background

The focus of this section will be on the way bitcoin transaction are communicated over the network, the network protocol used by the bitcoin ecosystem. We will also describe the bitcoin network protocol in detail.

### 2.1 Bitcoin Transaction Communication

When it comes to the official definitions of different network protocols, like the internet protocols HTTP (web browsing), POP and SMTP (receiving/sending e-mail), the Internet Engineering Task Force (IETF) plays a central role of keeping track of the official definitions of the protocols. The bitcoin protocol is not among the protocols tracked by the IETF or similar organisation, this means there is in theory no official authority keeping track of changes and formal registration of the network protocol.

The bitcoin community rather manages their own bitcoin protocol documentation. The commonly referred as the official documentation of the bitcoin ecosystem is the documentation of bitcoin client Bitcoind, maintained at the bitcoin.org website in the form of a Wiki [10]. Due to the nature of a Wiki anyone can commit proposals for improvements. These proposals are called 'Bitcoin Improvement Proposal' and are almost always abbreviated to BIP.

The first BIP [11] described how the workflow of the BIP process should be implemented (Fig. 1).

Since the implementation of the BIP workflow a total of nine BIP's are accepted (by consensus). This has led to the bitcoin ecosystem as documented on the bitcoin.org website. The definition of the network protocol as stated on the bitcoin.org

**Fig. 1** Workflow of a bitcoin improvement proposal

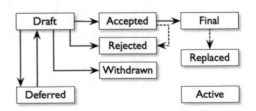

website will be seen as the official definition of the bitcoin network protocol. The bitcoin and its network is continuously under development, the version of the network protocol used for this research is version 70014 (released on the 23 August 2016).

For understanding where the opportunities lie for law enforcement within the bitcoin network protocol, a basic level of understanding of the relevant parts of the network protocol is required. The bitcoin network protocol is an extensive protocol due to the network management that is required in a peer-to-peer network. These parts of the network are interesting by itself but outside the scope of knowledge required to answer the research question. For sake of simplicity only the parts of the protocol required for sending and receiving a transaction will be explained. For a full explanation of the bitcoin protocol, the reader is encouraged to read the network protocol part of developer guide on the website of bitcoin.org (https://bitcoin.org/en/developer-reference).

As in many documents regarding encryption or communication we introduce the following characters to illustrate the different roles different parties have within the network communication:

- [Alice] A party that can send and/or receive bitcoin to other a user (Bob) of bitcoin
- [Bob] The counterpart of Alice. Just like Alice, Bob can receive and send bitcoin.

To illustrate how bitcoin transaction works an example of a transfer of bitcoin between Alice to Bob is described. Alice has a total of one bitcoin in her wallet. A wallet is the collection of all her bitcoin address of which she possess the private key. For sake of simplicity we assume she has one bitcoin address where the bitcoin resides on.

Alice has received the public bitcoin address of Bob at some point in the past during interpersonal communication.

Bob gave Alice a product of the value of ฿ 0.5 and Alice would like to settle this bill by transferring ฿ 0.5 from her wallet to Bob's public bitcoin address.

Both Alice and Bob have a fully functional network connection to the internet and are freely to access the internet without interference of any network appliances. Their bitcoin clients are up to date and communicating with use of the bitcoin network protocol.

The bitcoin software of Alice collects the amount of bitcoin she needs for the transaction, in this example they will all come from the address which has one bitcoin assigned to them. Her bitcoin wallet uses this bitcoin as input for the transaction to Bob.

The client of Alice creates a script that looks a bit like the riddle: "I have ฿ 0.5 for the one who represents a signature derived from the private key that belongs to this (Bob's) public key". Because Bob is the only one who is able to answer the question he can solve this riddle by presenting the signature signed by his private key.

From a bitcoin perspective it's not possible to transfer ฿ 0.5 and have the remainder left in the same bitcoin address. Because Alice doesn't want to lose her 0.5 bitcoin the bitcoin client will create a second transaction with almost the same riddle: "I have ฿ 0.49999995 for the one who represents a signature derived from the private key that belongs to this (Alice's) public key". Alice can solve this riddle by herself and will receive the ฿ 0.49999995. This is called the change of the transaction.

During the transaction Alice has 'lost' some bitcoin (฿ 0.00000005 to be exact), this is a transaction fee for the network. The parties within the bitcoin ecosystem that invest a lot of computational power will receive this fee when they are the one who process the transaction of Alice. When they have created a new block for the ledger (blockchain) they will receive the fee and the transaction is permanently recorded in the ledger.

This describes the high level steps of a bitcoin transfer between two individuals, this is backed up by a lot of cryptographic functions and a network protocol to facilitate the communication. The focus of this chapter is the network part of the bitcoin ecosystem therefore we will not go into the cryptographic details required for the transaction other than those who are used on by the network. Based on the example we know that there are few pieces of information are required from the network to conduct the transfer of the bitcoin. The following steps are required to send bitcoin between Alice and Bob:

- Alice needs to send her riddle to the bitcoin network so it can be picked up by someone to solve the riddle.
- Bob needs to receive the riddle from the bitcoin network so it knows it can send an answer to the riddle.
- The new block needs to be communicated to the clients.

## 2.2   Bitcoin Network Protocol

While Alice and Bob used their client to transfer bitcoin as explained in the previous section the network protocol made sure their actions were published to the network. This way the entire bitcoin community has been made aware of the transaction and Bob knows he received the bitcoin.

The messages that are used by the bitcoin ecosystem are divided in two categories; data massages and control messages. For the ease of reference in the documentation at later time, the same division is kept within this section.

Protocol Documentation [11] gives an excellent in-depth explanation of the messages. To understand the bitcoin network protocol that is required for forensic analysis of the network data for each message a short explanation is included.

### 2.2.1   Control Messages

A control message is used to manage the connection between two nodes, for example a client talks to a full node can use this set of messages to make sure they can still communicate with each other by sending a ping message and replying with a pong message. These messages are not meant to transfer content to each other, for the transport of content data messages are used.

### alert

This message is a deprecated feature of the bitcoin network protocol, this type of messages was retired on the 19th of January 2017 (bitcoin-alert-system-retired). Clients who receive this message can safely ignore this message.

### addr

The preliminary examination of network data showed that the addr message is a very common message on the bitcoin network. It's used to notify bitcoin clients of the network address of other bitcoin clients (also referred to as nodes) that accept network connections. This helps bitcoin clients to keep an update list of nodes to connect to. In the case of Alice and Bob, their clients do not accept external connection. Therefore she only receives addr messages.

### filterload

The *filterload* message is sent to a full-node (also referred to as peer). This message contains a probabilistic data structure. This probabilistic filter is calculated by the client of Alice to tell the full node in which transactions Alice might be interested in. Clients add extra bitcoin address to gain plausible deniability, but the use of a Bloom filter will always reduce the privacy Alice while using bitcoin [12].

### Ping and Pong

Ping and its counterpart pong are used to test if the network is still functional and the peer it was talking to is still around. If there is no pong after a ping the peer is marked as down and is being discarded by the client.

### Version and Verack

A version messages are used for the version handshake, this is a mutual process of the client and peer to tell each other what client they are using. It's interesting to note that it includes a string of the software it uses as well as a timestamp including time zone.

### 2.2.2 Data Message

The data messages are the messages who do the heavy lifting. These messages contain the actual content regarding information that needs to be shared between two nodes. They rely on the control message to make sure the connection is active.

### inv

The inv message is sent unsolicited by the peer to the client or if the client has requested for a data (for example the client has sent a *getblocks* message).

The message contains information about the knowledge of the peer (transaction information for example), not the actual information itself.

## getdata

The *getdata* message requests more information from another node. When a client has received what kind of information is available through an inv message received earlier, it can use the *getdata* message to receive the actual content.

The information that can be requested are: a *tx* message, block message, *merkleblock* message. If there is no answer for the *getdata* message a *notfound* message will be received.

## getblocks

If a client has interest in a block or multiple blocks it can send a **getblocks** message to a peer to retrieve them. After a peer has received the **getblocks** message it will reply with an inv message. The client can then use the **getdata** to learn more about the blocks.

## mempool

Mempool requests are sent to a peer by the client to ask information about a transaction that has been verified (the cryptography regarding the transaction is correct) but not yet confirmed (registered in the blockchain).

## merkleblock

The **merkleblock** message is another possible response on a **getdata** message, if there is a matching transaction a separate *tx* message is sent to the client.

Each block in the bitcoin blockchain contains a merkle tree of all the transactions it contains. A merkletree is a cryptographic way to summarise all the transactions within a block of the blockchain reducing the amount of data that is needs to be transferred to a client.

If the client has previously sent the *filterload* message, the **merkleblock** messages contains only the transaction that match the Bloom filter as well as the header of the merkletree, so the client can confirm the proof of work.

## tx

This is a transaction message which can be send and received to transfer bitcoin or to make a client aware that it receives bitcoin. A transaction message consist out one or more transaction inputs and one or more transaction outputs. The transaction output shows the amount of bitcoin and a script that contains information regarding the transaction.

A script is essentially a list of instructions recorded with each transaction that describes how the next person wanting to spend the bitcoin being transferred can gain access to them. The pay-to-pubkey-hash (P2PKH) is the script that used to pay an amount of bitcoin to a public bitcoin address. The following program code is an example of a P2PKH script (Fig. 2):

| 76 A9 14 524D3FDA6331C448C3758E8B0137EDD292150931 88 AC |

**Fig. 2** Example of P2PKH script

The first 2 bytes and last 2 bytes ($0 \times 76$, $0 \times A9$, $0 \times 88$ and $0 \times AC$) are the instructions (OP_DUP, OP_HASH160, OP_EQUALVERIFY and OP_CHECKSIG) telling the client what kind of data has been sent and how to handle the data. After the first two instructions there is one byte telling the client the size of data that needs to be handled, in this case it's 20 bytes ($0 \times 14$), the byte size of a hash-160 string. The example above translates to: *"Somebody who can proof ownership of the hash-160 in this message is allowed to spend the amount of bitcoin that is transferred by this message"*.

## 3 Related Work

To define the current state of research this chapter can be broken down in two parts, the current state of network forensic as a science by itself and the current state of research of the bitcoin with a focus on the network protocol.

### 3.1 Current State of Network Forensic

Network forensics is the science that deals with capture, recording, and analysis of network traffic for detecting intrusions and investigating them [13].

Network forensics is one of the many science fields a digital forensic investigator has to understand to some degree. The science of network forensics is, similar to everything else related with computers, a science that changes daily due to the release of new hardware and computer programs.

When it comes to the digital forensics there are a few forensic principles: minimising the contamination of original scene and evidence, maintaining the integrity of digital evidence, maintaining the chain of custody of evidence, and complying with rules of evidence for admissibility at the court [14]. To support a forensic investigator with the different types of sciences and the forensic principles different models are developed to break down the entire process into clear phases or stages.

For the network forensics, the model proposed in [13] (named GPMNF model) specifies nine phases of network forensics preparation, collection, preservation, examination, analysis, investigation and preparation, incident response and detection phase. Other scientific research have developed frameworks or methods they use similar phases. Even when not referred specifically they use a similar model, for example:

- A proposed architecture for network forensic system in large-scale networks [15]. In this paper a framework is described that utilises collecting, sfiltering, indexing, analysing (including malware analysis), reporting, alerting and visualisation. Many parts of the proposed architecture correspond to the GPMNF.
- A proactive approach in network forensic investigation process (Rasmi2015). The research proposes a more proactive network forensic investigation process. It suggests three modules: (1) Evidence collection (Preparation and evidence collection and preservation), (2) Online and offline alert correlation and optimisation (Analysis and examination) and (3) Evidence presentation as dissemination (presenting and reporting, disseminating and documenting).

The online and offline alert correlation as suggested by the second research is an innovative way to add value to the network data that was collected. Adding online available information to the collected network data is also applicable to bitcoin network traffic, while the user receives information regarding the bitcoin data a lot of other information is processed by others, for example statistics about the blockchain and its transactions [16].

Based on the papers mentioned above the GPMNF is one the abstract models that are suitable for conducting forensic network investigation. During research of forensic models no other model was found that was suggested as successor of the GPMNF.

While the phases of the GPMNF help investigator to collect and investigate network information, it does not specify how the phases of GPMNF should be applied in practice. This chapter focus on finding valuable information within the bitcoin network protocol therefore; previous work on (methods) analysing network data is studied. A part of computer forensics is the malware analysis or malware forensics, therefore information about analysing unknown network protocols can be found from research within cybersecurity science.

The following papers show approaches of how network information can be investigated:

- Automatic Network Protocol Analysis [17]. An approach in which network messages specifications are automatically created by monitoring the binary application that produces the messages. Within the researched protocols (ex. HTTP, DNS and IRC) the researchers were successful in extracting the format specification of message types used by these protocols.
- Network activity analysis of CryptoWall ransomware [18]. A high-level analysis of the network protocol used by the CryptoWall malware. A detailed explanation is given regarding the set-up of the lab and network manipulation. Regarding the analysis of the network packages, after the decryption, the types are described, but the content of the messages was only partly described.
- Highly Resilient Peer-to-Peer Botnets Are Here: An Analysis of Gameover Zeus [19]. The papers provide technical details on the topology and communication protocol of P2P Zeus malware. By reverse engineering the malware samples, collected from the Sandnet malware analysis environment (no longer in operation). By applying unspecified methods of reverse engineering the P2P-protocol of de

Gameover Zeus malware were defined. There was no information within the research paper about (methods of) analysing network data.

- PeerDigger: Digging Stealthy P2P Hosts through Traffic Analysis in Real-time [20]. The researchers propose, a system called PeerDigger, which is capable of detecting P2P bots in real-time. PeerDigger consists out of two phases: a P2P host detection phase and a P2P bot identification phase. The analysis is done by looking at the behavioural characteristics the network. Without analysing the payload of the packages, PeerDigger draws conclusions that the P2P is generated by a bot by analysing the connection patterns.

- Large-scale network packet analysis for intelligent DDoS attack detection development [21]. By applying machine learning techniques, the researchers were able to detect DDOS attacks based on characteristics of the network traffic. Similar as with Peerdigger, the content within the packages was not examined. If future research, methods of DDOS amplification could be taken in consideration. The application of machine learning could be relevant to bitcoin network research, for example to train the algorithms to detect transaction between known participants based on behavioural analysis.

There are several approaches for network information analysis, from the research papers mentioned above: from a malware analysis point of view researchers are likely to analyse network traffic by monitoring the malware execution and network traffic at the same time. This seems to be a good approach when analysing an unknown or undocumented piece of software, it's less applicable to bitcoin network analysis where everything is open source and well documented. This method is applicable to bitcoin network research when investigating a specific implementation of a bitcoin client or when trying to understand specific bitcoin implementation. Based on the malware analysis approach it seems advisable to stay as close to the implementation as possible, which can be done by using the same libraries as the bitcoin clients do.

The last research was done with a very different approach to analysing network data. This approach was based on automatically investigating the network data. While the need to analyse network traffic automatically is seen in papers about P2P botnets and DDOS attacks. The methods applied to this kind of network analysis are more high-level in nature. The analysing of packets is mostly irrelevant to find the bots or stopping/identifying the DDOS attack. From the perspective of the bitcoin analysis this might only be relevant within large network investigations for example where the investigator is called to scene and tries to find the computer system involved with the bitcoin P2P network.

## 3.2 Current State of Bitcoin Network Research

With the increasing interest of the public and by criminals in the Bitcoin ecosystem, an increase of interest in Bitcoin is noticeable in the academic world. There are many academic papers written on various topics regarding Bitcoin. While studying the papers, websites and articles that were published, a division in the following topics was noticed: blockchain research [22–24], bitcoin improvement [25, 26], bitcoin clients [12, 27], bitcoin privacy concerns [12] [28], bitcoin forensics for Windows systems [29] and privacy-oriented cryptocurrencies [30].

It seems that the blockchain is the most popular research topic, two reasons for this can be: In the first place the fact that the blockchain is a gigantic public source of information of all the transactions [31]. Secondly the blockchain technology by itself is one the promising developments for the future. Therefore, blockchain application, issues and improvements are being explored [32, 33].

While there is plenty of interest in researching the blockchain, the bitcoin network protocol got almost no attention of academic researchers. A reason for this low interest can be, that researching the bitcoin network protocol is a very limited domain and implementation specific. The study of the network traffic of bitcoin is mostly relevant to the bitcoin ecosystem where the blockchain technology can be used as solution to many problems.

Researches that were published regarding the bitcoin network protocol (sorted by year of publication):

- Deanonymisation of clients in Bitcoin P2P network [34]. With the help of functions within the bitcoin network protocol, this research found a method to deanonymize a significant fraction of Bitcoin users and correlate their pseudonyms with their public IP addresses.
- Analysing the deployment of Bitcoin's P2P network [35]. The development of a framework that traverses Bitcoin's P2P network and generates statistics regarding its size and distribution among autonomous systems.
- An analysis of anonymity in Bitcoin using P2P Network traffic [36]. Partially successfully developed heuristics for identifying ownership relationships between Bitcoin addresses and IP addresses.
- Bitcoin over Tor isn't a Good Idea [37]. Researchers found an exploit to force bitcoin participants to connect to a specific full node and a method to fingerprint bitcoin users.
- Anonymity Properties of the Bitcoin P2P Network [38]. Recent research shows that the bitcoin network has poor anonymity properties, and the change to a "better" way to spread information did not increase the anonymity.

When reviewing the research papers regarding the network communication of the bitcoin protocol the anonymity, or the lack of anonymity, of the network stands out. These researches show methods to relate an IP address to a bitcoin address, solving the cut between bitcoin addresses and their identity. There is an understandable need to find such methods for privacy concerns as well as for police cases. From the perspective of this chapter, the connection between IP address and bitcoin address is less relevant because when connecting a wiretap the IP address of interest is already identified by the methods mentioned above or through the follow-up of other leads.

The fourth listed research describes the use of an eavesdropper, to gain knowledge of the bitcoin network. The eavesdropper suggested in the research differs from a wiretap, the eavesdropping methods suggested in the paper is a modified version of bitcoin node that does not relay bitcoin network messages but only receive and analyse them. This require active engagement in the bitcoin network where a wiretap is fully passive.

To the knowledge of the author, no paper has been published where there was an attempt to analyse the bitcoin network traffic of the client, and no paper discusses the bitcoin network traffic from a forensic investigator perspective.

# 4 Methodology

Before going into the details of the methodology of the experiment, a motivation of the types of network data gathering is given and is suitable for the experiment. Working with network data requires knowledge about computer network protocols, therefore the required information on the fundamental network protocols in relation to the bitcoin network protocol is given.

After explaining the required network basics, the methodology of network analysis within the experiment is described. This methodology explains how the information is created, collected, analysed, investigated and presented.

Finally, the scope of the experiment is described. Since there are many bitcoin clients, the bitcoin network is extensive and there are many ways to create a dataset.

The bitcoin protocol uses both the TCP and UDP protocols. A bitcoin client needs to be able to communicate with other participants of the network, therefore it needs to find these participants before it can start interchanging messages. To discover participants it uses the DNS protocol, which by its turn relies on the UDP protocol, to retrieve a list of IP addresses of bitcoin participants.

After a bitcoin client has found other participants it start, using the TCP transport protocol for its network communication, which ensures a reliable network connection between its peers for communication.

We adapt the network forensics process mentioned in [13] in our forensic analysis of Bitcoin network. More details are described in the following sub-sections.

The model specifies nine phases of network forensics as shown in Fig. 3. The following parts of the model will be used during the research: the preparation, collection, preservation, examination, analysis, investigation and preparation phases. From the model the incident response and detection phase are left out, these phases are more applicable in cases where there was an intrusion or cyber-attack. Since this is not the case these two phases where omitted from the experiment.

The model is focused on finding the attacker of a computer network, this is different from focus of this chapter. Therefore, the investigation phase is adjusted from attribution of the attacker to finding answers to the research question.

**Fig. 3** Generic process
model for network forensics
[13]

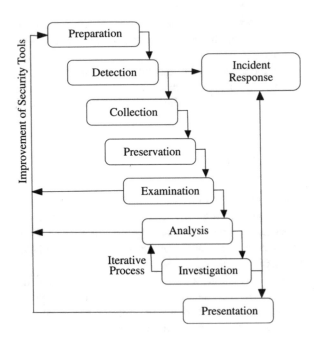

## 4.1   Preparation

To create a dataset containing bitcoin network traffic, a controlled environment is
prepared with the appropriate Network Forensic Analysis Tools (NAFT). Within
the environment an internet connection will be provided to the bitcoin client. The
provided internet connection enables the client to reach out to bitcoin network to
send and/or receive transactions. All the network traffic from and to the bitcoin
client will be collected for further analysis. A detailed description about the research
environment can be found in Sect. 5.

## 4.2   Collection

As shown in the section Fundamental network protocols a single IP address expresses
a single node in a network. In the consumer market this is the IP address of the modem
that resides in the home of an internet service provider (ISP) customer. The IP address
was given to the modem by the ISP during power on and network initialisation of
the device, the modem communicates with a router of the ISP, called the gateway, to
connect to the rest of the internet.

　　When intercepting network information between the two nodes (modem and gateway) there can be two types of network information intercepted and used for further
analysis.

- Netflow. The collection of network flows. A flow is defined in RFC3954 as a uni-directional sequence of packets with some common properties that pass through a network device (RFC3954). This gives insight in the metadata of the network connections. A flow as specified in the RFC does not contain payload of the packages.
- Package capture. The collection of network packages. A network package contains the complete information that was sent, this includes all the metadata and complete payload.

For the research regarding the bitcoin network protocol, a package capture is used. Netflow data will only provide the connection details when it comes to the bitcoin protocol. These connection details do not reveal much because other peers in the network are publicly available.

Within the experiment the analysis of the content of the data messages is most likely to hold information to find an answer to the research question. A precise insight of what was happening during the bitcoin connection, the full content of the package payload and header must be available for analysis. This can only be done with a package capture. With the help of the payload it's possible to determine if it's really bitcoin traffic and extract forensic artefacts from the package capture file.

To collect network information, a wiretap will be placed between the clients and the gateway computer. This will create a PCAP file that contains all the network communication from one bitcoin user per file.

## 4.3 Preservation

When network data is intercepted it needs to be preserved for further analysis. For preserving network information, a variation of file formats are available (wireshark-fileformatreference). The most common one is PCAP.

## 4.4 Examination

The captured network data as preserved in the PCAP is a complete collection of all the network data. For the purpose of this research of bitcoin network traffic only bitcoin network data is required, and even then, the amount of data of bitcoin network information can be a lot to analyse. Therefore the preserved data is reduced, filtered and duplicate messages are removed prior the analysis.

To reduce, filter and remove duplicates from the collected data a combination of tools and techniques are used. To find methods ways to reduce data NAFTs with a graphical user interface are used, to filter and remove duplicates a combination of NAFTs, custom scripts and other off the shelf Linux commands are used.

## 4.5   Analysis

To analyse PCAP files there are a lot of graphical NAFT available, for example: Wireshark [39], Network Miner [40] and Moloch [41]. To work with PCAP files programmers have developed libraries to ease the use of PCAP files in the program. For example for Python programming language the library Scapy (ScapyWebsite2017).

We use both the graphical and programmable approach to analyse PCAP data. For most of the rough work of finding the right packet NAFT Wireshark will be used. To find, select and export the right packages for the more detailed analysis the programmable approach will be taken.

## 4.6   Investigation

The investigation phase is an iterative process with the analysis phase. The goal of the investigation phase is to find answers that contribute to the research question. In this phase results of the analysis can be combined with other open source intelligence to enrich the analysed information.

## 4.7   Presentation

After analysis the bitcoin traffic knowledge is gained regarding the bitcoin network traffic. To be able to leverage this knowledge in the future or share this knowledge with other investigators simple tools will be developed. Besides that other investigators can benefit from the analysis it creates a reproducible outcome. Anybody with the same data set would get the same results, allowing evidence presented to the court to be examined by an independent third party.

## 4.8   Limitation and Scope

This research limits itself to the analysis of network connection between the consumer and the ISP or to put in different terms: The analysed wiretap PCAP files were created by intercepting the information of one single IP address.

The reason for the limit of analysing one IP address at a time is that the amount of network traffic is reduced, multiple simultaneous bitcoin participants can lead to confusion, the data creation and collection can be done within in a controllable environment.

Within the experiment, for each client the network data will be captured separately. This will result in two PCAP files each containing all the network data generated by the client. This method of capturing data simulates separated modems connecting to an ISP.

After the collection of the network data, not all the messages will be analysed for the purpose of this research. Only the relevant messages of which might answer the research question. These are the following:

- Control message—version. This message tells what client is used by both parties
- Control message—filterload. The Bloom filter that can be queried to see which bitcoin address or transaction result in a match
- Data message—**tx** The transaction message itself, there might be false positives due to use of the Bloom filter
- Data message—merkleblock The merkleblock might show what transaction matches the Bloom filter.

These are the messages this chapter will have its focus on, other bitcoin messages are less likely to contain information that will contribute to find the answer to the research question.

There are many bitcoin clients available on the internet, when searching for an appropriate client to use within the research, the following criteria had to be met:

- Must implement the bitcoin protocol as stated on the bitcoin's documentation page
- Must be open source to be able to study the inner workings where needed
- Must use filterload messages when retrieving content
- Must be a well-known bitcoin client
- Must supported the experiment operation system Linux.

When searching for a bitcoin client among the bitcoin clients suggested at website of bitcoin.org that met all these criteria. Based on the information provided on the website two wallets where examined: Electrum and Mulitbit HD. After installation the client showed to be using a different way to connect to the bitcoin network and using SSL for the encryption. After installation of the Multibit HD client an examination of the client and its produced network traffic showed it met the criteria as stated above.

## 5　Experiment

To be able to wiretap an internet connection of an internet service provider (ISP) customer, the police is required to have a signed court order. This court order is required to request the ISP to create a duplicate of its customers network traffic. To create a replica of a police wiretap would require the involvement of an ISP, a dedicated customer connection and involvement of the department of the police responsible for wiretaps. For this experiment the creation of such environment for experiments is not feasible.

Therefore a research environment for the purpose of conducting experiments is developed in which the creation of the dataset can be controlled so that anyone can redo the creation of the datasets and will create similar results as a police wiretap. The research environment has been created in such way that no external parties like an ISP are required.

With this environment it's possible to create datasets of users that use a bitcoin client. The datasets are created in such way that it can be used for forensic research and contain all the messages with the minimum amount of clutter or mixture of irrelevant data.

## 5.1  Experimental Environment

The environment is created with the help of multiple virtual machines (VMs). The software used to create the virtual environment was VirtualBox (version 5.0.30 r112061). The VMs are all running the Linux operation system, the publicly available Kali Rolling Light (2016.2) (Kali2016) ×64 image was used to install the VMs. This image was chosen because of the ease of use and completeness in pre-installed software.

On the client VMs the bitcoin software and its dependencies were installed according to the guidelines on their website (https://multibit.org/help/hd0.3/how-to-install-linux.html). One VM was configured as a network gateway providing internet access to the other VMs as well act as the point of interception (POI). This VM did not have the bitcoin software installed, instead is has the required software to create network captures. The client VMs, were all virtually connected among each other in a separate network from the host computer, hence the reason why one of the VMs is required to act as a network address translation gateway.

The point of interception has a dedicated network interface for each of the clients, to separate the network traffic during interception. This way an uncluttered dataset is created per client and by separating their subnets the client networks act independently.

The network as explained is shown in Fig. 4. The clients are represented as Alice and Bob, the Interception Gateway in the middle has two functions; it connects Alice and Bob to the Internet and at the same time functions as the POI as it records all network traffic of Alice and Bob.

## 5.2  The Dataset

The datasets of intercepted network data are created on the point of interception as described above. The tool used to create the datasets is TCPDump, this tool is a common tool to create immutable packet capture (PCAP) files. Before the start of interception of network data, the clock of the VM is synchronised with the help of

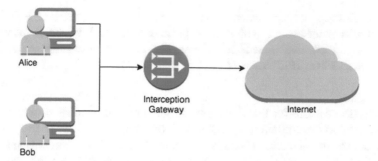

**Fig. 4** Network diagram of the experiment

Network Time Protocol (NTP). This way it's certain that when information has to be gathered from the blockchain both of the timestamps are synchronised.

The creation of the data sets is done by stepping trough the following steps:

- Synchronize the time with help of the NTP protocol
- Creating the output directories to store the captures
- Start capturing network traffic from Alice
- Start capturing network traffic from Bob.

The steps above regarding the creation of the PCAP files and synchronisation are combined into an automated script to be able to create multiple data sets in the same consistent way. The script that was used to create the datasets that were investigated Executing the script will result in two PCAP files, for each client one PCAP file containing all its network traffic.

This script was run multiple times to test if it was successful in recoding the network traffic.

## 5.3 Bitcoin Messages of Interests

As stated in Sect. 4, four bitcoin messages were identified that are likely contributing to the research question. This section goes into detail of these messages to select, extract, reduce and finally analyse them. After analysis, a method will be developed to help answer to the research question.

### 5.3.1 Bitcoin Message: Version Messages

The version message contains a user agent (a user agent is a piece of information that describes the software that communicate on the behalf of the user) field that helps to identify which client is used for the communication. Due to limitations within the Wireshark application an overview of Bitcoin client user agents cannot be made. A script was developed to create a list of the user agents of Bob and the nodes it was connected to. Within the experiment the result of the script was that Bob was using only one single bitcoin client. The following user agent was retrieved: *bitcoinj:0.13.3/MultiBitHD:0.4.1/*.

**Examination**
The selection criteria for the version messages are communicated over the TCP network protocol according to specifications of the bitcoin network protocol. The header of the bitcoin message contains the value 'version', Wireshark is able to use this to find all the version messages. Since Wireshark is able to select the correct bitcoin messages, Wireshark is also able to extract the messages from the captured data. The messages are extracted in a separate PCAP file to make sure the correct focus is given in the analysis and it's the author's experience that working with large network capture files can be time-consuming due to processing time.
**Analysis**. To analyse the information a small script is developed to support the analysis of the version message. This script extracts the content of the version message which is the user agent and prints it on the screen.
**Examination**. To reduce the amount, message duplicates and version messages from other then Bob's version messages need to be filtered out. Therefore the result of the script mentioned before can be combined with default commands of the Linux operating system. These commands sort and remove duplicates from the output of the script.
**Investigation**. After the reduction the version message analysis can be investigated. For this experiment the result was that there was only one bitcoin user agent used by Bob, which corresponds to the experiment environment (shown above).
**Presentation**. The result of the steps above is a single user-agent string of the bitcoin client that was used. This will likely means that the user made use of one single bitcoin client.

The ability to identify the bitcoin user-agent, and the used bitcoin client, from a forensic perspective is very valuable. Identifying the used client can be the first piece of the puzzle because it tells the abilities of the bitcoin client in use. Also, it gives the forensic investigator indicators which forensic artefacts to look for when conducting computer forensics if the computer is ever seized.

Within this experiment the user-agent of the Multibit HD Bitcoin client was found.

### 5.3.2   Bitcoin Message: Filterload

The filterload message contains a Bloom filter, this message are *send* to full nodes of the bitcoin network. It is used to tell a full node which information of the bitcoin network the client is interested in. This message is used to reduce the amount of data to be sent to the client.

Wireshark does not have the ability to test a value, like a bitcoin address, against a Bloom filter, therefore a script was developed.

As described in the current state of research it's advisable to stay close to original implementation of software that generates the network traffic. For that reason the same implementation of the Bloom filter in the Multibit HD client is used for the development of the script.

Multibit HD uses the source code framework BitcoinJ [42]. Therefore the same framework is used to process and evaluate the Bloom filter.

**Examination**. The selection criteria for the filterload messages are that they are communicated over the TCP network protocol according to specifications of the bitcoin network protocol. The header of the bitcoin message contains the value 'filterload', Wireshark is able to use the 'filterload' value to find all the filterload messages. In the captured network data many duplicate filterload messages are found. The number of duplicates can be reduced by the help of the checksum of the payload. This value can be found in the header and is a hash (Bitcoin uses the first 4 bytes of the SHA256 hash of the payload as checksum value) of the payload.

**Analysis**. From the examination three messages remained, analysis showed the messages have different 'Tweak value' (A random value to add to the pseudorandom number generator in the hash function used by the Bloom filter. (https://github.com/bitcoin/bips/blob/master/bip-0037.mediawiki) values, and different data values. Because the Tweak is different, a comparison or distance calculation cannot be made. Therefore, no further reduction of messages is possible.

**Investigation**. With the help of the script mentioned above to test a Bloom filter and the reduced dataset the bitcoin address of Bob was investigated. The script returned a positive match for the bitcoin address of Bob. To validate the result a limited group of arbitrary bitcoin address are tested and returned a negative result.

**Presentation**. The presentation of the Bloom filter is straight forward, the Bloom filter results in true or false. Which results in an outcome that if the filter tests is negative the client is not interested in this bitcoin address, if the test is positive the statement can be made that the client might be interested in the bitcoin address

During a wiretap or using other types of information gathering methods, bitcoin addresses can show up during the investigation. In such cases the ability to test a bitcoin address against a Bloom filter can be of value to the investigation. It gives the investigator insight of the possible interest of bitcoin client in a bitcoin address. Because of the probabilistic nature of the bloom filter, a positive match does not guarantee the interest in a bitcoin address. A negative match guarantees the investigator a bitcoin client is not interested in the bitcoin address.

Within this experiment the bitcoin address of Bob was known, by testing Bob's address against the Bloom filter and received a positive result confirmed that the Bloom filter likely contain the bitcoin address of the wallet.

In cases where the bitcoin address is not be known to the investigator other bitcoin network messages may help to answer such question.

### 5.3.3 Bitcoin Message: tx

The **tx** messages are used to notify the bitcoin client about a single transaction, the transaction details like the amount of bitcoin, the source and destination address are stored within the tx message. Because these messages contain the transaction information they are likely to contain relevant information for the investigator. This is a messages in which both are involved:

- Alice sends a tx message notifying the bitcoin network that she has made a transaction
- Bob receives a tx message to be notified that he was received an amount of bitcoin from Alice.

Because both parties contain relevant information regarding the research question, the collected network information of both parties is examined.

**Investigating tx messages of Alice**

**Examination**. The selection criteria for the tx messages are that they are communicated over the TCP network protocol according to the specifications of the bitcoin network protocol. The header of the bitcoin message contains the value '**tx**'. Wireshark is able to use this value to find the **tx** messages.

**Analysis**. Within the collected network information a limited amount **tx** messages was found, investigating the checksum field by hand shows that there is only one unique payload.

**Investigation**. To determine if this tx message is the transaction between Alice and Bob the transaction identifier (TXID) can be calculated. To calculate the TXID the bytes of the tx messages must be hashed twice with the sha256 algorithm followed by an endianness swap (so the result of the last sha256 algorithm is reversed). With the calculated TXID (shown below), open source investigation showed that this TXID was in fact the transaction between Alice and Bob.

**Presentation**. With the calculated TXID additional information can be presented about the transaction, for example: is this transaction known in the blockchain, which bitcoin addresses were involved, how much money was involved and which IP address was the first to relay this transaction.

The TXID that was calculated, and later determined to be the transaction between Alice and Bob:

895d34885e6e0d067c91d0087da9486755b3bc0d7c1b6eb317f0a7e1b81dd888

**Investigating tx Messages of Bob**

**Examination**. The selection criteria for the tx messages are the same as with Alice. In the captured network data of Bob, in contrary to the network data of Alice, a lot of duplicate **tx** messages were found. The number of duplicates can be reduced the same way as explained with filterload messages with the help of the checksum value of the payload. This reduction is the first step of preparing for the information for further analysis, the relevant data within the **tx** message needs to be extracted.

The tx messages have one or more P2PHK-scripts, these scripts contain the bitcoin addresses.

**Analysis**. The extracted bitcoin address needs to be further processed. In the experiment more than 600 bitcoin address were retrieved, the relevance of bitcoin addresses can be determined by testing the addresses against the Bloom filter set by the client. After analysing the addresses with the help of Bloom filter, only 1% of the bitcoin addresses remained.

**Investigation**. Publicly available resources to examine the blockchain can be used to get more info about this small set of bitcoin addresses. In the case of the experiment, open source information showed only one bitcoin address was involved in a transaction during period of the data capture. This was the transaction made by Alice to transfer bitcoin to Bob.

**Presentation**. The presentation of these results is that it was possible to determine one transaction that was committed during the time of network capture. There is always is a risk of a false positive due to the nature of a Bloom filter.

The **tx** messages are the most relevant message to analyse because it involves the transfer of the bitcoin. The analysis of Alice required some insight of how to extract the correct information and calculate the transaction identifier. The result of this calculation was the transaction between Alice and Bob.

The analysis of the messages of Bob was less straight forward due to large number of messages his client received, it's up to the investigator to reduce the amount of information. With the process described above it was possible to reduce the amount of relevant messages to one single address, which was involved with the transaction between Alice and Bob.

When a police investigation would have a wiretap on both parties, similar to the experiment, the information can be correlated. Alice is involved with one transaction which also matches with the Bloom filter and reduced information of Bob. From a police perspective a statement with the probability verging on certainty, that both parties are involved with the transaction can be made.

### 5.3.4 Bitcoin Message: Merkleblock

Merkleblock messages contain the transaction identifiers (TXID) in the requested block. The merkleblock messages contain transactions that will match the Bloom filter set by the client. This was the motivation to add this to relevant messages.

**Examination**. The selection criteria for the merkleblock messages are that they are communicated over the TCP network protocol according to specifications of the bitcoin network protocol. The header of the bitcoin message contains the value merkleblock, Wireshark is able to use this value to find all the merkleblock messages. The messages can be extracted the same way as version, **tx** and filterload messages with Wireshark.

**Analysis**. The extracted information needs to be further processed with the help of a script that retrieves the transaction identifiers from the message. Analysis of the retrieved data showed there was no relation possible between the transaction identifiers.

**Investigation.** The investigation between the transaction identifiers and the transaction no relationship could be established.

**Presentation.** No results could be presented.

During the experiment it showed that analysing the merkleblock message does not reveal any new information. The reason for this can be found in the time that it takes the bitcoin network to add a new block to the blockchain. The block that includes the transaction of Bob and Alice was added two days later. At the time of the block creation there was no data collection.

## 5.4 Discussion

From the results of the experiment as described above regarding the four different types of messages: *version*, *filterload, tx* and *merkleblock*. Three types of message contained relevant information that reveal information regarding information of the client.

The following information could be retrieved from the messages:

**Bitcoin message: version**. The client that was used during the transaction could be determined.

**Bitcoin message: filterload**. The filterload allows a bitcoin address to be tested against a Bloom filter, this gives the opportunity to test if a client might have interest in a certain bitcoin address.

**Bitcoin message: tx**. From the transaction messages received from the sender it was possible to retrieve the transaction information, through open sources, from the blockchain. The receiving party of the transaction required the application of the Bloom filter to reduce the number of transactions down to one transaction that fit within the experiment

Earlier research stated Bloom filters used by bitcoin clients induce the risk of leaking information about the addresses of Bitcoin users [12]. This experiment, confirms that the use of Bloom filter leaks information, in this case the information leakage is beneficial to the police investigator.

Given the scope and limitations, of the experiment. The experiment results showed that with the help of NAFTs, custom scripts and the documentation of the bitcoin protocol, a forensic investigator is able to determine the transaction that is sent and/or received.

The results of the experiment have shown that there are multiple parts to the bitcoin network protocol that are worth investigating during a forensic investigation. The research within this chapter has identified two messages that contain information regarding transactions sent or received by a bitcoin address. Besides the messages containing information regarding the transaction, one type of bitcoin messages

**Table 1** Forensic value of the investigated messages

| Message | Forensic value |
|---|---|
| Version | The version information of the client used during communication |
| Filterload | Contains the probabilistic Bloom filter that can be used to test if a client might be interested in a bitcoin address or have no interest in the bitcoin address at all |
| Tx | Contains the transaction bitcoins, which can be used to calculate the transaction identifier and/or identify the bitcoin addresses involved in the transaction |

showed its relevance because it contained the software version used by the client and by that revealed the capabilities of the bitcoin client.

With the two messages that contain information regarding the transaction received or sent it was, within the boundaries of the experiment, possible to determine the bitcoin addresses that were used and the amount of bitcoin transferred. In Table 1 an overview is given of the messages and their forensic value.

The information that was used during the experiment came from the network traffic captured combined with open source intelligence. The bitcoin ledger (the block chain) itself was not part of the experiment, this might have left out opportunities. An approach could be to determine if a TX message contains the information about a transaction in the past or happened during the network capture.

## 6 Conclusion and Future Work

This chapter puts the bitcoin network protocol under a forensic magnifier by looking at the inner parts of the bitcoin network protocol that might have relevance to a forensic investigation. Because there was only a limited amount of information available regarding the network protocol and the increase of use of bitcoin in police cases, this chapter tries to fill this void in information.

The fulfilment of the information gap is achieved in three ways: First, by providing a detailed background of the bitcoin ecosystem where the network part was centre of the background research. Second, the current state of research regarding network forensics in general is described, followed by a description of the current state of bitcoin network research. Third, an experiment was done to bring this theoretical background knowledge to a more practical situation.

While looking at the state-of-the-art bitcoin research, very little information was found in academic literature. With regards to the analysis focused on the bitcoin messages, no previous research was found.

When searching for bitcoin from a forensic perspective, only a limited number of articles regarding the blockchain were available. Therefore, this chapter can be seen as one of the first attempts to show the possibilities for forensics within the network data of bitcoin.

The boundaries of the research were the limitation in messages investigated, the duration of network capture, the limit of one transaction, the clean state of the clients. These boundaries might have caused the investigated dataset to be cleaner than a wiretap of an internet connection used by a suspect. In real cases the amount of transactions might cause an overflow of information requiring a time-consuming analysis.

The ability to determine the used bitcoin address of the receiving client depends on the use of the Bloom filter. The Bloom filter is a probabilistic filter, which means that the positive matches from this filter cannot be trusted completely and the result of false positives has always to be taken into account. This might require another investigative strategy to confirm the use of the bitcoin address. The Bloom filter has shown to be helpful to eliminate a suspected bitcoin address, when a Bloom filter tests negative the bitcoin address is definitely not in the filter.

The dataset within the experiment is limited to one transaction conducted by two newly installed bitcoin clients in a clean state. The findings of the experiment could have been confirmed if there were more datasets available. Furthermore an experiment with datasets collected from heavily used bitcoin clients could have confirmed if the methods found within the experiment would also worked on more polluted datasets.

As shown by the experiment, investigating the bitcoin network protocol can reveal relevant results, which means that clients using bitcoin network protocol as defined on the website of bitcoin.org will implement their network communication in the same way for interoperability with other clients. The findings within this chapter can apply to other bitcoin clients that are following the definition of bitcoin.org, and not only the investigated client Multibit HD.

This chapter is not a wholistic approach to the bitcoin network in the sense of investigating every bitcoin network message in every way possible. It shows that there is a lot of relevant information to be retrieved from the messages that were selected for this chapter. More research can be done to other message within the network protocol.

Secondly the difference between the clean state, as used in the experiment, of the bitcoin client and a heavily used bitcoin client will introduce new challenges to be investigated. For example, clients will create multiple addresses to be part of a transaction and Bloom filters can be updated to increase the number of false positives. Additional research can be done in the future to test if the findings and methods can be applied against heavily used bitcoin clients.

There are still a lot of investigate values within the communication protocol of the bitcoin network, therefore this chapter can be observed as invitation to explore the undiscovered information which is likely to reside within the bitcoin messages.

# References

1. National Cyber Security Centre: Cyber Security Assessment Netherlands 2016. National Cyber Security Centre, The Hague, Tech. Rep. (2015)
2. WhatsApp Inc.: WhatsApp Encryption Overview. WhatsApp Inc, Tech. Rep. [Online] (2016). Available: https://www.whatsapp.com/security/WhatsApp-Security-Whitepaper.pdf
3. Herzberg, A., Leibowitz, H.: Can Johnny finally encrypt? evaluating E2E-encryption in popular IM applications. In: STAST '16: Proceedings of the 6th Workshop on Socio-Technical Aspects in Security and TrustDecember, pp. 17–28 (2016). https://doi.org/10.1145/3046055.3046059
4. Viber: Viber Encryption Overview [Online] (2019). https://www.viber.com/en/security-overview. Accessed: 2019-06-02
5. Sgaras, C., Kechadi, T., Le-Khac, N.-A.: Forensics acquisition and analysis of instant messaging and VoIP applications. In: Lecture Notes in Computer Science, vol. 8915, pp. 188–199 (2015). https://doi.org/10.1007/978-3-319-20125-2_16
6. Lewis, J.A., Zheng, D.E., Carter, W.A.: The effect of encryption on lawful access to communications and data. CSIS, Tech. Rep., [Online] (2017). Available: https://csis-prod.s3.amazonaws.com/s3fs-public/publication/170203_Lewis_EffectOfEncrytion_Web.pdf
7. Ryder, S., Le-Khac, N.-A.: The end of effective law enforcement in the cloud?—to encrypt, or not to encrypt. In: The 9th IEEE International Conference on Cloud Computing, San Francisco, CA USA (2016). https://doi.org/10.1109/CLOUD.2016.0133
8. Le-Khac, N.-A., Markos, S., Kechadi, M.-T.: Towards a new data mining-based approach for anti money laundering in an international investment bank. In: International Conference on Digital Forensics and Cyber Crime (ICDF2C 2009), Springer, Berlin LNICST 31, 30 Sept–2 Oct, Albany, New York, USA (2009)
9. Nakamoto, S.: Bitcoin: A Peer-to-Peer Electronic Cash System, p. 9 (2008). Www.Bitcoin.Org [Online]. Available: https://bitcoin.org/bitcoin.pdf
10. Bonneau, J., Miller, A., Clark, J., Narayanan, A., Kroll, J.A., Felten, E.W.: SoK: research perspectives and challenges for bitcoin and cryptocurrencies. In: 2015 IEEE Symposium on Security and Privacy, pp. 104–121. IEEE (2015)
11. Protocol Documentation: Protocol Documentation—Bitcoin Wiki [Online]. Available: https://en.bitcoin.it/wiki/Protocol_documentation. Accessed: 2019-03-17
12. Gervais, A., Capkun, S., Karame, G.O., Gruber, D.: On the privacy provisions of Bloom filters in lightweight bitcoin clients. In: Proceedings of the 30th Annual Computer Security Applications Conference on—ACSAC'14, pp. 326–335. ACM Press, New York, USA (2014)
13. Pilli, E.S., Joshi, R.C., Niyogi, R.: Network forensic frameworks: survey and research challenges. Dig. Invest. 7(1–2), 14–27 (2010)
14. Rogers, M., Goldman, J., Mislan, R., Wedge, T., Debrota, S.: Computer forensics field triage process model. J. Dig. Forensics Secur. Law 1(2), 19–38 (2006)
15. Tafazzoli, T., Salahi, E., Gharaee, H.: A proposed architecture for network forensic system in large-scale networks. Interface (2015) [Online]. Available: https://arxiv.org/ftp/arxiv/papers/1508/1508
16. Blockchain Luxembourg S.A.: Bitcoin Block Explorer—Blockchain [Online]. Available: https://blockchain.info/. Accessed: 2019-07-08
17. Wondracek, G., Comparetti, P.: Automatic network protocol analysis. In: Proceedings of the 15th Network and Distributed System Security Symposium (NDSS) (2008)
18. Cabaj, K.: Network activity analysis of CryptoWall ransomware. PRZEGLA̧D ELEKTROTECHNICZNY 1(11), 203–206 (2015)
19. Andriesse, D., Rossow, C., Stone-Gross, B., Plohmann, D., Bos, H.: Highly resilient peer-to-peer botnets are here: an analysis of Gameover Zeus. In: Proceedings of the 2013 8th International Conference on Malicious and Unwanted Software: The Americas, MALWARE 2013, pp. 116–123 (2013)
20. He, J., Yang, Y., Wang, X., Tang, C., Zeng, Y.: PeerDigger: Digging stealthy P2P hosts through traffic analysis in real-time. In: Proceedings—17th IEEE International Conference on Computational Science and Engineering, CSE 2014, Jointly with 13th IEEE International Conference

on Ubiquitous Computing and Communications, IUCC 2014, 13th International Symposium on Pervasive Systems, pp. 1528–1535 (2015)

21. Kato, K., Klyuev, V.: Large-scale network packet analysis for intelligent DDoS attack detection development. In: The 9th International Conference for Internet Technology and Secured Transactions (ICITST-2014), London UK (2014)
22. Spagnuolo, M., Maggi, F., Zanero, S.: BitIodine: extracting intelligence from the Bitcoin network. In: Lecture Notes in Computer Science (including subseries Lecture Notes in Artificial Intelligence and Lecture Notes in Bioinformatics), vol. 8437, pp. 457–468 (2014)
23. Meiklejohn, S., Pomarole, M., Jordan, G., Levchenko, K., McCoy, D., Voelker, G.M., Savage, S.: A fistful of bitcoins. In: Proceedings of the 2013 conference on Internet measurement conference—IMC'13, pp. 127–140. ACM Press, New York, USA (2013)
24. Moser, M., Bohme, R., Breuker, D.: An inquiry into money laundering tools in the Bitcoin ecosystem. In: 2013 APWG eCrime Researchers Summit, pp. 1–14. IEEE (2013)
25. Barber, S., Boyen, X., Shi, E., Uzun, E.: Bitter to Better—How to Make Bitcoin a Better Currency. In: Lecture Notes in Computer Science (including subseries Lecture Notes in Artificial Intelligence and Lecture Notes in Bioinformatics), vol. 7397, pp. 399–414. LNCS (2012)
26. Miers, I., Garman, C., Green, M., Rubin, A.D.: Zerocoin: anonymous distributed E-cash from Bitcoin. In: 2013 IEEE Symposium on Security and Privacy, pp. 397–411. IEEE (2013)
27. Van der Horst, L., Choo, K.K.R., Le-Khac, N.-A.: Process memory investigation of the Bitcoin clients electrum and bitcoin core. IEEE Access 5(1) (2017). https://doi.org/10.1109/ACCESS. 2017.2759766
28. Androulaki, E., Karame, G.O., Roeschlin, M., Scherer, T., Capkun, S.: Evaluating user privacy in Bitcoin. IACR Cryptol. ePrint Arch. 7859, 34–51 (2013)
29. Zollner, S., Choo, K.K.R., Le-Khac, N.-A.: An automated live forensic and postmortem analysis tool for bitcoin on windows systems. IEEE Access 7 (2019). https://doi.org/10.1109/ACCESS. 2019.2948774
30. Koerhuis, W., Kechadi, T., Le-Khac, N.-A.: Forensic analysis of privacy-oriented cryptocurrencies. Elsevier (2020). https://doi.org/10.1016/j.fsidi.2019.200891
31. Fantazzini, D., Nigmatullin, E., Sukhanovskaya, V., Ivliev, S.: Everything you always wanted to know about bitcoin modelling but were afraid to ask, p. 49 (2016). [Online]. Available: https://mpra.ub.uni-muenchen.de/71946/
32. Hurlburt, G.: Might the blockchain outlive Bitcoin? IT Prof. 18(2), 12–16 (2016) [Online]. Available: http://ieeexplore.ieee.org/document/7436669/
33. Dorri, A., Steger, M., Kanhere, S.S>, Jurdak, R.: BlockChain: A Distributed Solution to Automotive Security and Privacy (2017) [Online]. Available: http://arxiv.org/abs/1704.00073
34. Biryukov, A., Khovratovich, D., Pustogarov, I.: Deanonymisation of clients in Bitcoin P2P network. In: Proceedings of the 2014 ACM SIGSAC Conference on Computer and Communications Security—CCS '14, pp. 15–29. ACM Press, New York, USA (2014)
35. Feld, S., Schönfeld, M., Werner, M.: Analyzing the deployment of Bitcoin's P2P network under an AS-level perspective. Procedia Comput. Sci. 32, 1121–1126 (2014)
36. Koshy, P., Koshy, D., McDaniel, P.: An analysis of anonymity in Bitcoin using P2P network traffic. In: Lecture Notes in Computer Science (including subseries Lecture Notes in Artificial Intelligence and Lecture Notes in Bioinformatics), vol. 8437, pp. 469–485 (2014)
37. Biryukov, A., Pustogarov, I.: Bitcoin over Tor isn't a good idea. In: 2015 IEEE Symposium on Security and Privacy, pp. 122–134. IEEE (2015)
38. Fanti, G., Viswanath, P.: Anonymity Properties of the Bitcoin P2P Network (2017) [Online]. Available: https://arxiv.org/pdf/1703.08761.pdf
39. Wireshark Foundation: Wireshark Go deep, 27.05.2010 (2010). [Online]. Available: https://www.wireshark.org/
40. NETRESEC: "NetworkMiner—The NSM and Network Forensics Analysis Tool (2017) [Online]. Available: http://www.netresec.com/?page=NetworkMiner
41. AOL: Moloch (2017) [Online]. Available: http://molo.ch/
42. BitcoinJ: Bitcoinj [Online]. Available: https://bitcoinj.github.io/

**Cornelis Leendert (Eelco) van Veldhuizen** received his Master's Degree, with first class honors, from University College Dublin, Ireland in 2017. He is currently Technical Coordinator at the National High Tech Crime Unit of the Netherlands. His research interests include cryptocurrencies, cybersecurity, bulletproof hosting and Digital Forensics.

**Madhusanka Liyanage** received his Ph.D in communication engineering from the University of Oulu, Oulu, Finland. He is currently working as an Ad Astra Fellow/Assistant Professor at School of Computer Science, University College Dublin, Ireland. He is also an adjunct professor at the University of Oulu, Finland. He is also a recipient of Marie Skłodowska-Curie Actions-Individual Fellowships. In 2011-2012, he was a research scientist at I3S Laboratory and Inria, Sophia Antipolis, France. Also, he was a visiting research fellow at and Computer Science and Engineering, The University of Oxford, Data61, CSIRO, Sydney Australia, Infolabs21, Lancaster University, UK and Computer Science and Engineering, The University of New South Wales during 2016-2018. He is a Member of IEEE and ICT. Madhusanka is a co-author of over 70 publications including three edited book with Wiley. He is also a management committee member of EU COST Action IC1301, IC1303, CA15107, CA15127, CA 16226 projects. Dr. Liyanage's research interests are SDN, IoT, Blockchain, MEC, mobile and virtual network security. More info at https://www.madhusanka.com/

**Kim-Kwang Raymond Choo** received the Ph.D. in Information Security in 2006 from the Queensland University of Technology, Australia. He currently holds the Cloud Technology Endowed Professorship at the University of Texas at San Antonio (UTSA). In 2015, he and his team won the Digital Forensics Research Challenge organized by the Germany's University of Erlangen-Nuremberg. He is the recipient of the 2019 IEEE Technical Committee on Scalable Computing (TCSC) Award for Excellence in Scalable Computing (Middle Career Researcher), 2018 UTSA College of Business Col. Jean Piccione and Lt. Col. Philip Piccione Endowed Research Award for Tenured Faculty, British Computer Society's 2019 Wilkes Award Runner-up, 2019 EURASIP Journal on Wireless Communications and Networking (JWCN) Best Paper Award, Korea Information Processing Society's Journal of Information Processing Systems (JIPS) Survey Paper Award (Gold) 2019, IEEE Blockchain 2019 Outstanding Paper Award, Inscrypt 2019 Best Student Paper Award, IEEE TrustCom 2018 Best Paper Award, ESORICS 2015 Best Research Paper Award, 2014 Highly Commended Award by the Australia New Zealand Policing Advisory Agency, Fulbright Scholarship in 2009, 2008 Australia Day Achievement Medallion, and British Computer Society's Wilkes Award in 2008. He is also a fellow of the Australian Computer Society, an IEEE senior member, and co-chair of IEEE Multimedia Communications Technical Committee's Digital Rights Management for Multimedia Interest Group.

**Nhien-An Le-Khac** is a lecturer at the School of Computer Science (CS), University College Dublin (UCD), Ireland. He is currently the program director of MSc program in Forensic Computing and Cybercrime Investigation (FCCI)—an international program for the law enforcement officers specializing in cybercrime investigations. He is also the co-founder of UCD-GNECB Postgraduate Certificate in fraud and e-crime investigation. Since 2008, he is a research fellow in Citibank, Ireland (Citi). He obtained his Ph.D. in Computer Science in 2006 at the Institut National Polytechnique de Grenoble (INPG), France. His research interest spans the area of Cybersecurity and Digital Forensics, Data Mining/Distributed Data Mining for Security, and Fraud and Criminal Detection. Since 2013, he has collaborated on many research projects as a principal/co-PI/funded investigator. He has published more than 150 scientific papers in peer-reviewed journal and conferences in related research fields, and his recent edited book has been listed the Best New Digital Forensics Book according to the Book Authority.

# So, What's Next?

Nhien-An Le-Khac and Kim-Kwang Raymond Choo

**Abstract** In this book, we reported the findings on a broad range of topics, ranging from Internet of Things (IoT) malware analysis to IoT testbed setup to malware
analysis (e.g. ransomware) to CCTV forensics to financial investigations (e.g. Pay-
Pal accounts and Bitcoins), cloud forensics to network and ToR forensics, which
were carried out mainly by students enrolled in the MSc in Forensic Computing
and Cybercrime Investigation program at University College Dublin, Ireland. These
students are also employed in law enforcement and government agencies in Canada,
Ireland, Germany, The Netherlands, and United Kingdom.

**Keywords** Digital forensics · Digital forensic education · Digital investigations ·
Criminal investigations · Civil litigations · National security investigations

## 1 Concluding Remarks

We hope the findings reported in this book will benefit the digital forensics and
investigation community, although the topics explored here are only a (very) small
fraction of what needs to be focused on.

There is a continuing need to keep a watchful brief on the constantly evolving
cyberthreat landscape, and how criminals and those with malicious intent can abuse
technologies to facilitate their nefarious activities. Future research topics include:

- Forensic, security and/or vulnerability analysis of different technologies, particularly emerging technologies such as smart lens, Internet of Things (IoT) devices
  embedded in human bodies, and other IoT devices in complex, adversarial settings.
- Related policy and regulatory topics, such as those relating to data and user privacy.
- Education and knowledge sharing activities so that the wider community can be
  kept abreast of the potential risks and cyber hygiene best practices.

N.-A. Le-Khac (✉)
University College Dublin, Dublin, Ireland
e-mail: an.lekhac@ucd.ie

K.-K. R. Choo
University of Texas at San Antonio, San Antonio, USA

N.-A. Le-Khac and K.-K. R. Choo (eds.), *Cyber and Digital Forensic Investigations*,
Studies in Big Data 74, https://doi.org/10.1007/978-3-030-47131-6_12

**Nhien-An Le-Khac** is a lecturer at the School of Computer Science (CS), University College Dublin (UCD), Ireland. He is currently the program director of MSc program in Forensic Computing and Cybercrime Investigation (FCCI)—an international program for the law enforcement officers specializing in cybercrime investigations. He is also the co-founder of UCD-GNECB Postgraduate Certificate in fraud and e-crime investigation. Since 2008, he is a research fellow in Citibank, Ireland (Citi). He obtained his Ph.D. in Computer Science in 2006 at the Institut National Polytechnique de Grenoble (INPG), France. His research interest spans the area of Cybersecurity and Digital Forensics, Data Mining/Distributed Data Mining for Security, and Fraud and Criminal Detection. Since 2013, he has collaborated on many research projects as a principal/co-PI/funded investigator. He has published more than 150 scientific papers in peer-reviewed journal and conferences in related research fields, and his recent edited book has been listed the Best New Digital Forensics Book according to the Book Authority.

**Kim-Kwang Raymond Choo** received the Ph.D. in Information Security in 2006 from the Queensland University of Technology, Australia. He currently holds the Cloud Technology Endowed Professorship at the University of Texas at San Antonio (UTSA). In 2015, he and his team won the Digital Forensics Research Challenge organized by the Germany's University of Erlangen-Nuremberg. He is the recipient of the 2019 IEEE Technical Committee on Scalable Computing (TCSC) Award for Excellence in Scalable Computing (Middle Career Researcher), 2018 UTSA College of Business Col. Jean Piccione and Lt. Col. Philip Piccione Endowed Research Award for Tenured Faculty, British Computer Society's 2019 Wilkes Award Runner-up, 2019 EURASIP Journal on Wireless Communications and Networking (JWCN) Best Paper Award, Korea Information Processing Society's Journal of Information Processing Systems (JIPS) Survey Paper Award (Gold) 2019, IEEE Blockchain 2019 Outstanding Paper Award, Inscrypt 2019 Best Student Paper Award, IEEE TrustCom 2018 Best Paper Award, ESORICS 2015 Best Research Paper Award, 2014 Highly Commended Award by the Australia New Zealand Policing Advisory Agency, Fulbright Scholarship in 2009, 2008 Australia Day Achievement Medallion, and British Computer Society's Wilkes Award in 2008. He is also a fellow of the Australian Computer Society, an IEEE senior member, and co-chair of IEEE Multimedia Communications Technical Committee's Digital Rights Management for Multimedia Interest Group.

Printed in the United States
by Baker & Taylor Publisher Services